Lecture Notes in Medical Informatics

Edited by D. A. B. Lindberg and P. L. Reichertz

23

Selected Topics in Image Science

Edited by O. Nalcioglu and Z.-H. Cho

Springer-Verlag
Berlin Heidelberg New York Tokyo 1984

Editors

O. Nalcioglu
Depts. of Radiological Sciences, Electrical Engineering
and Medicine (Cardiology), University of California
Irvine, CA 92717, USA

Z.-H. Cho
Dept. of Electrical Science, Advanced Institute of Science,
Seoul, Korea
Dept. of Radiology, Columbia University
New York, N.Y. 10032, USA

ISBN-13:978-3-642-93255-7 e-ISBN-13:978-3-642-93253-3
DOI: 10.1007/978-3-642-93253-3

Softcover reprint of the hardcover 1st edition 1984

2145/3140-543210

PREFACE

The continuing growth of computed tomography (CT) and other imaging techniques motivated us to bring together a comprehensive review of the state of the art in diagnostic imaging.

Twelve years after the first appearance of x-ray CT, computerized diagnostic imaging has grown so rapidly in sophistication that it is difficult to follow current developments in this diversified field.

In this book, we have attempted to cover the basic developments in several areas. The first part includes some of the fundamental aspects of computerized diagnostic imaging such as algorithms and detectors. Specific applications in emission tomography, digital radiography, ultrasound and nuclear magnetic resonance imaging are dealt with in the second part. The contributed papers are by experts currently in the field, whom we feel would certainly enlighten the subject matter and possibly suggest directions for future development.

We would like to express our sincere thanks to those who have contributed to this volume. We are sure that their original papers will be beneficial for readers and will also remain as an important reference for researchers in the years to come.

We would also like to thank Betty Trent for her expert and patient typing of the entire book. Finally, special thanks are due to Mrs. Ingeborg Mayer of Springer-Verlag for her encouragements, support and patience throughout the preparation of this book.

Irvine, California
November 1983

Orhan Nalcioglu
Zang-Hee Cho

<u>CONTENTS</u>

Methods and Algorithms Toward 3-D Volume
Image Reconstruction with Projections
by Z.H. Cho and J.B. Ra

Direct Fourier Reconstruction Techniques
in NMR Tomography
by Z.H. Cho, O. Nalcioglu, J.C. Jeong and H.B. Song

Radiation Detectors for CT Instrumentation
by M.R. Farukhi

Page

Positron Emission Tomography - Basic Principles,
Corrections and Camera Design
by C. Bohm, L. Eriksson and M. Bergström

Single Photon Emission Computed Tomography:
Potentials and Limitations
by C.B. Lim and S. Gottschalk

Page

Energy Selective Digital Radiography
by R.E. Alvarez

Matched Filtering for Digital Subtraction Angiography
by R.A. Kruger

Functional Analysis of Angiograms
by Digital Image Processing Techniques
by R. Brennecke and J.H. Bürsch

Acoustical Imaging: History, Applications,
Principles and Limitations
by G. Wade, H. Lee and C. Schueler

Ultrasound Tomography by Galerkin or Moment Methods
by S.A. Johnson and F. Stenger

NMR Imaging: Principles, Algorithms and Systems
by Z.H. Cho, H.S. Kim, C.H. Oh, H.W. Park and S.W. Lee

METHODS AND ALGORITHMS TOWARD 3-D VOLUME IMAGE RECONSTRUCTION WITH PROJECTIONS

Z.H. Cho and J.B. Ra

Department of Electrical Sciences,
Korea Advanced Institute of Science, Seoul, Korea and
Department of Radiology, Columbia University, New York

1. INTRODUCTION

Since the introduction of computerized tomography (1), many reconstruction algorithms (2-4) have been suggested using various mathematical techniques. Especially, in 3-dimensional volume imaging, the Fourier transform approach (5) is usually preferred. This Fourier technique can be categorized into conventional 2-D slice by slice technique (6-8) or direct true 3-D image reconstruction (9-11) and also depending on the mode of data collection, i.e., parallel (6,7,9,10) or fan (cone for 3-D) beam (8,12-16). Reconstruction procedure can also be classified depending on the mode of processing; namely filtered backprojection (FB) (7, 10), backprojection filtering (BF) (17,51), or direct Fourier transform reconstruction (DFR) (18-20). Each method has its own characteristics and preference depends on the kind of tomographic system being employed and computational hardware involved.

Conventional two-dimensional image reconstruction using one-dimensional projection data with a sufficient number of views provides high quality images. By stacking a series of 2-D images reconstructed by this method, one can obtain a true 3-D volume image. Recently, however, interest has been focused on reconstruction of 3-D volume image directly from 2-D projection data since this has an advantage, especially in statistical image quality improvement (9,10). The latter is an important question usually raised in emission computerized tomography (ECT) where the data statistics are usually poor. In this paper, we will review and develop 2-D plane and 3-D volume image reconstruction algorithms using a complete spherical detection scheme. These algorithms have been investigated by many authors, we will, therefore, focus our attention to the direct 3-D volume image reconstruction algorithm and propose the generalized true three-dimensional reconstruction (GTTR) algorithm (21-23) which is applicable to any geometry from a complete to any truncated spherical geometry. The algorithm is applicable to both elongated human body as well as a large spherical object and provides high quality images near the system center by a focusing concept. The algorithm proposed is basically a filtered backprojection approach which has computational advantages as well as high image quality. Filtering was performed by polar angle dependent composite filter functions which we will describe in detail. Consideration was also given to the practical implementation of 3-D volume image reconstruction where 2-D projection data processing requires some care, i.e., slope discontinuities at the boundary which tend to produce ring artifacts in the high density resolution image has been eliminated using a new derivative weighting function.

In section 2, 2-D and 3-D reconstruction algorithms of various types are briefly reviewed. In section 3 and 4, the direct volume image, i.e., true 3-D reconstruction algorithm and its extended version are presented.

2. RECONSTRUCTION ALGORITHMS

2.1 Preliminary

The image reconstruction using measured projection data is the basis of computerized tomography and various 3-D image reconstruction algorithms have been investigated during the last ten years such as the direct matrix inversion technique (24), maximum entropy reconstruction (25), iterative method (26,27), and Fourier reconstruction (5). Among others, the Fourier reconstruction method is the most widely used technique in computerized tomographic applications such as X-ray, nuclear medicine (single photon or positron) (27-33), ultrasound (34-36), and NMR (nuclear magnetic resonance) (37-39) computerized tomography. Fourier reconstruction techniques may be categorized into three types, namely, filtered backprojection (FB) (7), backprojection filtering (BF) (17), and direct Fourier transform reconstruction (DFR) (19). Theoretical aspects and characteristics of these algorithms are discussed for 2-D plane image reconstruction. The need for direct three-dimensional reconstruction has increased recently, especially in the area of emission CT and algorithms toward this end will be introduced. However, interest has been focused initially on the algorithms suitable for the complete spherical geometry.

2.2 Two-dimensional Image Reconstruction Algorithms

2.2.1 Filtered backprojection algorithm

The filtered or convolved backprojection algorithm is the most popular reconstruction method in CT scanning. For the mathematical formulation of this convolution-backprojection algorithm, there are two basic forms in existence, namely parallel beam mode and fan beam scheme. We will consider these two basic techniques in some detail which is useful for the development of the necessary formulation of the true 3-D volume image reconstruction.

Parallel beam mode - Let's assume an object distribution function $f(x,y)$ in the coordinates (x,y) and their rotated coordinates (x',y').

The data to reconstruct the image are projections which represent sets of line integrals of the object in various directions. Projection data $p_\phi(x')$ is the line integral taken along the y' direction and may be expressed as

$$p_\phi(x') = \int_{-\infty}^{\infty} f(x',y')\,dy' \tag{1}$$

$$= \int_{-\infty}^{\infty}\int_{-\infty}^{\infty} f(x,y)\,\delta(x\cos\phi + y\sin\phi - x')\,dxdy, \tag{2}$$

where
$$\begin{pmatrix} x' \\ y' \end{pmatrix} = \begin{pmatrix} \cos\phi & \sin\phi \\ -\sin\phi & \cos\phi \end{pmatrix} \begin{pmatrix} x \\ y \end{pmatrix}. \tag{3}$$

Since, in practice, the object is space-limited, $p_\phi(x')$ will also be limited to a finite extent. Fourier transform of projection data $p_\phi(\omega)$, can be related to projection data $p_\phi(x')$ as

$$p_\phi(\omega) = \int_{-\infty}^{\infty} p_\phi(x')e^{-i\omega x'}\,dx' \tag{4}$$

$$= \int_{-\infty}^{\infty}\int_{-\infty}^{\infty} f(x,y)e^{-i\omega(x\cos\phi+y\sin\phi)}\,dxdy. \tag{5}$$

Eq. (5) can also be written as

$$p_\phi(\omega) = \int_{-\infty}^{\infty}\int_{-\infty}^{\infty} f(x,y)e^{-i(\omega_x x+\omega_y y)}\,dxdy$$

$$= F(\omega_x,\omega_y)\Big|_\phi = F(\omega,\phi), \tag{6}$$

where $\omega_x = \omega\cos\phi$ and $\omega_y = \omega\sin\phi$ when (ω_x, ω_y) and (ω, ϕ) represent the Cartesian and polar coordinates of (x,y) in the spatial frequency domain, respectively. Eq. (6) states that a one-dimensional Fourier transform of projection data at a given angle ϕ represents the 2-D Fourier transform values of the object function $f(x,y)$ in spatial frequency domain along the radial frequency with a given angle ϕ. This is the well known projection theorem which plays a central role in the 2-D and 3-D image reconstruction (19). From the above theorem, it can easily be shown that the object function $f(x,y)$ can be recovered as

$$f(x,y) = F_2^{-1}\left[P_\phi(\omega)\right] \tag{7}$$

$$= \frac{1}{4\pi^2} \int_{-\infty}^{\infty} \int_{-\infty}^{\infty} P_\phi(\omega) \exp\left[i(x\omega_x + y\omega_y)\right] d\omega_x d\omega_y \tag{8}$$

$$= \frac{1}{4\pi^2} \int_{0}^{\pi} \int_{-\infty}^{\infty} P_\phi(\omega) \exp\left[i\omega(x\cos\phi + y\sin\phi)\right] |\omega| \, d\omega d\phi, \tag{9}$$

where $F_2^{-1}[\cdot]$ denotes 2-D inverse Fourier transform operator and Eq.(9) is a polar coordinate representation of Eq.(8) in the spatial frequency domain. By changing the order of integration, Eq. (9) becomes

$$f(x,y) = \frac{1}{2\pi} \int_{0}^{\pi} d\phi \left(\frac{1}{2\pi} \int_{-\infty}^{\infty} d\omega \, P_\phi(\omega) |\omega| \exp\left[i\omega(x\cos\phi + y\sin\phi)\right] \right) \tag{10}$$

$$= \frac{1}{2\pi} \int_{0}^{\pi} d\phi \int_{-\infty}^{\infty} dx' \, p_\phi(x') h(x\cos\phi + y\sin\phi - x') \quad , \tag{11}$$

where

$$h(t) = F_1^{-1}\left[|\omega|\right] , \tag{12}$$

and $F_1^{-1}[\cdot]$ is the 1-D inverse Fourier transform operation. Eq. (11) is well known convolution backprojection operation. The convolution kernel $h(\cdot)$ is an inverse Fourier transform of $|\omega|$ and its exact form is not realizable in practice. Several modified filter functions, therefore, have been suggested (7,40,41,53), and the selection of a particular filter function will affect the characteristics of the reconstructed image such as the desired image resolution and smoothness.

<u>Fan beam mode</u> - Although the parallel beam reconstruction algorithm has

been a fundamental tool for image reconstruction, the fan beam reconstruction algorithm is, nevertheless, widely used, for example in X-ray CT mainly because basic data collection procedure involved is in fan beam mode. In addition, the image reconstruction with the fan beam algorithm often provides better resolution with same dimension of sampled data as with the parallel case due to improved sampling at the center region (45,46). This resolution improvement with fan beam mode reconstruction is found to be especially important, for example, in the positron emission tomography where the intrinsic image resolution is restricted by the detector size.

Among the various approaches to the derivation of the fan beam reconstruction algorithm (8,42), use of the coordinate transformation (43) from the parallel beam algorithm is found to be useful. If the fan beam projection data set is represented by $P_{\alpha 1}(\beta_1)$, where β_1 and α_1 represent detector position and rotation angle of the beam, as shown in Fig. 1, the relation between parallel and fan beam projection data with coordinates (x',ϕ) and (α_1,β_1) can be given as

$$\phi = \alpha_1 + \beta_1, \qquad x' = R_d \sin\beta_1, \tag{13}$$

where R_d is the distance between the center point and the apex of the fan. Since the reconstruction formula derived for the parallel projection data set given in Eq. (11) is equally valid for $0 - 2\pi$, it can be rewritten as

$$f(x,y) = \frac{1}{4\pi}\int_0^{2\pi} d\phi \int_{-\infty}^{\infty} dx'\; p_\phi(x')h(x\cos\phi+y\sin\phi-x'). \tag{14}$$

By use of Eq. (13) and Eq. (14) the fan beam analogy can be derived as

$$f(x,y) = \frac{1}{4\pi}\int_0^{2\pi} d\alpha_1 \int_{-\beta_{1m}}^{\beta_{1m}} d\beta_1 p_{\alpha_1}(\beta_1)h\left[x\cos(\alpha_1+\beta_1)+y\sin(\alpha_1+\beta_1)-R_d\sin\beta_1\right]J(\beta_1), \tag{15}$$

where $J(\beta_1)$ is a Jacobian of the transformation which is given by

$$J(\beta_1) = \left|\frac{\partial(x',\phi)}{\partial(\beta_1,\alpha_1)}\right| = R_d\cos\beta_1. \tag{16}$$

Eq. (15) can be rewritten as

$$f(x,y) = \frac{1}{4\pi}\int_0^{2\pi} d\alpha_1 \int_{-\beta_{1m}}^{\beta_{1m}} d\beta_1\; p_{\alpha_1}(\beta_1)J(\beta_1)h\left[v_1\sin(\beta_1'-\beta_1)\right], \tag{17}$$

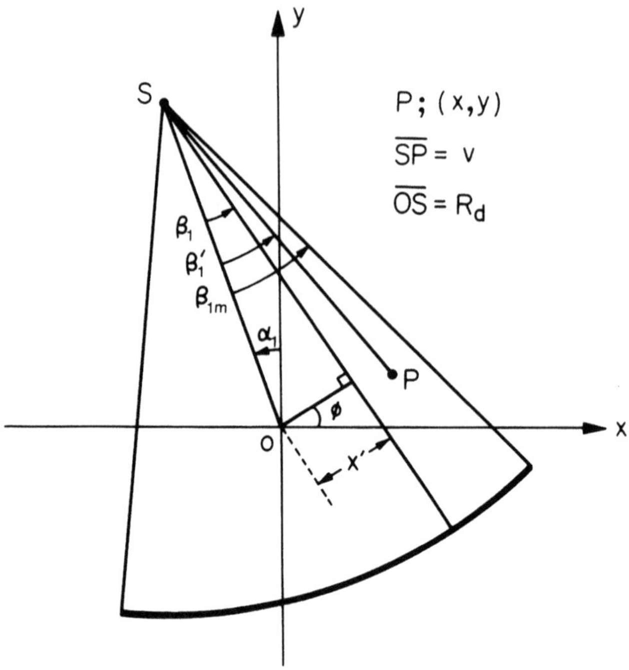

Fig. 1. Geometry for 2-D fan beam reconstruction algorithm.

where

$$v_1 = \sqrt{(x\cos\alpha_1+y\sin\alpha_1)^2+(x\sin\alpha_1-y\cos\alpha_1+R_d)^2}, \tag{18}$$

and

$$\tan\beta_1' = \frac{x\cos\alpha_1+y\sin\alpha_1}{x\sin\alpha_1-y\cos\alpha_1+R_d} . \tag{19}$$

Substituting $t = v_1\sin\xi$ and using the transformation $\omega' = \omega v_1\sin\xi/\xi$ in Eq. (12), it can be shown that

$$h(v_1\sin\xi) = \left(\frac{\xi}{v_1\sin\xi}\right)^2 \; h(\xi) \quad . \tag{20}$$

By using the relation given in Eq. (20) and substituting $\xi = \beta_1'-\beta_1$, Eq. (17) can be written as

$$f(x,y) = \frac{1}{4\pi} \int_0^{2\pi} d\alpha_1 \int_{-\beta_{1m}}^{\beta_{1m}} d\beta_1 \; P_{\alpha_1}(\beta_1)J(\beta_1)g(\beta_1'-\beta_1)\frac{1}{v_1^2} \; , \tag{21}$$

where

$$g(\xi) = \left(\frac{\xi}{\sin\xi}\right)^2 h(\xi) \ . \tag{22}$$

The fan beam reconstruction algorithm given in Eq. (21) consists of three parts:

(a) Weighting the projection data with $J(\beta_1)$
(b) convolution with $g(\beta_1)$ and
(c) weighted backprojection with weight $1/v_1^2$

Eq. (21) may, therefore, be written in a simplified form as

$$f(x,y) = B_{w_2} \left[\{p_{\alpha_1}(\beta_1)J(\beta_1)\} * g(\beta_1)\right] \ , \tag{23}$$

where B_{w_2} is a 2-D weighted backprojection operator.

2.2.2 Backprojection-filtering algorithm

As was mentioned earlier, the filtered backprojection algorithm provides a high quality image as well as computational efficiency and is by far the most popular algorithm in computed tomography (7). The use of this algorithm is limited, however, to equi-sampled parallel or fan mode of the straight line integral projection data. Hence a rebinning and interpolation processes are required for arbitrarily oriented straight rays in the more general case. It is also not possible to apply this algorithm to line integral data of curved rays occurring in ultrasound reflection tomography.

An alternative backprojection-filtering (BF) algorithm is proposed to overcome the problems and found to be more general as long as the point spread function (PSF) of a backprojected image is taken as $1/r$ where r is the distance from point source. Although the method appears attractive and general (17,44) it is rarely used because the resultant images were usually poorer than those obtained by the filtered backprojection algorithm. The main reasons for poor image quality may be due to:

(a) The spatial extend of the backprojected image is unlimited although the projection data from the finite size object are limited. Truncation of the 2-D backprojected image data for the digital processing will, therefore, result in image degradation.

(b) 2-D filter function to be used in the rectangular Cartesian coordinate has slope discontinuities at the boundary (near cutoff frequency), resulting in a ring artifact. Use of an overweighted window function such as the Hanning or Hamming window in an attempt to avoid the ring artifacts necessarily degrades the image resolution.

In this section we will show a BF algorithm with which high quality images equal to the filtered backprojection images can be obtained.

If the point spread function (PSF) of the backprojected image is $1/r$, the relation between the object density function $f(x,y)$ and the backprojected or blurred image $b(x,y)$ is as follows:

$$b(x,y) = f(x,y) ** \frac{1}{r} , \tag{24}$$

where r is the distance from the source point at origin and ** represents the 2-D convolution. Fourier transform of both sides of Eq. (24) is given by

$$B(\omega_x,\omega_y) = F(\omega_x,\omega_y) \frac{1}{\omega} , \tag{25}$$

where $B = F_2[b]$, $F = F_2[f]$ and ω is the radial spatial frequency. From Eq. (25), we obtain the object function $f(x,y)$ by inverse Fourier transform operation.

$$f(x,y) = F_2^{-1}\left\{\omega B(\omega_x,\omega_y)\right\} \tag{26}$$

$$= F_2^{-1}\left\{\omega\right\} ** b(x,y) , \tag{27}$$

where

$$\omega B(\omega_x,\omega_y) = F(\omega_x,\omega_y) . \tag{28}$$

Generally, the computation time of convolution operation in the space domain is larger than the processing time required in the spatial frequency domain. Relative computational time gain of spatial frequency domain operation increases as the number of pixel points increases, especially in 2-D data processing. Let us now consider Eq. (26) as the basic form of the BF algorithm. In the implementation of this algorithm we have to carefully consider two previously mentioned aspects. First is the size and form of the backprojected image and second is the shape of the filter function.

In conventional 3-D image processing, an image of matrix size N x N is modified to 2N x 2N data in which outside of the N x N center array is filled with zeros to avoid the aliasing effect arising in circular convolution (52). Similar procedure can not be applied for the backprojected data array since the backprojection data decays slowly and values outside the N x N array boundary are still found to be significant. Truncation of backprojected image data, therefore, will result in a severe truncation artifact. To avoid this artifact we expand sampling data twice, i.e., take full 2N x 2N backprojected image data, instead of taking only N x N array from the backprojected image and appending the rest of 2N x 2N array with zeros. The use of these full data reduces the artifact significantly.

Selection and formulation of the 2-D filter function is another important factor which determines the image quality in 2-D BF image reconstruction since the slope discontinuity at the cutoff frequency causes ring artifact. We have selected, therefore, a modified 2-D version of the one-dimensional Shepp-Logan filter function (7).

$$H(\omega_x,\omega_y)\Big|_{|\omega_x| \le \omega_N,\ |\omega_y| \le \omega_N} = H(\omega,\phi) = \begin{cases} \dfrac{2\omega_N}{\pi} \sin \dfrac{\pi\omega}{2\omega_N}, & \omega < \omega_N \\ \dfrac{2\omega_N}{\pi}, & \omega > \omega_N \end{cases} \tag{29}$$

where ω_N is the Nyquist frequency corresponding to a sampled interval of the backprojection image. This filter function, therefore, acts as ω filter for small ω but is flat at all high frequency boundaries including the four corners of 2-D zonal filter. This 2-D zonal filter was found to be an excellent filter for BF image reconstruction algorithm. The overall reconstruction procedures in this case are as follows:

(a) Backprojection to get 2N x 2N blurred image.
(b) 2-D FFT of the full 2N x 2N data and filtering with the filter given in Eq. (29).
(c) 2-D inverse FFT and adopt N x N image in the central region.
(d) Finally normalization of the image with precalculated reference data such as uniform disc image.

2.2.3 Direct Fourier transform reconstruction (DFR)

Even though the FB algorithm is widely used because of its high image quality, it requires considerable computation, especially for the back-projection operation. One alternative algorithm is direct Fourier domain mapping of the Fourier transform of each projection data in polar coordinate and then two-dimensional inverse Fourier transform of the resultant Fourier domain data. This follows from the projection theorem which was presented in Eq. (6). Since the Fourier transform of the projection data set obtained by transaxial scanning will provide a radial data set of the 2-D Fourier domain object function, it is possible to form a set of polar rasters which represent the Fourier transform of the object function obtained by 1-D Fourier transform of the projection data. In DFR technique, therefore, the reconstruction procedures involved are the interpolation problem which is the conversion of data from polar form to Cartesian form and 2-D inverse Fourier transform of the interpolated data with FFT. This method is much faster than the FB algorithm, it requires, however, an accurate and fast polar to Cartesian coordinate interpolation technique and the accuracy of this interpolation is of critical importance to the quality of the images. Several authors have studied the interpolation scheme in order to obtain an image of comparable quality to that of the FB algorithm and have been successful (18-20).

Another important point to be considered in DFR technique is the computation time involved in the interpolation since 2-D interpolation usually requires a large amount of computational time. Solution to this problem has been suggested and its modified versions were proposed (18, 19). These interpolation schemes based on concentric square raster have an advantage that it can be operated in one-dimensional form to obtain the Cartesian raster suitable for the FFT. Although DFR technique with concentric square raster sampling appears attractive, it is not easily applicable to many practical systems such as XCT and PCT (positron CT) mainly due to the fixed detector spacing and sampling distance. It is found useful, however, in NMR (nuclear magnetic resonance) tomography where data appears in the Fourier domain and the sampling interval is adjustable by simply varying the magnitude of the x-y gradient fields.

2.3 Three-dimensional Image Reconstruction Algorithms for Complete Sphere

2.3.1 Line integral projection data

In the conventional reconstruction algorithm, projections of an object are taken transaxially and are used to reconstruct the object slice by slice. The stacked slices constitute the three-dimensional volume image of the object.

In emission tomography, where the radionuclides are injected into the patient, the photons are emitted isotropically in 4π directions. This slice stacking method for a volume object does not make full utilization of the photons since this scheme captures photons emitted only in the direction perpendicular to the (long) axis of the body which constitute only a small fraction of the entire emitted photons. Especially in emission tomography, all the emitted photons should preferably be collected using, e.g. complete spherical geometry. For image reconstruction with the data collected in 4π geometry, the direct true three-dimensional reconstruction (TTR) algorithm will be required for the maximum utilization of all the available photons.

Parallel beam mode (9,10) - Let us consider a spherical geometry which completely surrounds the object and assume that all the emitted photons are captured and rearranged into 2-D parallel data sets in 4π directions. The point spread function of this system is $1/r^2$ and the backprojected blurred function is given by

$$b(x,y,z) = f(x,y,z) *** \frac{1}{r^2}, \tag{30}$$

where *** indicates a three-dimensional convolution operator.

By 3-D Fourier transform of Eq. (30), blurred image function in spatial frequency domain will be obtained, i.e.,

$$\begin{aligned} B(\omega_x,\omega_y,\omega_z) &= F(\omega_x,\omega_y,\omega_z) \cdot F_3\left(\frac{1}{r^2}\right) \\ &= F(\omega_x,\omega_y,\omega_z) \cdot \frac{1}{\rho}, \end{aligned} \tag{31}$$

where $B = F_3\left(b\right)$, $F = F_3\left(f\right)$, $\rho = (\omega_x^2+\omega_y^2+\omega_z^2)^{\frac{1}{2}}$ and $F_3\left(\cdot\right)$ is the 3-D Fourier transform operator. Eq. (31) may be rewritten as

$$F(\omega_x,\omega_y,\omega_z) = \rho B(\omega_x,\omega_y,\omega_z) . \tag{32}$$

Finally the object function can be obtained by inverse Fourier transform of Eq. (31), i.e.,

$$f(x,y,z) = F_3^{-1}\left[\rho\right]*** \, b(x,y,z). \tag{33}$$

Eq. (4) (the three-dimensional version of Eq. (27) is essentially a 3-D convolution (deconvolution) of simple backprojected (blurred) image and this 3-D convolution process as well as backprojection operation would take an unusually long computation time.

As an alternative approach, the filtered backprojection method has been proposed. The 3-D version of the projection theorem states that the Fourier transform of the 2-D projection represents the plane data passing through the center in the frequency domain in the same direction as the projected plane of the object. It may be written as

$$\begin{aligned} F_2\left[p_{\theta,\phi}(x',z')\right] &= P_{\theta,\phi}(\omega_{x'},\omega_{z'}) \\ &= F(\omega_{x'},\omega_{z'};\theta,\phi), \end{aligned} \tag{34}$$

where ($\omega_{x'}$, $\omega_{z'}$) is the spatial frequency domain coordinates of (x',z') in the direction of (θ,ϕ). If the uniform planes which pass through the center are superimposed from all possible direction in 3-D space, the overlapped density function becomes $1/\rho$. Therefore, to obtain the 3-D object function in the Fourier domain, 2-D Fourier transform of each projection data set should be compensated with the factor ω before the superposition for the normalization. Thus image function in spatial frequency domain appears as,

$$F(\omega_x,\omega_y,\omega_z) = \frac{1}{4\pi}\int_0^{\pi} d\theta\sin\theta \int_0^{\pi} d\phi\left[\omega P_{\theta,\phi}(\omega_{x'},\omega_{z'})\delta(\omega_{y'})\right], \tag{35}$$

where $\omega = (\omega_{x'}{}^2+\omega_{z'}{}^2)^{1/2}$.

Image function can now be obtained by

$$\begin{aligned} f(x,y,z) &= F_3^{-1}\left[\frac{1}{2\pi}\int_0^{\pi} d\theta\sin\theta \int_0^{2\pi} d\phi\ \omega P_{\theta,\phi}(\omega_{x'},\omega_{z'})\delta(\omega_{y'})\right] \\ &= \frac{1}{4\pi^2}\int_0^{\pi} d\theta\sin\theta \int_0^{2\pi} d\phi\ F_2^{-1}\left[\omega P_{\theta,\phi}(\omega_{x'},\omega_{z'})\right] \\ &= \frac{1}{4\pi^2}\int_0^{\pi} d\theta\sin\theta \int_0^{2\pi} d\phi\{F_2^{-1}\left[\omega\right]**P_{\theta,\phi}(x',z')\}. \end{aligned} \tag{36}$$

Eq. (36) shows a 3-D FB algorithm which is in fact the backprojection of the filtered 2-D projection data in 3-D space. In this case the 2-D filter function or kernel is simply $F_2^{-1}(\omega)$.

<u>Cone beam mode</u> - As an extension of the fan beam mode, a three-dimensional image reconstruction with cone beam data can be formulated. A direct cone beam reconstruction algorithm has several advantages over the parallel beam mode, namely, improved sampling in the same detector dimension and elimination of the rebinning process in 3-D X-ray or positron tomography. There have been several attempts to provide 3-D image reconstruction with the cone beam mode (13,15), a practically implementable algorithm has not yet been developed, however. Even though the cone beam algorithm using filtered backprojection has been considered (13), the method requires space variant filtering which is practically impossible to implement due to the extremely lengthy computation time. In this section, we will introduce a volume image reconstruction approach (14) which is based on the fan beam algorithm.

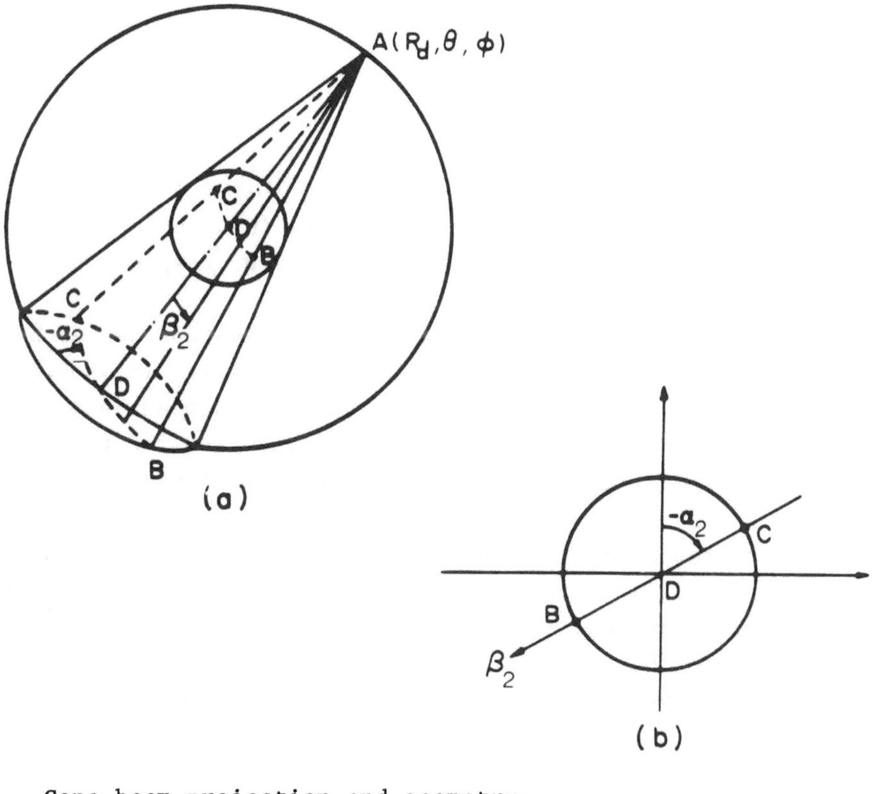

Fig. 2. Cone beam projection and geometry: (a) geometry and notations, (b) coordinates of projection data seen from cone apex toward center or detector space.

Cone beam scanning for a spherical geometry is shown in Fig. 2(a). Here, the cone apex is represented by (R_d, θ, ϕ) in conventional polar coordinates with ranges of ϕ and θ given by

$$0 \leq \phi < 2\pi$$

and

$$0 \leq \theta < \pi \quad . \tag{37}$$

Fig. 2(b) is a projection data plane of the cone seen from the cone apex A toward the center of the sphere and α_2 and β_2 are angular and radial components of the cone projection which represent detector positions. The ranges of α_2 and β_2 are

$$-\frac{\pi}{2} \leq \alpha_2 < \frac{\pi}{2} \tag{38}$$

and

$$-\beta_{2m} \leq \beta_2 \leq \beta_{2m},$$

where

$$\beta_{2m} = \sin^{-1} \frac{R_o}{R_d} \ . \tag{39}$$

R_o and R_d are the radius of the inner sphere in which an object is to be contained and the radius of the outer sphere, respectively.

A projection data set at a given view can now be represented by $P_{\theta,\phi}(\alpha_2, \beta_2)$ where θ and ϕ denote direction of cone axis. It should be noted that a data set in β_2-direction is one of the fan beam data sets in a given 2-D cone shown in Fig. 2(a) which is required for the reconstruction of a particular plane where the given fan belongs.

If planes are reconstructed and summed around an arbitrary vector $\underline{\Omega}_i$ as shown in Fig. 3(a) and (b), a 3-D volume image $f(r;\underline{\Omega}_i)$ can be constructed. Similarly, another 3-D volume image can be constructed for another vector $\underline{\Omega}_j$ and so on. Hence a 3-D volume image can be constructed by summation of the volume images over the entire 4π solid angle, i.e.,

$$f(\vec{r}) = \frac{\int_\Omega f(\vec{r};\underline{\Omega})\, d\underline{\Omega}}{\int_\Omega d\underline{\Omega}} \ , \tag{40}$$

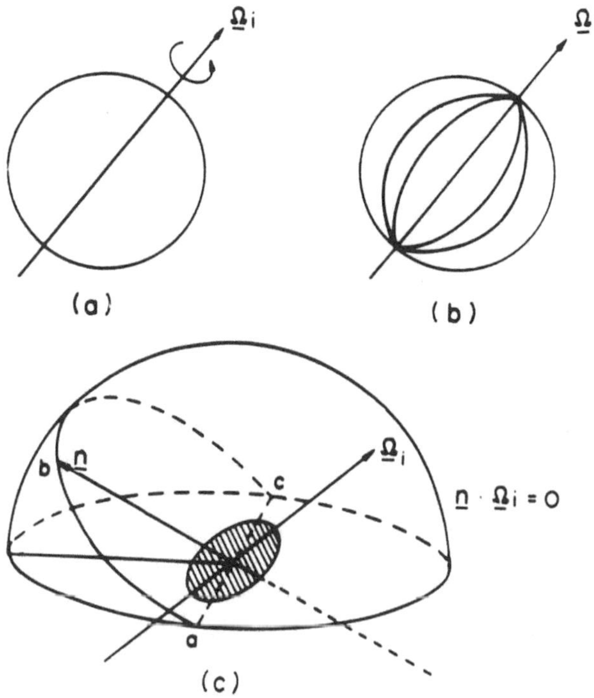

Fig. 3. Volume image construction for cone beam geometry: (a) a plane, (b) a sphere formed by many planes, (c) a plane (hatched area) among the planes of the sphere and its normal vector $\underline{n}$.

where $\int_{\Omega} d\underline{\Omega} = \int_0^{\frac{\pi}{2}} d\theta \sin\theta \int_0^{2\pi} d\phi = 2\pi$ is a normalization factor and Ω represents the hemisphere of the orientation vector.

A 3-D volume image $f(\vec{r};\underline{\Omega}_i)$, given above, is equivalent to a volume which is the summation of planes represented by all the normal vectors which are perpendicular to $\underline{\Omega}_i$ as shown in Fig. 3(b). The tips of these vectors are on the path $a \rightarrow b \rightarrow c$ as shown in Fig. 3(c). The final 3-D volume image $f(\vec{r})$ can be obtained by summation of planes represented by all normal vectors whose tips are on the hemisphere as shown in Fig.3(c). The 3-D volume image, then, can be written as

$$f(\vec{r}) = \frac{1}{2\pi} \int f_2(\vec{r};\underline{n})\, d\underline{n}\ , \tag{41}$$

where $f_2(\vec{r};\underline{n})$ is a 2-D image represented by a normal vector n whose tip is on the hemisphere. Then a 3-D volume can be constructed by summation of all 2-D planes and each 2-D plane $f_2(\vec{r};\underline{n})$ can be constructed with the fan beam algorithm.

Only one fan in each cone beam is used for the reconstruction of a plane, the other fan projection data of the cone will, later, contribute to the other 2-D planes which have a common vector $\underline{\Omega_i}$. Since only one view is provided from a cone for the reconstruction of a plane, other views necessary for the reconstruction of that plane have to be obtained by backprojection of the convolved and weighted backprojection of $P_{\theta,\phi}(\alpha_2,\beta_2)$ around α_2 with respect to β_2 (similar to the procedure used in the fan beam reconstruction algorithm).

Based on the above principles, a reconstruction formula for cone beam geometry can be derived. The fan data (view) from a cone is now processed in exactly the same way as the fan beam reconstruction procedure, i.e., weighted according to β_2 and convolved as

$$P^*_{\theta,\phi}(\alpha_2,\beta_2') = \int_{-\beta_{2m}}^{\beta_{2m}} P_{\theta,\phi}(\alpha_2,\beta_2)J(\beta_2)g(\beta_2'-\beta_2)d\beta_2, \tag{42}$$

where $g(\beta_2)$ and $J(\beta_2)$ are the same as the ones used in the fan beam reconstruction algorithm. If $P^*_{\theta,\phi}(\alpha_2, \beta_2)$ is backprojected with weight $\frac{1}{v_2{}^2}$ for all θ and ϕ, the result is a 3-D volume iamge, which consists of planes with each view having a unique normal vector for θ and ϕ. A 3-D volume image is then reconstructed by

$$f(\vec{r}) = \frac{1}{4\pi} \int_0^{2\pi} d\phi \int_0^{\pi} d\theta \sin\theta \; P^*_{\theta,\phi}(\alpha_2,\beta_2') \frac{1}{v_2{}^2} \tag{43}$$

where $P^*_{\theta,\phi}(\alpha_2,\beta_2')$ is given by Eq. (42), v_2 is the distance between cone apex (fan apex for each fan projection) and reconstruction point, and $\int_0^{2\pi} d\phi \int_0^{\pi} d\theta \sin\theta$ represents the backprojection with uniform solid angle sampling.

Combining Eq. (42) and Eq. (43), the reconstruction formula can be expressed as,

$$f(\vec{r}) = \frac{1}{4\pi} \int_0^{2\pi} d\phi \int_0^{\pi} d\theta \sin\theta \left(\int_{-\beta_{2m}}^{\beta_{2m}} P_{\theta,\phi}(\alpha_2,\beta_2)J(\beta_2)g(\beta_2'-\beta_2)\frac{1}{v_2^2}d\beta_2\right). \tag{44}$$

Eq. (44) is identical to Eq. (21) except for the addition of the back-projection operation over an angle θ Eq. (44) can also be written in in a simplified form as

$$f(\vec{r}) = B_{w_3} \left[\{p_{\theta,\phi}(\alpha_2,\beta_2)J(\beta_2)\} *g(\beta_2)\right] , \tag{45}$$

where B_{w_3} is a 3-dimensional weighted backprojection operator.

2.3.2 Plane integral projection reconstruction

Fourier NMR (nuclear magnetic resonance) imaging techniques (38-40) suggested the possibility of exciting the entire volume of an object and thereby obtaining planar integral data (11,49) with which an efficient whole volume image reconstruction can be made. Although this total volume excitation and planar data collection method is one of several existing methods in NMR imaging (48), it appears worth examining the algorithm for general 3-D image reconstruction. Another possible application of this algorithm is the 3-D ultrasound reflection tomography (35) where the reflection data as a function of time appears as the plane integrals of the reflectivity at the same distance from the source point.

One-dimensional plane integral projection data of the x'-y' plane at angles of θ and ϕ of a 3-D object function $f(x',y',z')$ can be given as

$$p_{\theta,\phi}(z') = \int_{-\infty}^{\infty}\int_{-\infty}^{\infty} f(x',y',z')\, dx'dy' \tag{46}$$

$$= \int_{-\infty}^{\infty}\int_{-\infty}^{\infty}\int_{-\infty}^{\infty} f(x,y,z)\delta(t-z')\, dxdydz, \tag{47}$$

where

$$t = x\sin\theta\cos\phi+y\sin\theta\sin\phi+z\cos\theta, \tag{48}$$

and (x',y',z') are the rotated Cartesian coordinates. By taking 1-D Fourier transform of $p_{\theta,\phi}(z')$ and using Eq. (47) and Eq. (48), one obtains

$$P_{\theta,\phi}(\rho) = \int_{-\infty}^{\infty} p_{\theta,\phi}(z')\exp\left[-i\rho z'\right]dz' \tag{49}$$

$$= \int_{-\infty}^{\infty}\int_{-\infty}^{\infty}\int_{-\infty}^{\infty} f(x,y,z)\exp\left[-i\rho(x\sin\theta\cos\phi+y\sin\theta\sin\phi+z\cos\theta)\right]dxdydz \tag{50}$$

$$= \int_{-\infty}^{\infty}\int_{-\infty}^{\infty}\int_{-\infty}^{\infty} f(x,y,z)\exp\left[-i(x\omega_x+y\omega_y+z\omega_z)\right] dxdydz \tag{51}$$

$$= F(\omega_x,\omega_y,\omega_z)\Big|_{\theta,\phi} , \tag{52}$$

where $F(\omega_x,\omega_y,\omega_z)$ is the image function in the spatial frequency domain and ρ is the radial spatial frequency. Eq. (52) is another 3-D projection theorem which states that the 1-D Fourier transform of the plane integral data at an angle θ and ϕ gives the one-dimensional radial frequency data in 3-D Fourier space. It is, therefore, possible to obtain spatial domain image function by direct inverse Fourier transform analogous to the method shown in section 2.3. We will, however, direct our attention to the filtered backprojection method. Recovery of the original volume image function $f(x,y,z)$ is then obtained by inverse Fourier transform of Eq. (52) as

$$f(x,y,z) = \frac{1}{8\pi^2}\int_{-\infty}^{\infty}\int_{-\infty}^{\infty}\int_{-\infty}^{\infty} F(\omega_x,\omega_y,\omega_z)\exp\left[i(\omega_x x+\omega_y y+\omega_z z)\right]d\omega_x d\omega_y d\omega_z \tag{53}$$

$$= \frac{1}{8\pi^2}\int_0^{\pi} d\theta\sin\theta \int_0^{2\pi} d\phi \left\{\tfrac{1}{2}\int_{-\infty}^{\infty} d\rho\ \rho^2 P_{\theta,\phi}(\rho)\exp\left[i\rho t\right]\right\}. \tag{54}$$

If the inverse Fourier transform of ρ^2 is denoted as $h_p(\cdot)$, Eq.(54) can be written as

$$f(x,y,z) = \frac{1}{8\pi^2}\int_0^{\pi} d\theta\sin\theta\int_0^{2\pi} d\phi\left(\int_{-\infty}^{\infty} dz'\ p_{\theta,\phi}(z')h_p(t-z')\right) . \tag{55}$$

The problem then becomes the selection of the filter kernel $h(z')$. In the 2-D image reconstruction, Shepp and Logan suggested a filter function for $|\omega|$ as

$$H(\omega) = \left|\frac{2}{a}\sin\frac{a\omega}{2}\right| , \tag{56}$$

which satisfies $|\omega|$ for small ω and assures the slope continuity at the boundary. Its spatial domain kernel is given by

$$h(ka) = -\frac{4}{\pi a^2(4k^2-1)} . \tag{57}$$

A similar approach can be taken for the filter function ρ^2, i.e.,

$$H_p(\rho) = \left(\frac{2}{a}\right)^2 \sin^2 \frac{a\rho}{2} . \tag{58}$$

Resultant space domain kernel $h_p(\cdot)$ is then given by

$$h_p(ka) = \begin{cases} \frac{2}{a^3} & ; \ k = 0 \\ -\frac{1}{a^3}; & |k| = 1 \\ 0 & ; \ k \neq 0 \text{ and } |k| \neq 1. \end{cases} \tag{59}$$

It is interesting to note that Eq. (59) and the second order differentiation approximation given by Shepp (11) are identical.

Although it appears awkward, this planar integral projection reconstruction concept can be applied to the 3-D line integral data by simply forming the planar integral data from the available line integral data sets.

3. ALGORITHM FOR GENERALIZED VOLUME IMAGE RECONSTRUCTION

3.1 Preliminary

The ideal detector configuration for emission tomography is a complex sphere. In this case, because of its spherical symmetry, the reconstruction slices (planes) can be oriented in any direction over a 4π solid angle, therefore, all the photons emitted in 4π directions can be fully utilized. Algorithms for this complete spherical geometry have been considered in section 2.3. Practical system design (22,51,54,55), however, prohibits the construction of such a configuration when one considers the elongated shape of the human body.

As a practical alternative, a truncated spherical configuration as shown in Fig. 4 can be considered. Parts of the surface of the sphere are removed along the body axis to adapt to the human shape. Although the sphere is truncated, this geometry still retains spherical symmetry in the remaining spherical surfaces and also retains possiblity of reconstructing true 3-D volume image even though it requires more complicated reconstruction algorithms than the ideal sphere. This 3-D reconstruction algorithm with limited solid angle has been examined by several authors, for example with two rotating opposite planar detector systems. Pelc considered this problem using a concept in which the measurement density in frequency space is made uniform on the basis of the projection theorem presented in Eq. (34) (51), while Colsher proposed a 3-D filter derived from a 3-D point spread function Eq.(52). These algorithms, however, are found to be insufficient to obtain quantitatively accurate images in some circumstances. Another drawback of these algorithms is the difficulty of reconstructing the image in a limited region due to the fact that these algorithms are based on the frequency domain processing which tends to treat the problem a in more general way.

To overcome these problems, a new true three-dimensional reconstruction algorithm, based on the superposition of all the 2-dimensionally reconstructed images in the space domain, has been considered. As will be discussed in the following sections, this algorithm is implemented by the convolution backprojection method with a new 2-D filter which varies according to the polar angle θ_p and results appear promising.

3.2 Theoretical Background

Let the radius of the detector sphere be R_d and the truncation angle be Θ_m in Fig. 4. Θ_m is the angle between the horizonal center and the truncated ring. An object which is to be reconstructed is contained within a sphere of radius R_o. In conventional multi-slice reconstruction the reconstructed volume image is obtained by the collection of 2-D projection data over a 2π azimuthal angle and the subsequent filtered backprojection of the data into the reconstruction space. In this case, all the projections are normal to the detector axis; thus, in this technique, only the annihilation photons normal to the detector axis are utilized for reconstruction. The image slice planes are, therefore, also normal to the detector axis. However, in the proposed spherical geometry, other orientations of slice planes can also be used for reconstruction since the spherical symmetry provides complete sets of 2-D projection data naturally oriented toward the center of the sphere as far as the detector surface covers the object. In these orientations, projection is parallel to the slice planes and thus annihilation photons emitted at angles which are not perpendicular to the detector axis are also utilized for reconstruction. The final 3-D image within the reconstruction sphere is obtained by summing all images reconstructed at each slice orientation in which a complete set of 2-D projection data is provided by the detected annihilation events and normalized by dividing by the number of slice orientations of the reconstructed image. Essentially, this is the basis of the reconstruction algorithm which has been developed for truncated spherical geometry. The derivation of the analytical reconstruction algorithm thus amounts to a complete enumeration of all possible slice orientations and the removal of computational redundancies in their numerous 2-D filtering operations. We will first derive a basic formulation for the continuous case, and later a practically implementable formula.

Let us set up some conventions and introduce some analysis necessary for derivation of algorithms for a truncated spherical configuration. Each slice orientation will be represented by a unit vector originating from the center of the sphere and normal to the slice planes, which is one set constituting a volume object. It should be remembered that reconstruction from these orientations utilizes projection data normal to these slice orientations which have complete sets of 2-D projection data are represented by a spherical cone as shown in Fig. 4. The polar angle limit of this cone is determined by the radii of both the detector and reconstruction spheres, i.e., R_d, R_o, and Θ_m, the truncation angle. If we denote this polar angle by Θ_u, it is given by

$$\Theta_u = \Theta_m - \cos^{-1}\frac{(R_d^{\,2}-R_o^{\,2})^{1/2}}{R_d} . \tag{60}$$

Angle Θ_m determines the emission solid angle utilized for reconstruction. This solid angle becomes an azimuthal strip bounded by $\pm\Theta_u$ around $\pi/2$ in polar angle since the photons utilized are perpendicular to the slice orientation vectors and there is azimuthal symmetry in the cone. The useful emission solid angle Ω_u is then given by

$$\Omega_u = 2\pi \int_{\pi/2-\Theta_u}^{\pi/2+\Theta_u} \sin\theta d\theta = 4\pi\sin\Theta_u \simeq 4\pi\sin n\Delta\Theta_u , \tag{61}$$

where $\Delta\Theta_u$ is the angle determined by the height of the single detector ring and n is the number of detector ring layers.

This solid angle should be compared with the solid angle subtended by the strip of single detector ring to appreciate the increase of photon

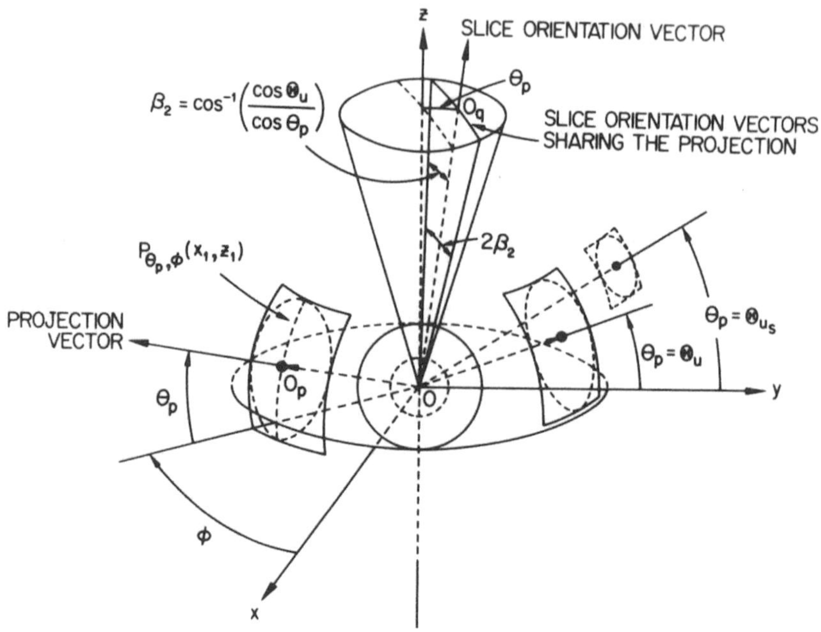

Fig. 4. Basic geometry of the truncated spherical positron camera system from which the original TTR algorithm is derived.

utilization associated with the proposed spherical geometry.

It would be worth noting a few words on the emission solid angle strip utilized for reconstruction in relation to the solid angle for annihilation photon detection at each point within the reconstruction sphere. In Fig. 4, the detection solid angle for annihilation line events from a point source is largest at the spherical center with $4\pi \sin\Theta_m$, and gradually decreases to a minimum of $4\pi \sin\Theta_u$ along the radius within the reconstruction sphere. Thus, the utilization solid angle strip bounded by $\pm\Theta_u$ around the polar angle $\pi/2$ is a minimum common solid angle or universal to every point in the reconstruction sphere. Because of this fact, it is termed as universal solid angle strip. Note also that this same solid angle is used in the space-invariancy requirement of a point source response function in the backprojection and deconvolution algorithm given in reference (51). The fact that we can utilize only the universal solid angle portion of the detection solid angle at each point is a direct consequence of object size and spherical truncation. The object size in a given truncation restricts the region of slice orientation and thereby the annihilation solid angle Θ_u useful for reconstruction. The impact of this can be better appreciated in Fig. 4 by comparing Θ_u with Θ_{u_s} for a smaller object.

3.3 Basic Formulation

We can now embark on the derivation of reconstruction algorithm for truncated spherical geometry. As stated previously, the final 3-D image is obtained by summation of 3-D images reconstructed at each slice orientation and normalizing it with number of slice orientations. All possible slice orientations for reconstruction are contained in the spherical cone or orientation vectors shown in Fig. 4. If we represent an orientation vector within a given spherical cone by $\underline{\Omega}_p = (\theta, \phi)$ and the 3-D image reconstructed at that orientation by $f(\vec{r};\underline{\Omega}_p)$, the final 3-D image, $f(x,y,z)$ or $f(\vec{r})$, can be represented by

$$f(\vec{r}) = \frac{\int_{\Omega_p} d\underline{\Omega}_p f(\vec{r};\underline{\Omega}_p)}{\int_{\Omega_p} d\underline{\Omega}_p} = \frac{\int_0^{2\pi} d\phi \int_0^{\Theta_u} d\theta f(\vec{r};\underline{\Omega}_p)\sin\theta}{2\pi(1-\cos\Theta_u)} , \tag{62}$$

where Ω_p represents the spherical cone of the slice orientation vector. The 3-D image reconstructed at slice orientation $\underline{\Omega}_p$ is obtained by filtered backprojection of the 2-dimensional projection data projected normal to the orientation vector $\underline{\Omega}_p$ and represented by

$$f(\vec{r};\underline{\Omega}_p) = \frac{1}{2\pi} \int_0^{\pi} d\phi' \left[p_{\phi}'(x',z';\underline{\Omega}_p) ** h(x',z')\right] , \tag{63}$$

where (x',z') is a rotated coordinate system in which the z'-axis is the direction of $\underline{\Omega}_p$ and the x'-axis is rotated by ϕ' in the plane normal to $\underline{\Omega}_p$, $p_{\phi'}(x',z';\underline{\Omega}_p)$ is a projection at azimuthal angle ϕ' normal to $\underline{\Omega}_p$, $h(x',z')$ is a 2-D convolution kernel and $**$ represents the two-dimensional convolution operation. The convolution kernel is an extended 2-D ramp filter given in the frequency domain by

$$H(\omega_{x'},\omega_{z'}) = |\omega_{x'}| . \tag{64}$$

One can easily see that

$$h(x',z') = F_2^{-1}\left(|\omega_{x'}|\right) \delta(z') ,$$

where F_2^{-1} and F_1^{-1} are the inverse Fourier transform operators of two and one-dimension, respectively.

Eq. (62) is then given by

$$f(\vec{r}) = \frac{\int_{\Omega_p} d\underline{\Omega}_p \frac{1}{2\pi} \int_0^{\pi} d\phi' p_{\phi'}(x',z';\underline{\Omega}_p)**h(x',z')}{2\pi(1-\cos\Theta_u)}$$

$$= \frac{\int_{\Omega_p} d\underline{\Omega}_p \int d\phi' \; p^*_{\phi'}\;(x',z':\underline{\Omega}_p)}{4\pi^2(1-\cos\Theta_u)} \tag{65}$$

where $p^*_{\phi'}(x',z':\underline{\Omega}_p)$ is the 2-D filtered projection data.

According to Eq. (65), spatial filtering and backprojection can be performed for image reconstruction at each orientation and then finally summed up for all possible slice orientations, i,e., over the entire spherical cone Ω_p as the equation suggests. This computational route requires approximately $N_\phi \times N_\phi \times N_\theta$ time of two-dimensional convolution and backprojection where N_ϕ and N_θ are the number of azimuthal and polar samplings. This method of reconstruction also requires additional 2-D linear interpolation of projection data. With careful observation, however, one notices that some projection data shared by other slice orientations and thus the filtering process of the projection data for these slice orientations can be combined. This is possible because integration is a linear operation. This fact allows us to carry out all filtering operations for a given 2-D projection data set in a single step and thus cuts the computing time by N_ϕ.

3.4 Formulation of a Practically Implementable Algorithm

For the implementation of the algorithm discussed previously, it is necessary to identify all the slice orientations involved in each set of projection data and generate the corresponding composite filter function. In fact, each 2-D parallel projection data set corresponding to a certain object size has a fixed number of slice orientations. This allows us to treat a certain projection data set in a unified way, i.e., 2-D projection data set can be processed with one filter function. Let us consider a 2-D projection set projected in one direction parallel to the line $\overrightarrow{OO_p}$ in Fig. 4. Here O_p is the center of the projection plane. As it is shown, this projection data is shared by slice plane orientations rotated around the line $\overrightarrow{OO_p}$ from $-\beta$ to β. The corresponding orientation vectors are represented by the intersection line shown on the surface of the spherical cone of orientation vectors shown in Fig. 4. This intersection line, therefore, indicates the number of slice orientations, e.g., the number of slice orientations is maximum at $\theta_p = 0$, and minimum at $\theta_p = \Theta_u$. The limit of rotation angle is, therefore, a function of projection direction or polar angle θ_p since the slice orientation vectors generated by rotation should stay within the spherical cone of orientation vectors for reconstruction.

The rotation angle limit β for a general projection inclined at an angle θ_p from the equatorial plane as shown in Fig. 4 is given by

$$\beta = \cos^{-1}(\cos\Theta_u/\cos\theta_p) . \tag{66}$$

The number of slice orientations which commonly share the projection data set $p_{\theta_p,\phi}$ projected in parallel to projection vector $\overrightarrow{OO}_p$ at angle ϕ and θ_p are fixed for a given geometry, i.e., a given object size and detector sphere. By modification of the projection data $p_{\phi'}(x',z':\underline{\Omega}_p)$ to $p_{\theta_p,\phi}(x_1,z_1)$ by rearrangement of the data set and introduction of the composite filter function concept, Eq. (65) can be rewritten as

$$f(\vec{r}) = \frac{\int_0^{\pi} d\phi \int_{-\Theta_u}^{\Theta_u} d\theta_p \cos\theta_p \int_{\beta}^{\beta} d\alpha \left[p_{\theta_p,\phi}(x_1,z_1) ** h(x_1,z_1:\alpha)\right]}{\int_0^{2\pi} d\phi \int_{-\Theta_u}^{\Theta_u} d\theta_p \cos\theta_p \int_{-\beta}^{\beta} d\alpha} , \tag{67}$$

where $p_{\theta_p,\phi}(x_1,z_1)$ is the parallel projection data set at polar angle θ_p and azimuthal angle of ϕ and $h(x_1,z_1:\alpha)$ is the filter function of projection data $p_{\theta_p,\phi}$ at a slice orientation corresponding to the rotation angle variable α and rotated form of conventional 2-D filter with angle α. Here the z_1 axis is in the direction of the line $\overrightarrow{OO}_q$ and the x_1 axis is normal to $\overrightarrow{OO}_p$ and $\overrightarrow{OO}_q$ as shown in Fig. 4.

The filter kernel $h(x_1,z_1:\alpha)$ in Eq. (67) is then given as

$$h(x_1,z_1:\alpha) = F_2^{-1}\left[H(\omega_{x_1},\omega_{z_1}:\alpha)\right] ,$$

and

$$H(\omega_{x_1},\omega_{z_1}:\alpha) = |\omega_{x_2}| = \omega|\cos(\xi-\alpha)| , \tag{68}$$

where ω_{x_2} is the rotated axis from ω_{x_1} with angle α and (ω,ξ) are the polar coördinates of $(\omega_{x_1},\omega_{z_1})$.

The denominator of Eq. (67) is the normalizing factor which represents the sum of all the weighting coefficients of projection data sets. It is interesting to note that this normalizing factor is equal to $4\pi^2(1-\cos\Theta_u)$ and is proportional to the solid angle of the spherical cone of the slice orientation vector.

The projection data set $P_{\theta_p,\phi}(x_1,z_1)$ is independent of α and convolution is a linear operation, therefore, Eq. (67) can be further simplified to

$$f(\vec{r}) = \frac{\int_0^{\pi} d\phi \int_{-\Theta_u}^{\Theta_u} d\theta_p \cos\theta_p \left[P_{\theta_p,\phi}(x_1,z_1)** \int_{-\beta}^{\beta} d\alpha\, h(x_1,z_1:\alpha)\right]}{4\pi^2(1-\cos\Theta_u)}$$

$$= \frac{\int_0^{\pi} d\phi \int_{-\Theta_u}^{\Theta_u} d\theta_p \cos\theta_p\, F_2^{-1}\left(P_{\theta_p,\phi}(\omega_{x_1},\omega_{z_1})H_{\theta_p}(\omega_{x_1},\omega_{z_1})\right)}{4\pi^2(1-\cos\Theta_u)} \tag{69}$$

The composite filter function $H_{\theta_p}(\omega_{x_1},\omega_{z_1})$ is given by

$$\begin{aligned} H_{\theta_p}(\omega_{x_1},\omega_{z_1}) &= F_2\left(\int_{-\beta}^{\beta} d\alpha\, h(x_1,z_1:\alpha)\right) \\ &= \int_{-\beta}^{\beta} d\alpha\, F_2\left(h(x_1,z_1:\alpha)\right) \\ &= \int_{-\beta}^{\beta} d\alpha\, \omega|\cos(\xi-\alpha)| \\ &= H_{\theta_p}(\omega,\xi)\;. \end{aligned} \tag{70}$$

Further examination reveals that the composite filter function $H_{\theta_p}(\omega,\xi)$ in $0 \leq \xi \leq \pi/2$ is given by

$$H_{\theta_p}(\omega,\xi) = \begin{cases} 2\omega\cos\xi\sin\beta & 0 \le \xi \le \frac{\pi}{2} - \beta \\ 2\omega(1-\sin\xi\cos\beta) & \frac{\pi}{2} - \beta < \xi \le \frac{\pi}{2} . \end{cases} \tag{71}$$

This composite filter function $H_{\theta_p}(\omega,\xi)$ is symmetrical about ω_{x_1} and ω_{z_1} axes. It is interesting to note that the reconstruction formula given in Eq. (69) leads to the conventional 2-D reconstruction algorithm as well as to 3-D volume image reconstruction algorithm for complete sphere as will be discussed in the next section.

3.5 Generality of the TTR Algorithm

Let us first consider the conventional 2-D convolution reconstruction algorithm for multiple slices which are normal to the long axis. This can be derived from Eq. (69) by assuming $\Theta_u \to 0$. Eq. (69) then becomes

$$f(\vec{r}) = \frac{\int_0^{\pi} d\phi \int_{-\Theta}^{\Theta_u} d\theta_p \cos\theta_p\, F_2^{-1}\left(P_{\theta_p,\phi}(\omega_{x_1},\omega_{z_1})2\beta|\omega_{x_1}|\right)}{2\pi\int_{-\Theta_u}^{\Theta_u} d\theta_p 2\beta\cos\theta_p} \tag{72}$$

where $2\beta|\omega_{x_1}|$ is derived from Eq. (68), i.e.,

$$\begin{aligned} H_{\theta_p}(\omega_{x_1},\omega_{z_1}) &\simeq 2\omega\sin\beta\cos\xi \\ &= 2|\omega_{x_1}|\sin\beta \\ &= 2|\omega_{x_1}|\beta, \text{ for small } \beta. \end{aligned} \tag{73}$$

Since $\theta_p \to 0$ and $P_{\theta_p,\phi}(\omega_{x_1},\omega_{z_1})$ is no longer a function of θ_p, by changing the order of integration in Eq. (72) it can be rewritten as

$$f(\vec{r}) = \frac{\int_0^{\pi} d\phi F_2^{-1}\left(P_{\theta_p=0,\phi}(\omega_{x_1},\omega_{z_1})\ |\omega_{x_1}|\right) \int_{-\Theta_u}^{\Theta_u} d\theta_p 2\beta\cos\theta_p}{2\pi \int_{-\Theta_u}^{\Theta_u} d\theta_p 2\beta\cos\theta_p}$$

$$= \frac{1}{2\pi} \int_0^{\pi} d\phi F_2^{-1} \left(P_{\theta_p=0,\phi}(\omega_{x_1},\omega_{z_1})|\omega_{x_1}|\right) . \tag{74}$$

Eq. (74) is a well known convolution or filtered backprojection with 2-D projection data $P_{\theta_p=0,\phi}(\omega_{x_1},\omega_{z_1})$. Eq. (74) can also be written in convolution form as

$$f(\vec{r}) = \frac{1}{2\pi} \int_0^{\pi} d\phi \left(P_{\theta_p=0,\phi}(x_1,z_1)**h(x_1)\delta(z_1)\right)$$

$$= \frac{1}{2\pi} \int_{-\infty}^{\infty} dz_1\ \delta(z_1-z) \int_0^{\pi} d\phi \left(P_{\theta_p=0,\phi}(x_1,z_1)*h(x_1)\right) \tag{75}$$

where

$$h(x_1) = F_1^{-1}\left(|\omega_{x_1}|\right). \tag{76}$$

Here * and ** denote one and two-dimensional convolution operations, respectively. Eq. (75) is essentially a 1-D convolution process stacked in the z direction.

It can also be shown that the Eq. (69) would lead to the true three-dimensional volume reconstruction algorithm for an ideal spherical detector array. In this case, we simply replace the Θ_u with $\frac{\pi}{2}$ so that

$$\int_{-\Theta_u}^{\Theta_u} d\theta_p \cos\theta_p = \int_0^{\pi} d\theta \sin\theta \tag{77}$$

and

$$\int_{-\Theta_u}^{\Theta_u} d\theta_p \left(2\beta\cos\theta_p\right) = 2\pi . \tag{78}$$

From Eq. (68) it can be shown that $H_{\theta_p}(\omega_{x_1},\omega_{z_1})$ becomes 2ω. Eq. (68) then can be written as

$$f(\vec{r}) = \frac{1}{2\pi^2} \int_{-\frac{\pi}{2}}^{\frac{\pi}{2}} d\theta_p \cos\theta_p \int_0^{\pi} d\phi F_2^{-1} \left(P_{\theta_p,\phi}(\omega_{x_1},\omega_{z_1})\omega\right) , \tag{79}$$

or

$$f(\vec{r}) = \frac{1}{4\pi^2} \int_0^{\pi} d\theta \sin\theta \int_0^{2\pi} d\phi F_2^{-1} \left(P_{\theta,\phi}(\omega_{x_1},\omega_{z_1})\omega\right) . \tag{80}$$

Eq. (80) is equivalent to Eq. (36) which is given by the convolution form, i.e.

$$f(\vec{r}) = \frac{1}{4\pi^2} \int_0^{\pi} d\theta \sin\theta \int_0^{2\pi} d\phi \{P_{\theta,\phi}(x_1,z_1) ** F_2^{-1}\left(\omega\right)\} . \tag{81}$$

4. EXTENDED TTR ALGORITHM FOR VOLUME IMAGING

4.1 Preliminary

In the previous section, we derived a generalized TTR (True Three-dimensional Reconstruction) algorithm which is applicable to direct true three-dimensional volume image reconstruction.

In the development of the previous TTR algorithm, attempts have been made to utilize as many oblique rays as possible to enhance the statistical quality of the reconstructed image. Limitation of the method, however, was that the maximum size of the object which can be constructed was limited by the universal angle Θ_u or band of the detector array.

Rays obtained at an angle larger than Θ_u are, therefore, discarded since they cannot be used to form the 2-dimensional parallel ray set which covers the entire object. Oblique rays with angles larger than Θ_u therefore, cannot be utilized and this loss becomes larger as the object size increases as shown in Fig. 4. In this section, we will attempt to utilize more oblique rays by use of an image superposition concept. This new algorithm, which we have termed the extended TTR (ETTR), is found to be useful in many applications where finite photon statistics and large object size, such as the elongated shape of the human body, are the limiting factors in the reconstruction of images (22,50,51) (e.g. the case with a spherical shaped positron tomograph (54,55).

4.2 Theoretical Analysis

Let us now consider a cylindrical object which extends above and below the truncated sphere detector array as shown in Fig. 5. In this situation, all the horizontal rays can be used for conventional 2-dimensional transaxial image reconstruction, therefore, the reconstructed region of maximum size is the region of the cylinder bounded by the two planes, tangent to the respective two circular apertures of the truncated spherical array. For this situation, according to the previously discussed TTR algorithm, only the horizontal rays, excluding all the oblique rays, were useful for the reconstruction.

It should be remembered, however, that the underlying basic concept of TTR algorithm in section 3 is based on the plane reconstruction principles which are the same as conventional 2-D transaxial image reconstruction from 1-D projections. In fact, with this TTR algorithm all the oblique rays with an angle as large as Θ_m (see Fig. 5) can be used for reconstruction of planes, e.g. even the line integral $\int f dL$ shown in Fig. 5 can be used for the construction of the plane which is tilted with an angle Θ_m. This is the most plausible conclusion one can draw from this TTR algorithm. Since the inclined planes can be perfectly reconstructed, image reconstruction can be extended up to the limit where the reconstructed image is virtually a single slice. The region of the cylinder within the geometry, then, can be reconstructed by several different steps, with each reconstruction step being a different region depending on the degree of oblique ray utilization. The shapes of these regions are similar to the folded cones and the volume decreases gradually as the number of used oblique rays increases, since the size of the 2-D projection data decreases as the polar angle Θ_p increases (see Fig. 5). Obviously, the region reconstructed by use of the oblique rays of the largest Θ_u is a point at the center and it is the result of the superposition of all the reconstructed planes having orientation angle from $-\Theta_m$ to Θ_m.

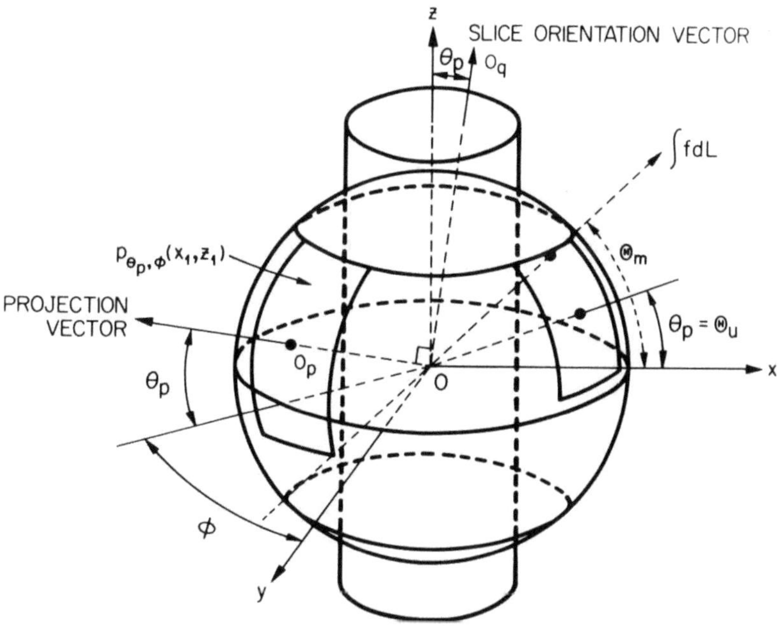

Fig. 5. Basic geometry of the truncated spherical positron camera system for the extended TTR algorithms.

Image reconstruction, therefore, may start from the rays with the smallest value of Θ_u, i.e., $\Theta_u = 0^o$ (which is equivalent to the conventional 2-D transaxial image reconstruction) and proceeds toward larger tilted angles by steps. After completion of the image reconstruction by several steps up to the largest Θ_u, the regions of poorer statistics are replaced with better ones to form volume image which have improved statistics toward the center. It should be noted, however, that all of the regional images utilizing different angular ranges of oblique rays are all correctly reconstructed images which differ only in their statistical quality of image and in sizes. In practice a two-step image reconstruction may suffice, i.e., reconstruction of an outer region which has relatively poor S/N but is acceptably large in size and then reconstruction of a smaller region near the central part with higher S/N. The high S/N region (image) will be the region of interest while the other poor S/N region (image) could be used as an auxiliary information. Synthesis of two or more different S/N images for the visualization of the quantitatively superior image at the interested region is the central theme of the extended TTR algorithm which allows us to exploit maximum sensitivity of the system, e.g. in the spherical positron tomograph (S-PET) where a large detector solid angle is provided for the central region of the object.

In this way, depending on the size of the region of interest, all or almost all of the oblique rays are effectively used for the reconstruction of the image and hence the statistically superior image can be obtained without loss of available projection data in the truncated spherical detector array.

For example, a simple two-step reconstruction can be written as,

$$f(\vec{r}) = \frac{1}{2\pi} \int_0^{\pi} d\phi F_2^{-1} \left. P_{\theta_p=0,\phi}(\omega_{x_1},\omega_{z_1}) |\omega_{x_1}| \right) \Big|_{\text{outer part of region of interest}}$$

$$+ \left. \frac{\int_0^{\pi} d\phi \int_{\Theta_u}^{\Theta_u} d\theta_p \cos\theta_p F_2^{-1} \left(P_{\theta_p,\phi}(\omega_{x_1},\omega_{z_1}) H_{\theta_p}(\omega_{x_1},\omega_{z_1}) \right)}{\int_0^{2\pi} d\phi \int_{-\Theta_u}^{\Theta_u} d\theta_p \, 2\cos\theta_p} \right|_{\text{region of interest}} , \qquad (82)$$

where Θ_u is the maximum angle of used rays determined by the region of interest.

The first part of Eq. (82) represents the low S/N ratio image or region reconstructed by stacking parallel image planes where each plane is obtained from conventional 2-D transaxial reconstruction, while the second part represents the high S/N ratio image region using all the available oblique rays. Since the volume of interest can sometimes be a small limited region, we can locate the region of interest at the center of the system and then only the second term in Eq. (82) can be used. This "focusing" concept used for the high sensitivity imaging is an interesting and important property that the ETTR algorithm uniquely possesses.

The potential benefit of this type of extended TTR algorithm can not be overemphasized, for example, in positron tomography, where 4π or near 4π detection of annihilation photons is most desirable but the practical difficulties in building such an ideal spherical detection system limit the system to a form of truncated spherical shape similar to the one illustrated in Fig. 5.

5. CONCLUSION

2-D and 3-D image reconstruction algorithms of various types are reviewed and their merits and shortcomings are discussed.

Direct three-dimensional volume image reconstruction algorithms with projections, i.e., available line or plane integral projection data are discussed in contrast to the conventional 2-D slice stacking volume image reconstruction. After a brief discussion of the total 3-D volume image reconstruction of a simple case, i.e., ideal spherical geometry, we have introduced a new generalized true three-dimensional reconstruction (TTR) algorithm and its extended version ETTR.

This new approach to true three-dimensional reconstruction is applicable to a truncated spherical geometry, therefore, to an elongated object such as a human body. Computer simulations were performed to verify the usefulness of the algorithm in real situations. It is found that the proposed algorithm provides accurate image reconstruction as well as improved image quality resulting from the full utilization of all the available projection data or rays, e.g. in a positron tomograph system where positron annihilation photons are emitted in 4π space. It is also shown that the new TTR algorithm presented in this chapter appears, indeed, as a general three-dimensional reconstruction algorithm which could be applicable to the geometries ranging from a complete ideal sphere to a single slice.

Preliminary computer simulation results indicate the potential of the algorithm and further simulation studies of various geometries are under progress and will be reported elsewhere.

APPENDIX

Derivative Weighting Function

For image reconstruction with Eq. (69), the selection of a weighting function for the composite filter H_{θ_P} is found to be an important factor. In an attempt to reconstruct a 2-dimensional plane or 3-dimensional volume image using 2-D data processing, a weighting function which ensures slope continuity at the boundary of the filter function in two-dimensional frequency space is found to be necessary to avoid spurious ringing in the reconstructed image. Unlike the one-dimensional weighting function (or window function) the selection of the two-dimensional weighting function requires some attention. In reconstructive image processing with BF in section 2.2.2, 2-dimensional filtering usually results in boundary slope discontinuity in 2-D space. It is precisely this discontinuity which produces undesirable ringing. To avoid this ringing problem, it has been common practice to sacrifice the resolution by applying excess weighting in the high frequency region. In the previous backprojection filtering reconstruction we have introduced an optimally weighted 2-dimensional ω filter which enables us to reconstruct accurate 2-D image. Although slope continuity in the simple ω filter is easy to realize (see Eq.(29)), it was found that difficulty exists in the case of the composite 2-D filter where slope continuity at the boundaries is not always assured. For this purpose an optimum weighting function which ensures the continuity at the 2-D boundaries was sought. One way of producing a smooth and continuous composite filter function at the boundaries is to generate a weighting function whose multiplication with the unweighted composite filter function is continuous at all boundaries. A 2-dimensional weighting function which satisfies such a condition can be written as

$$\begin{aligned} W_2(\omega_x,\omega_z) &= W_1(\omega_x)\, W_1(\omega_z) \\ &= \left[\frac{1}{2}\{1+\cos a(\omega_z)\omega_x\}\right]\left[\frac{1}{2}\{1+\cos b(\omega_x)\omega_z\}\right] , \end{aligned} \tag{A-1}$$

where $a(\omega_z)$ and $b(\omega_x)$ are the coefficients which make the derivatives of the products $W_1(\omega_x)H_{\theta_p}(\omega_x,\omega_z)$ and $W_1(\omega_z)W_1(\omega_x)H_{\theta_p}(\omega_x,\omega_z)$ equal to zero, i.e., continuous at the boundaries. The coefficient $a(\omega_z)$ and $b(\omega_x)$ are found numerically for each $H_{\theta_p}(\omega_x,\omega_z)$ by taking partial derivatives, i.e.

$$\frac{\partial}{\partial\omega_x} W_1(\omega_x)\, H_{\theta_p}(\omega_x,\omega_z)\Bigg|_{\omega_z = \text{const}} = 0, \quad \text{for } a(\omega_z) \text{ at a given } \theta_p \tag{A-2}$$

and

$$\frac{\partial}{\partial\omega_z} W_1(\omega_x;\text{fixed a})\, W_1(\omega_z)\, H_{\theta_p}(\omega_x,\omega_z)\Bigg|_{\omega_x = \text{const}} = 0, \quad \text{for } b(\omega_x) \text{ at a given } \theta_p . \tag{A-3}$$

This derivative weighting function is found to be effective in reducing the ringing artifact in the reconstructed image without impairing the image resolution.

ACKNOWLEDGEMENTS

Present paper represents works of many students and colleagues at both Imaging System Science Laboratory at KAIS and Imaging Center at Columbia University. Authors would like to thank those who have contributed directly and indirectly to make this paper possible, especially to H.B. Min, J.C. Jeong and H.B. Song at KAIS, and C.B. Lim at Technicare. Finally authors would like to thank D.J.R. Cumming for critical review of the paper.

REFERENCES

1. Hounsfield GN (1972). A method of and apparatus for examination of a body by radiation such as X-ray of gamma radiation. U.K. Patent 1283915.

2. Cho ZH (1974). General views on 3-D image reconstruction and computerized transverse axial tomography. IEEE Trans Nucl Sci NS-21: 44-71.

3. Budinger TF and Gullberg GT (1974). Three-dimensional reconstruction in nuclear medicine emission imaging. IEEE Trans Nucl Sci NS-21:2-20.

4. Herman GT (ed) (1979). Topics in applied physics. (In) Image Reconstruction from Projections, Vol. 32, Springer-Verlag, New York.

5. Barett HH and Swindell W (1977). Analog reconstruction methods for transaxial tomography. Proc IEEE 65:89-111.

6. Ramachandran GN and Lakshiminarayanan AV (1971). Three-dimensional reconstruction from radiographs and electron micrographs: Application of convolutions instead of Fourier transform. Proc Natl Acad Sci(US) 68:2236-2240.

7. Shepp LA and Logan BF (1974). The Fourier reconstruction of a head section. IEEE Trans Nucl Sci NS-21:21-43.

8. Herman GT, Lakshiminarayanan AV and Naparstek A. Convolution reconstruction technique for divergent beams. Comput Biol Med 6:259-271 (1976), and also Lakshiminarayanan AV: Reconstruction from divergent ray data. Technical Report 92, Department of Computer Science, State University of New York at Buffalo (1975).

9. Nalcioglu O and Cho ZH (1978). Reconstruction of 3-D objects from cone beam projections. Proc IEEE 66:1584-1586.

10. Pelc NJ and Chesler DA (1979). Utilization of cross-plane rays for three-dimensional reconstruction by filtered backprojection. J Comput Assist Tomogr 3:385-395.

11. Shepp LA (1980). Computerized tomography and nuclear magnetic resonance. J Comput Assist Tomogr 4:94-107.

12. Cho ZH, Tsai CM and Ahn I (1976). Pseudo fan beam reconstruction algorithms and their physical problems. IEEE Trans Nucl Sci NS-23: 560-567.

13. Denton RV, Friedlander B and Rockmore AJ (1979). Direct three-dimensional image reconstruction from divergent rays. IEEE Trans Nucl Sci NS-26:4695-4703.

14. Min HB (1982). Three-dimensional volume image reconstruction with cone beam projections. ISS LAB Report No. 3, KAIS, Korea.

15. Altschuler MD, Herman GT and Lent A (1978). Fully three-dimensional image reconstruction from cone-beam sources. IEEE Conference on Pattern Recognition and Image Processing, Chicago, Illinois.

16. Schlindwein M (1978). Iterative three-dimensional reconstruction from twin-cone beam projections. IEEE Trans Nucl Sci NS-25:1135-1143.

17. Budinger TF and Gullberg GT (1975). Reconstruction by 2-dimensional filtering of simple superposition transverse-section image. (In) Digest of the Optical Society Meeting on 2-D and 3-D Reconstruction: ThA 9-1-ThA 9-4 WB6, held August 4, 1975 at Stanford University, Palo Alto, California.

18. Jeong JC, Song HB and Cho ZH (1982). Direct Fourier transform 3-D image reconstruction with modified concentric square sampling technique. ISS LAB Report No. 3, KAIS, Korea, and submitted for publication in IEEE Trans Med Imaging.

19. Mersereau RM and Oppenheim AV (1974). Digital reconstruction of multi-dimensional signals from their projections. Proc IEEE 62: 1319-1337.

20. Stark H, Woods JW, Paul I and Hingorani R (1981). An investigation of computerized tomography by direct Fourier inversion and optimum interpolation. IEEE Trans Bio Eng BME-28:496-505.

21. Ra JB and Cho ZH (1981). Generalized true three-dimensional reconstruction algorithm. Proc IEEE 69:668-670.

22. Ra JB, Lim CB, Cho ZH, Hilal SK and Correll J (1982). A true three-dimensional reconstruction algorithm for the spherical positron emission tomograph. Phys Med Biol 27:37-50.

23. Cho ZH, Ra JB and Hilal SK. True three-dimensional reconstruction (TTR) - Application of algorithm toward full utilization of oblique rays. Paper invited by IEEE Trans on Medical Imaging.

24. Budinger TF and Gullberg GT (1974). Lawrence Berkeley Lab Report LBL-2146.

25. Wernecke SJ and D'addario LR (1977). Maximum entropy image reconstruction. IEEE Trans Comput C-26:351-364.

26. Herman GT and Lent A (1976). Iterative reconstruction algorithms. Comput Biol Med 6:273-294.

27. Ter-Pogossian MM, Phelps ME, Hoffman EJ, et al (1975). A positron emission transaxial tomograph for nuclear imaging (PETT). Radiology 114:89-98.

28. Phelps ME, Hoffman EJ, Huang SC and Kuhl DE (1978). ECAT: A new computerized tomographic imaging system for positron emitting radiopharmaceuticals. Nucl Med 19:625-647.

29. William C, Crabtree MC and Burgiss SG (1979). Design and performance characteristics of a positron emission computed axial tomograph ECAT-II. IEEE Trans Nucl Sci NS-26:619-627.

30. Cho ZH, Chan J and Eriksson L (1976). Circular ring transverse axial positron camera (CRTAPC) for 3-D reconstruction of radionuclides distribution. IEEE Trans Nucl Sci NS-23:613-622.

31. Budinger TF, Derenzo SE, Gullberg GT, et al (1977). Emission computer assisted tomography with single photon and positron annihilation photon emitters. J Comput Assist Tomogr 1:131-145.

32. Ter-Pogossian MM, Mullani NA, Hood JT, Higgins CS and Curie CM (1978). A multislice positron emission computerized tomograph (PETT IV) yielding transverse and longitudinal images. Radiology 28:477.

33. Ter-Pogossian MM, Mullani NA, Hood JT, Higgins CS and Picke DC (1978). Design consideration for a positron emission transverse tomograph (PETT V) for imaging of the brain. J Comput Assist Tomogr 2:539-544.

34. Norton SJ and Linzer M (1979). Ultrasonic reflectivity tomography: Reconstructin with circular transducer arrays. Ultrasonic Imaging 1:154-184.

35. Norton SJ and Linzer M (1979). Ultrasonic reflectivity imaging in three dimensions: Reconstruction with spherical transducer arrays. Ultrasonic Imaging 1:210-231.

36. Nahamoo D and Kak AC (1981). Ultrasonic echo imaging with pseudorandom and pulsed sources: A comparative study. Ultrasonic Imaging 3:1-36.

37. Lauterbur PC (1974). Magnetic resonance zeugmatography. Pure and App Chem 40:149-157.

38. Hinshaw WS, et al (1978). Display of cross-sectional anatomy by Nuclear Magnetic Resonance Imaging. Brit J Radiol 51:273-280.

39. Mansfield P and Maudsley A (1977). Planar spin imaging by NMR. J Mag Res 27:101-119.

40. Kwoh YS, Reed IS and Truong TK (1977). Generalized |w|-filter for 3-D reconstruction. IEEE Trans Nucl Sci NS-24:1990-1998.

41. Ra JB (1979). Critical review of reconstruction filters for convolution algorithm. ISS LAB Report No. 1:1-25, KAIS, Korea.

42. Lim CB, Chang LT and Jasaczak RJ (1980). Performance analysis of three camera configurations for single photon emission computed tomography. IEEE Trans Nucl Sci NS-27:559-568.

43. Nahamoo D, Crawford CR and Kak AC (1981). Design constraints and reconstruction algorithms for transverse-continuous-rotate CT scanners. IEEE Trans Bio Eng BME-28:79-98.

44. Lee HS (1979). Development of two-dimensional filter for the reconstructive imaging processing. ISS LAB Report No. 1:59-75, KAIS, Korea.

45. Herman GT (1979). Data collection for cross-sectional image reconstruction by a moving ring of positron annihilation detectors. J Comput Assist Tomogr 3:361-366.

46. Cho ZH, Hong KS, Ra JB and Lee SY (1981). A new sampling scheme for the ring positron camera. IEEE Trans Nucl Sci NS-28:94-98.

47. Shepp LA and Vardi Y (March 1982). Maximum likelihood reconstruction for emission tomography. International Workshop on Physics and Engineering in Medical Imaging.

48. (1980). ISS LAB Report No. 2, KAIS, Seoul, Korea.

49. Lai C-M and Lauterbur PC (1981). True three-dimensional image reconstruction by nuclear magnetic resonance zeugmatography. Phys Med Biol 26:851-856.

50. Pelc NJ (1979). Ph.D. Thesis, MIT.

51. Colsher JG (1980). Fully three-dimensional positron emission tomography. Phys Med Biol 25:103-115.

52. Crawford CR and Kak AC (1979). Aliasing artifacts in computerized tomography. Applied Optics 18:3704-3711.

53. Huang S, Phelps ME, Hoffman EJ and Kuhl DE (1980). Cubic splines for filter design in computed tomography. IEEE Trans Nucl Sci NS-27:1368-1374.

54. Cho ZH, Hilal SK and Correll JW (1980). High resolution spherical positron tomograph with bismuth germanate scintillation crystals. Abstract Med Phys 7.

55. Cho ZH (Oct 1982). Paper presented at IEEE Nuclear Science Symposium.

DIRECT FOURIER RECONSTRUCTION TECHNIQUES

IN NMR TOMOGRAPHY

Z.H. Cho*,**, O. Nalcioglu†, J.C. Jeong* and H.B. Song*

*Department of Electrical Sciences, Korea Advanced Institute of Science, Seoul, Korea,
**Department of Radiology, Columbia University, New York,
†Department of Radiological Sciences, School of Engineering(Electrical) and Department of Medicine(Cardiology), University of California, Irvine, California.

1. INTRODUCTION

Nuclear Magnetic Resonance (NMR) tomography is rapidly developing since its first inception by Lauterbur (1). NMR imaging appears to be a competing modality with X-ray CT in the study of anatomical structures. Like emission tomography, NMR may also provide biochemical and physiological image information (2-7). The most commonly used reconstruction technique in NMR is the filtered backprojection method, similar to X-ray CT (8). Although the filtered backprojection with Fourier convolution is well known and easy to implement, it is not the optimal technique for NMR tomography. This is because NMR projection data is obtained in the frequency domain, rather than the spatial domain as in X-ray CT. It is, therefore, not necessary to follow the conventional X-ray CT type of reconstruction procedure in NMR which would require transformation of data into the spatial domain (via Fourier transform) in order to perform convolution and backprojection. It appears that the direct Fourier domain treatment of the data may be more convenient in NMR tomography. In this chapter, we will discuss two different techniques which are geared toward working with the data in the Fourier domain so that time-consuming backprojection operation can be eliminated. These are the direct Fourier methods of Kumar-Welti-Ernst (KWE) (9) and Mersereau-Oppenheim (10). These two techniques will be treated with a unified format so that both can be understood within the framework of 2-D Fourier transform operation.

2. IMAGE FORMATION WITH DIRECT FOURIER TRANSFORMATION

The Fourier domain interpolation technique developed by Mersereau and Oppenheim (10) and direct Fourier transform method of Kumar-Welti-Ernst (9) are direct and simple ways of handling free induction decay (FID) data. While the later approach is fundamentally different from the former, in essence both methods provide a 2-D frequency domain representation of the object function. It will be shown that the Mersereau and Oppenheim technique corresponds to polar raster sampling while the KWE method corresponds to direct Cartesian sampling, both in the frequency space.

2.1 Direct Fourier Reconstruction Method of Mersereau and Oppenheim: Polar Raster Sampled Data

Although most of the reconstruction algorithms used in transaxial X-ray and emission CT are based on the Fourier convolution concept, the reconstruction has rarely been carried out in the Fourier domain. The main reason behind the preference of the spatial domain reconstruction is that the collected projection data is in the spatial domain.

The algorithm proposed by Mersereau and Oppenheim is based on the projection theorem and demonstrates that the reconstruction can be carried out in the frequency domain without the use of time-consuming backprojection operation. This is an attractive method, especially for the NMR application, since the data obtained is already in Fourier domain and 2-D Fourier transform operations can easily be performed by Fast Fourier Transform (FFT) using readily available array processors. The main reason why this technique has not been widely used is due to the sampling and interpolation problems as will be discussed below.

For completeness, let us first prove the projection theorem. Let us define (u,v) to be the coordinate system which is rotated by the projection angle θ with respect to the (x,y) frame. (See Figure 1).

The projection data (profile) $P_\theta(u)$ along the u-axis (i.e., line integral along the direction of v-axis) can be expressed as

$$\begin{aligned} P_\theta(u) &= \int f(u,v)dv \\ &= \iint f(x,y)\delta(x\cos\theta + y\sin\theta - u)dy\,dx. \end{aligned} \tag{1}$$

The 2-D FT of the object f, on the other hand, is defined as

$$\begin{aligned} \hat{F}(\omega_x,\omega_y) &= \iint f(x,y)\exp[-j(x\omega_x + y\omega_y)]\,dydx \\ &= \hat{F}(\rho,\phi), \end{aligned} \tag{2}$$

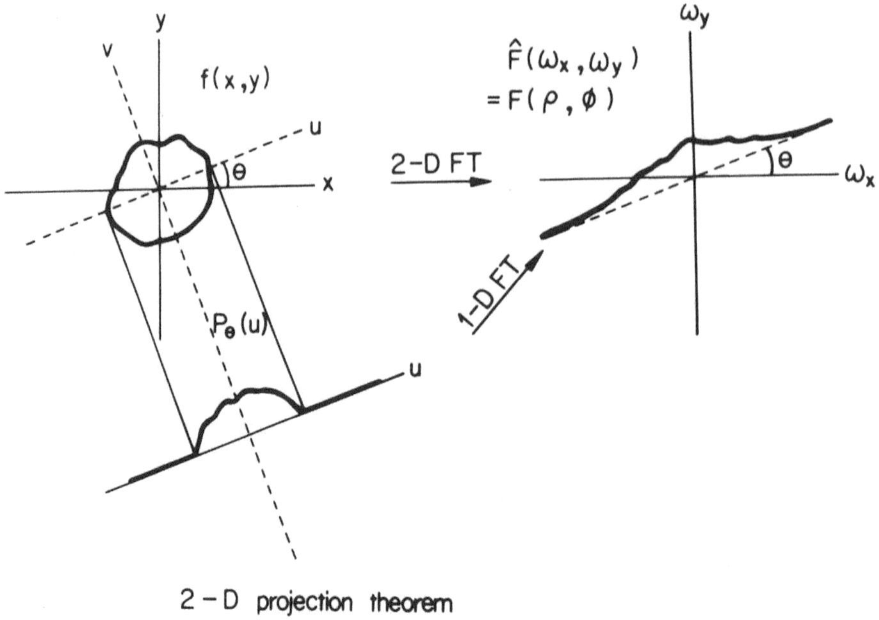

Fig. 1. Illustration of the projection theorem. Projection data at a given view corresponds to a frequency domain line through the origin. This basic theorem leads to the polar raster data in the frequency (Fourier) domain for the transaxial scanning.

where f(u,v) and f(x,y) are the object functions and $\hat{F}(\rho,\phi)$ is the polar coordinate representation of $\hat{F}(\omega_x,\omega_y)$, respectively. It can be shown that the 1-D FT of projection $P_\theta(u)$ is equal to the components of $\hat{F}(\rho,\phi)$ at $\phi = \theta$, i.e.,

$$\hat{F}(\rho,\phi)\Big|_{\phi=\theta} = \int P_\theta(u)\,\exp(-j\rho u)du. \tag{3}$$

2-D projection theorem can easily be extended to the 3-D case and is given as follows: Let the projection axes (u,v,w) be the rotated coordinates of (x,y,z) by projection angle (α,θ). The projection $P_{\alpha,\theta}(u)$ along u-axis (i.e., plane integral in the direction perpendicular to u-axis) can be calculated by,

$$\begin{aligned} P_{\alpha,\theta}(u) &= \iint f(u,v,w)dwdv \\ &= \iiint f(x,y,z)\delta(x\ \sin\theta\cos\alpha + y\ \sin\theta\sin\alpha + z\ \cos\theta - u)dz\ dy\ dx \end{aligned} \tag{4}$$

3-D FT of f(x,y,z) is defined as

$$\hat{F}(\omega_x,\omega_y,\omega_z) = \iiint f(x,y,z)\ \exp[-j(x\omega_x + y\omega_y + z\omega_z)]dzdydx$$
$$= \hat{F}(\rho,\theta_\rho,\phi_\rho), \qquad (5)$$

where $\hat{F}(\rho,\theta_\rho,\ \phi_\rho)$ is the $\hat{F}(\omega_x,\omega_y,\omega_z)$ represented in the spherical coordinates. It can easily be shown that the 1-D FT of $P_{\alpha,\theta}(u)$ and $\hat{F}(\rho,\theta_\rho,\phi_\rho)$ have the following relation,

$$\hat{F}(\rho,\theta_\rho,\phi_\rho)_{\substack{\theta_{\rho=\alpha}\\ \phi_{\rho=\theta}}} = \int P_{\alpha,\theta}(u)e^{-j\rho u}\ du. \qquad (6)$$

Thus, it is seen again that the 1-D FT of the projection $P_{\alpha,\theta}$ is equal to the 3-D FT of the object f along a cord passing through the origin and in the direction given by angles (α and θ). Eqs. (1)-(6) are the foundations of 2-D and 3-D direct Fourier transform image reconstruction.

Projection theorem leads directly to the Direct Fourier Reconstruction (DFR) algorithm in a straight-forward manner. Initially the projection data, $P_\theta(u)$ (or $P_{\alpha,\theta}(u)$ for 3-D case) for all the angular views from θ equals 0 to π (or $0 \leq \theta < \pi$, $0 \leq \phi < \pi$ for 3-D), are collected and transformed by using the 1-D FT. Finally, this data is used to produce the frequency domain image function. After appropriate interpolation, 2-D (or 3-D) inverse Fourier transform is performed for the retrieval of the final spatial domain image.

In the case of 2-D reconstruction, the 2-D Fourier transform of the object function, eq. (3), is sampled along radii which are separated by the angle between adjacent views. It is seen that, in the conventional transaxial parallel ray sampling with constant angular increments between views, the object function will be sampled in a polar raster in the frequency domain. If one wishes to use the 2-D FFT to transform the object function back into spatial domain, it is necessary to transform data from the polar raster into Cartesian raster. Cartesian samples can be obtained by interpolation of the given polar raster or grid samples.

In the DFR procedure optimal sampling of the polar raster data and conversion to Cartesian raster data, in Fourier domain, are the central theme of the entire technique described in the following.

i. Algorithm of the DFR method of Mersereau and Oppenheim

A description of the DFR algorithm and computational steps for the 2-D case are given as follows:

(a) Projection data for N views with M rays per view are obtained from eq. (1) and given by

$$P_{\theta_i}(u_j) = \iint f(x,y)\,\delta(x\cos\theta_i + y\sin\theta_i - u_j)\,dy\,dx \qquad \text{for } 1 \leq i < N,\ 1 \leq j < M. \tag{7}$$

(b) 1-D FFT of each projection data is performed to form a polar raster in frequency domain by discrete FFT;

$$\begin{aligned}\hat{F}(\rho_k,\theta_i) &= \sum_{\ell=1}^{M} P_{\theta_i}(u_\ell)\exp(-ju_\ell\rho_k)\\ &= \sum_{\ell=1}^{M} P_{\theta_i}(u_\ell)\exp[-j\Delta u(\ell-\tfrac{M}{2})\,(k-1)\,\Delta\rho]\\ &= \sum_{\ell=1}^{M} P_{\theta_i}(u_\ell)\exp[-j\,\frac{\pi(\ell-M/2)(k-1)}{M}]\ ,\ \text{for } k=1,2,\ldots,M.\end{aligned} \tag{8}$$

where $\Delta u\Delta\rho = \frac{\pi}{M}$ and $u_\ell = (\ell - \frac{M}{2})\ \Delta u$.

(c) The values of Cartesian points in the first and second quadrants of the given polar points are estimated by first order linear interpolation, which is given by:

$$\begin{aligned}\hat{F}(\omega_x,\omega_y) &= c_{m,n}\hat{F}(\rho_m,\theta_n) + c_{m,n+1}\hat{F}(\rho_m,\theta_{n+1}) + c_{m+1,n}\hat{F}(\rho_{m+1},\theta_n)\\ &\quad + c_{m+1,n+1}\hat{F}(\rho_{m+1},\theta_{n+1}),\end{aligned} \tag{9}$$

where $c_{k,\ell}$ are the weighting coefficients which can be calculated easily from the surrounding points (these values are saved in a look-up table for repeated fast calculation).

The values of the function in the third and fourth quadrants are obtained by using the complex conjugate property:

$$\begin{aligned}\hat{F}(-\omega_{x_i},-\omega_{y_j}) &= \hat{F}^*(\omega_{x_i},\omega_{y_j})\\ \hat{F}(\omega_{x_i},-\omega_{y_j}) &= \hat{F}^*(-\omega_{x_i},\omega_{y_j})\end{aligned} \tag{10}$$

(d) Finally inverse 2-D FFT is performed. The real part of the result will then be the desired image.

In the execution of the above technique, two important operations have to be performed; optimum angular sampling and interpolation. Both operations require polar to Cartesian raster transformation.

ii. Optimum angular sampling and interpolation

(a) Modified concentric square raster sampling schemes in the DFR method

To obtain the Cartesian form of object function $\hat{F}(\omega_x,\omega_y)$ from the polar coordinate object function $\hat{F}(\rho,\theta_\rho)$ requires special effort. The concentric Square Raster Sampling (CSRS) method proposed by Mersereau and Oppenheim seems to be efficient, elegant, and also provides relatively accurate estimation by simple low order interpolation. The original concept of CSRS is illustrated in Fig. 2, where frequency sampling interval is varied depending on the angle of each view. The largest sampling interval results at 45° and 135°, while the shortest appears at $\theta = 0^\circ$ and 90°. The resultant frequency domain sampling interval at the view of projection angle θ_i appears as

$$\Delta\omega(\theta_i) = \Delta\omega(0)\frac{1}{\max(|\cos\theta_i|,\ |\sin\theta_i|)} \tag{11}$$

With the number of sampling points being kept fixed, the spatial domain sampling interval ΔR and the full length of sampling array R should be maintained as

$$\Delta R(\theta_i) = \Delta R(0)\cdot\max(|\cos\theta_i|,\ |\sin\theta_i|), \tag{12}$$

$$R(\theta_i) = R(0)\cdot\max(|\cos\theta_i|,\ |\sin\theta_i|). \tag{13}$$

This $R(\theta_i)$ path is shown in Fig. 2(a). It consists of four semicircles with each center placed at $(\frac{R(0)}{4}, 0)$, $(-\frac{R(0)}{4}, 0)$, $(0, \frac{R(0)}{4})$, and $(0, \frac{R(0)}{4})$, respectively.

<u>Equal angular sampling ($\Delta\theta$ = constant)</u>

If we maintain the equal angular sampling interval $\Delta\theta$ between two adjacent views, the view angle will increase at a constant rate as follows,

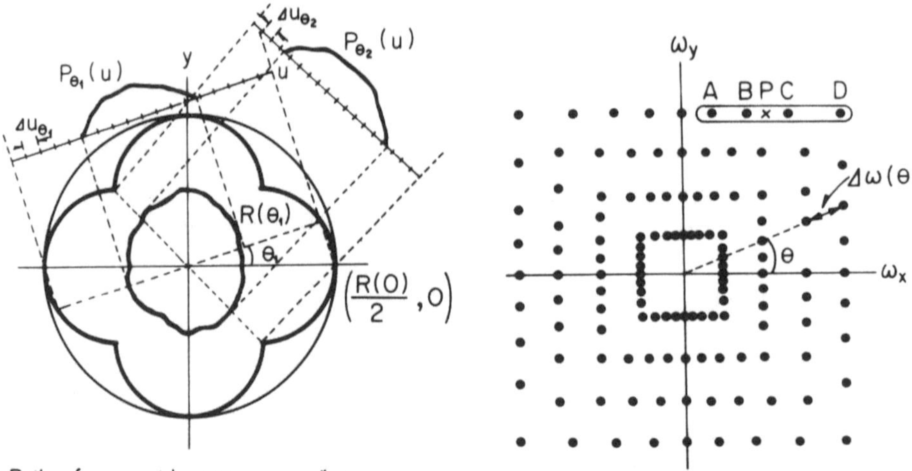

Path of concentric square sampling.

(a)

Sampled concentric square raster with equal $\Delta\theta$.

(b)

Fig. 2. Scan path and obtained polar raster concentric square data. Polar raster data which is similar to the Cartesian raster as shown in (b) can be obtained by following the scan path shown in (a). This path simply indicates the variable sampling interval with a fixed number of sample points, that is, largest interval appears at x axis or $\theta=0$ while smallest sampling interval appears at $0 = 45^\circ$. The polar raster data obtained in this way appears close to the Cartesian raster and suggests simple conversion of the polar raster to Cartesian raster. The later is a necessary requirement for the FFT operation with a digital computer.

$$\theta_i = \Delta\theta(i=1), \text{ for } i = 1,2,\ldots,N, \tag{14}$$

where $\Delta\theta = \frac{\pi}{N}$.

In this case, the sampling density becomes sparse towards 45° and 135° directions as illustrated in Fig. 2(b). This poor sampling and resulting interpolation errors near 45° and 135° detrimentally affect the reconstructed images, especially for high resolution images such as the X-ray or NMR CT images where contrast resolutions of 0.5% to 1.0% are common.

<u>Variable angular sampling ($\Delta\theta$ = variable)</u>

To avoid difficulties due to the constant $\Delta\theta$, the vertical (or horizontal) distance betweeen neighboring sample points $\Delta\omega_y(\Delta\omega_x)$ was held equal by increasing the number of views towards 45° and 135° directions. The view angle then may be expressed as,

$$\theta_i = \begin{cases} \tan^{-1}\{\frac{i-1}{N/4}\}, & \text{for } i = 1,2,\ \ldots,\ \frac{N}{4}+1, \\ \frac{\pi}{2} - \theta_{\frac{N}{2}+2-i}, & \text{for } i = \frac{N}{4}+2,\ \ldots,\ \frac{N}{2} \\ \frac{\pi}{2} + \theta_{i-\frac{N}{2}}, & \text{for } i = \frac{N}{2}+1,\ \ldots,\ N. \end{cases} \qquad (15)$$

Using this technique the interpolation error could be kept nearly the same for all the frequency space and some improvement in image quality were noticeable as will be seen in a later section. The improvements, however, do not match the resolution requirements of the new generation X-ray or NMR CT system due to several reasons, as will be discussed in conjunction with the optimized Concentric Square Raster Sampling (CSRS) technique.

Optimized angular sampling ($\Delta\theta$ = variable)

The new interpolation scheme in eq. (15) improved image quality, although some artifacts still remained and image quality was inferior compared to the Fourier convolution method. Poor image quality may be due to the fact that a 2-dimensional Fourier transform of a real object usually appears in frequency domain as concentric circles with the amplitude decaying along the radial direction from the origin (zero frequency point) to the outer rim. That is, variations along the radial direction are much more rapid than the variation in the angular direction. Thus rapid variations may result in the poor interpolation.

When the interpolation is performed from the original concentric square raster, the errors in the vicinity of 0° and 90° directions are much smaller than those of 45° and 135° directions, since the Cartesian and the polar coordinates ω_x and ω_y coincide much more closely along the 0° and 90° directions.

In order to equalize the errors over all views, it is required to increase the number of views towards 45° and 135° even more than the uniform distance ($\Delta\omega_x$ or $\Delta\omega_y$ = constant) concept introduced by Mersereau and Oppenheim as proposed in eq. 15.

The more systematic approach is the equalization of the mean radial distance of the sampling points along the direction for different angular views in which the interpolation is performed.

This approach leads to a new view angle distribution given as;

$$\theta_i = \begin{cases} \cos^{-1}\left\{\dfrac{N/4}{N/4 + (\sqrt{2}-1)\ (i-1)}\right\}, & \text{for } i=1,2,\ldots,N/4+1 \\[2ex] \dfrac{\pi}{2} - \theta_{N/2+2-i} & , \text{ for } i=N/4+2,\ldots,N/2 \\[2ex] \dfrac{\pi}{2} - \theta_{i-N/2} & , \text{ for } i=N/2+1,\ldots,N. \end{cases} \tag{16}$$

The resultant view angle distribution of the later scheme is shown in Fig. 3. In Fig. 4, the other two previously described view angle distributions together with the new one are plotted for comparison. Fig. 4(a), (b) and (c) corresponding to eqs. (14), (15), and (16), respectively. A more uniform radial distribution is provided by eq. (16) where equalization of the radial sampling distances are systematically sought.

The view angle distribution following eq. (16) was found to be near optimal as will be demonstrated later with the reconstructed images.

(b) Higher order interpolation - Lagrange interpolation

In addition to the optimized sampling technique discussed above, it was found that for high resolution imaging, a higher order interpolation was necessary.

In the present study, higher order interpolation was achieved by Lagrange interpolation. A four point Lagrange interpolation scheme was used and is given by:

$$F(P) = \frac{BP\cdot CP\cdot DP}{BA\cdot CA\cdot DA}F(A) + \frac{AP\cdot CP\cdot DP}{AB\cdot CB\cdot DB}F(B) + \frac{AP\cdot BP\cdot DP}{AC\cdot BC\cdot DC}F(C) + \frac{AP\cdot BP\cdot CP}{AD\cdot BD\cdot CD}F(D). \tag{17}$$

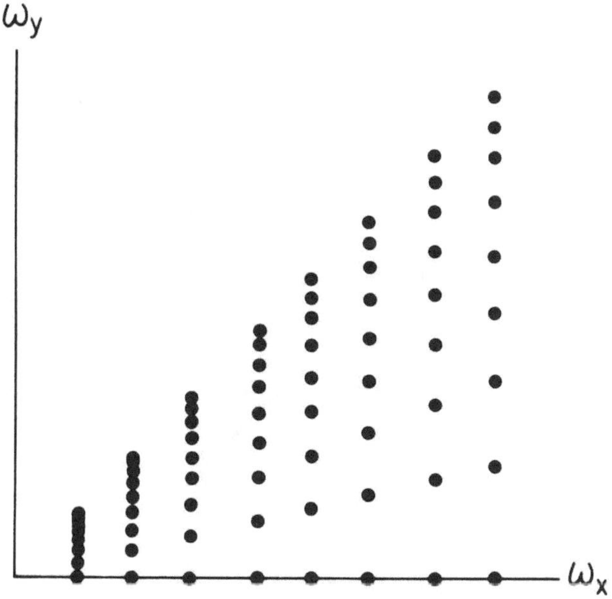

Fig. 3. Modified concentric square raster sampling (CSRS) obtained by following the scan path shown in Fig. 2(a) with variable angular interval. This method leads to finer sampling toward 45° or 135° to overcome the intrinsically poor sampling resulting from the circularly symmetric nature of the Fourier domain image data.

Where BP is the distance between point B and P for example.

We will demonstrate the effect of this high order interpolation on the reconstructed images in a later section.

iii. Relation of the spatially resolved NMR signals (FID) and object function in direct Fourier transform operation.

A most interesting part of NMR imaging is the Fourier transform operation of the FID which leads to a signal which represents the spatially resolved spin density function. As given by eq. (1), the two dimensional spin density function $\rho(x,y)$ and the line integral data (NMR Fourier transformed) obtained at a given angle can be written as (similar to eq. 1).

$$FT\left[P_\theta(x')\right] = \int d\omega_{x'} \cdot [\iint \rho(x,y)\delta(x\cos\theta+y\sin\theta-x')dydx] \exp(-jx'\omega_{x'}) \quad (18)$$

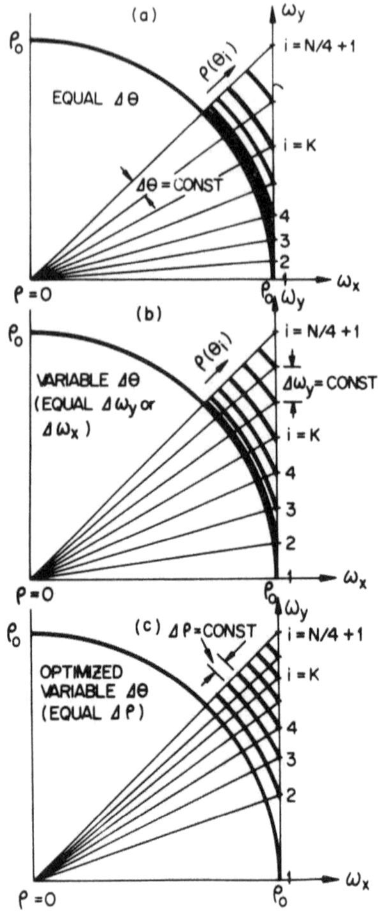

Fig. 4. Three different CSRS schemes: (a) Original CSRS proposed by Mersereau and Oppenheim, (b) Another CSRS scheme also proposed by Mersereau and Oppenheim partly to overcome the poor radial sampling near the 45° and 135°. This scheme is based on the equal sampling of $\Delta\omega_x$ and $\Delta\omega_x$ axis. This scheme is termed as variable $\Delta\omega$ sampling. (c) Proposed optimized CSRS scheme to overcome the shortcomings of the above two methods. In this scheme, equal radial sampling was sought to match the concentric circular behavior of Fourier domain spectra. This scheme is close to the CSRS data obtained in Fig. 3, but optimized in terms of equal

where $x' = u$. The NMR equivalent of eq. (18), that is the FID signal $s(t)$ is given by,

$$s(t) = \int d\omega_{x'} \int \rho(x',y')\, dy' \exp\left[-j\gamma(H_o+G_{x'}x')t\right] \tag{19}$$

where $\gamma H_o = \omega_o$, $\gamma G_{x'} x' = \omega_{x'}$ and γ is gyromagnetic ratio, $G_{x'}$ is field gradient, and $\omega_{x'}$ is the precession frequency. From eqs. (18) and (19), it is easy to see the following relationships;

$$FT\,[P_\theta(x')] = \hat{F}(\rho,\theta)$$

and

$$\hat{F}(\rho,\theta) \simeq s_\theta(t) \tag{20}$$

The relationships given in eq. (18), (19), and (20) are directly related to the 2-D projection theorem given in eqs. (1) to (3), and allow us to process directly in the Fourier domain.

2.2 Direct Fourier Reconstruction Methods of Kumar-Welti-Ernst and Hutchison - Cartesian Raster Sampled Data

Another interesting method for direct Fourier transform approach is the method derived by Kumar, Welti and Ernst (9,15) and a related technique by Hutchison (12). These two methods have several distinct advantages over other methods; namely, the simplicity of data collection, data handling, and fast image reconstruction capability. The latter is mainly due to the elimination of the time-consuming backprojection operation. Although the method itself is mathematically simple, it has to be explained in conjunction with the NMR imaging method.

i. Original Kumar-Welti-Ernst method (9)

In the original KWE method, FID signals are obtained for three time periods and the 3-D Fourier transform of this data set then gives the original 3-D spin density distribution function. In Fig. 5 the basic form of r.f. pulsing and field gradient sequences of the 2-D KWE method is illustrated. In 2-D case, the z-component of the local magnetic field and gradient pulse sequence are given by

$$H_z(x,y;z_o) = \begin{cases} H_o + G_y y, & \text{for } 0 < t \leqslant t_y \\ H_o + G_x x, & \text{for } t_y < t \end{cases} \tag{21}$$

where time scale t_y is varied according to preassigned sequences, i.e., $t_y = 0$ to t_a, where t_a is the optimal observation time of the FID. The FID signals observed within the t_x periods will form a full 2-D FID

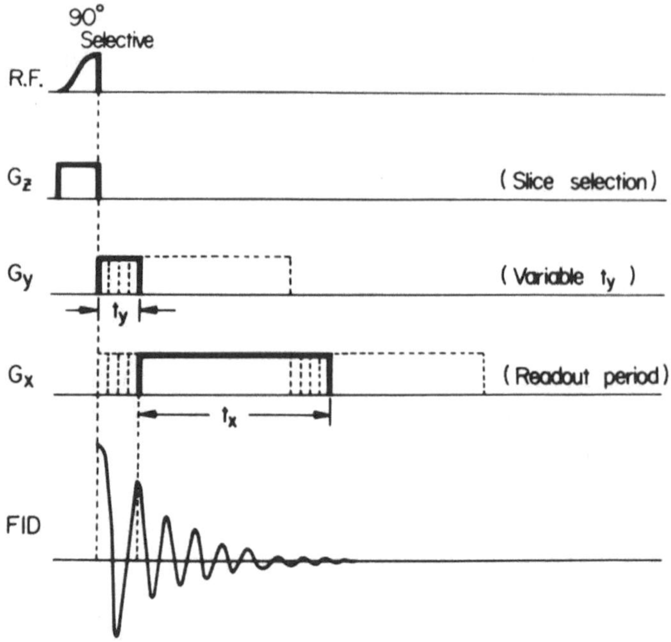

Fig. 5. The imaging sequences of original 2-D KWE direct Fourier imaging. Pulse length t_y of gradient G_y is varied from $t_y = 0$ to $t_{y_{max}}$ (or $t_{x_{max}}$) with intervals of $\Delta t_y \simeq \frac{t_{x_{max}}}{N}$ where N is the number of samples along ω_y coordinate.

signal set sufficient for representation of the entire plane spin density after 2-D Fourier transformation.

The observed FID signal $s(t_x,t_y;z_o)$ is given by

$$s(t_x,t_y;z_o) = M_o \iint \rho(x,y;z_o)\cos\{\gamma(G_x x t_x + G_y y t_y)\}\exp\{-(t_x+t_y)/T_2^*\}dydx, \tag{22}$$

where M_o is spin magnetization per unit volume, and T_2^* is the effective transverse relaxation time which is given by $1/T_2^* = 1/T_2 + \gamma\Delta H/2$, where T_2 is the spin-spin relaxation time and ΔH is the inhomogeneity of the magnetic field.

Inverse Fourier transformation of eq. (22) results in the spatial spin density function $\rho(x,y;z_o)$ or $\hat{S}(\omega_x,\omega_y;z_o)$, where

$$\hat{S}(\omega_x,\omega_y;z_o) = \int\int s(t_x,t_y;z_o)\exp\{j(\omega_x t_x+\omega_y t_y)\}dt_y dt_x$$

or $$\hat{S}(x,y;z_o) = \int\int s(t_x,t_y;z_o)\exp\{j\gamma(G_x x t_x+G_y y t_y)\}dt_y dt_x \quad (23)$$

Let us now consider the actual digital implementation of eq. (23). In the 2-D Fourier transform of an N x N image matrix, it is necessary to obtain 2N FID's and digitize each FID into 2N sample points. In the original KWE method, however, only N FID's are obtained and only the real part of transformed values are stored in a matrix. The second part of the Fourier transform operation is performed with a transposed matrix (real part only). During the latter operation N zeros are added to each row before the transformation. Absolute values of the N^2 complex Fourier coefficients are then used for mapping of a two-dimensional NMR section image or Zeugmatogram.

Images obtained by this method, however, were poor and show some artifacts. These drawbacks of the original KEW method can be overcome by the improved KWE direct Fourier imaging methods described in the following section.

ii. Improved (modified) KWE method (15,16)

Although this method can be extended to 3-dimensional form, the method described in this section is limited to 2-D imaging. This method overcomes the inherent shortcomings of sampling and other problems associated with data collection of the original KWE method mentioned in the previous section.

In this modified scheme, a single slice is first selected by the application of a narrow band 90° r.f. pulse and field gradient G_z. Two-dimensional FID sets are then obtained by varying t_y, the time for which G_y is applied. During the G_x period two spin echoes are produced by application of two broad band 180° r.f. pulses. In effect four FID's are obtained during this period, i.e., FID's labeled as A, B, C, and D. FID's A and B as well as C and D are complex conjugate pairs. Sampling intervals of t_x and t_y are determined by Nyquist sampling to avoid aliasing effects. A series of FID sets are obtained as a function of t_y to form a complete set of 2-D FID signals (N FID signals). This data set will fill all the necessary data from 0 to 2π both in the x and y directions. Note that in the case of the original KWE method only the quadrant A is obtained.

The spin echo technique employed in this scheme is based on the Meiboom-Gill spin echo technique rather than the Hahn spin echo. Within period A, the spins in the selected slice are dephased with field gradient G_x, as shown in Fig. 6. After the first 180° spin echo pulse in the B period, the spins are rotated about y' axis and begin to rephase.

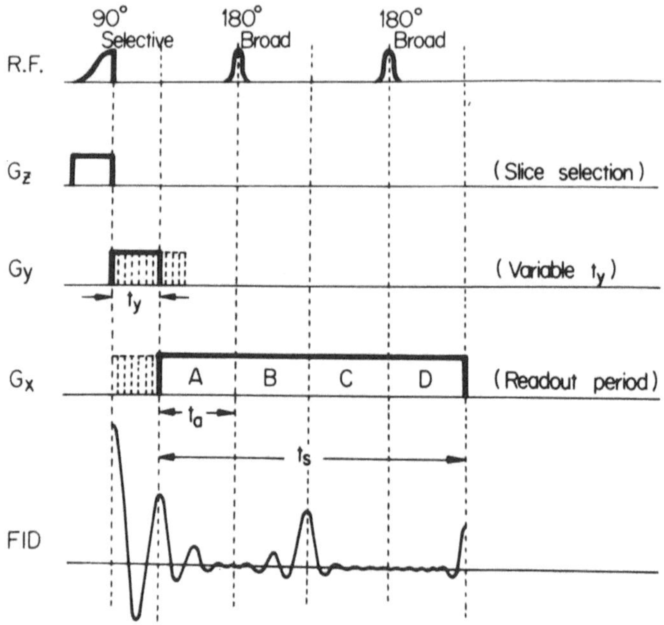

Fig. 6. The imaging sequences of single slice 2-D modified KWE direct Fourier imaging (MKWE).

The FID of period B has a conjugate relationship with that of period A. In period C, the spins are dephased in a different manner than in a period A. After the second 180° spin echo pulse, in period D, the spins are rotated about y' axis again and begin to rephase. The FID's of the period C and D also have a conjugate relationship.

When the FID's are measured with quadrature phase sensitive detection equipment, the resulting signals obtained in the periods A, B, C, and D are given by:

$$s(t_x,y_y;z_o) = M_o \int\int \rho(x,y;z_o)\ \exp\{-j\gamma(x\cdot G_x\cdot t_x + y\cdot G_y\cdot t_y)\}dydx \qquad (24)$$

$$s_A = s(t_x,t_y;z_o)A_1 \qquad (25)$$

$$s_B = s^*\ (t_x,t_y;z_o)A_1\ \exp\{-(\tfrac{1}{2}t_s - 2t_x)/T_2\} \qquad (26)$$

$$s_C = s(t_x,-t_y;z_o)A_1\ \exp\{-\tfrac{1}{2}t_s/T_2\} \qquad (27)$$

$$s_D = s(-t_x,t_y;z_o)A_1\ \exp\{-(t_s-2t_x)/T_2\} \qquad (28)$$

where $A_1 = \exp\{-t_x + t_y)/T_2^*\}$.

By simple attenuation correction, eqs. (26), (27), and (28) can now be placed to fill the required data sets. Conventional 2-D Fourier transform of the complete data set A, B, C, and D will result in a fully reconvered 2-D spin density image.

iii. Hutchison-KWE method (12)

The improved KWE method appears useful and enables one to obtain all the four quadrants of data unambiguously, however, it is difficult to apply in some situations.

First, the large 180° spin echo requires a large intensity broad band r.f. pulse which would be difficult to obtain and secondly the complex pulse sequencing of double spin echo method (15) might also be difficult to implement. In conjunction to the KWE method previously discussed, by use of Hutchison method which does not require varying pulse width (t_y) and use of the gradient pulse reversing spin echo technique, it is possible to obtain two quadrants of data which will be sufficient to form a complete Fourier domain image data. In addition, this spin echo technique using gradient reversal, provides imaging which is free from gradient pulse transient time dependent artifacts. This combined Hutchison-KWE method is illustrated in Fig. 7. In this scheme, a simple variation of the amplitude of gradient pulse G_y with fixed pulse length will be required. The advantage of this method is twofold. First, the pulse sequencing is easy since timing is not critical as the other methods,

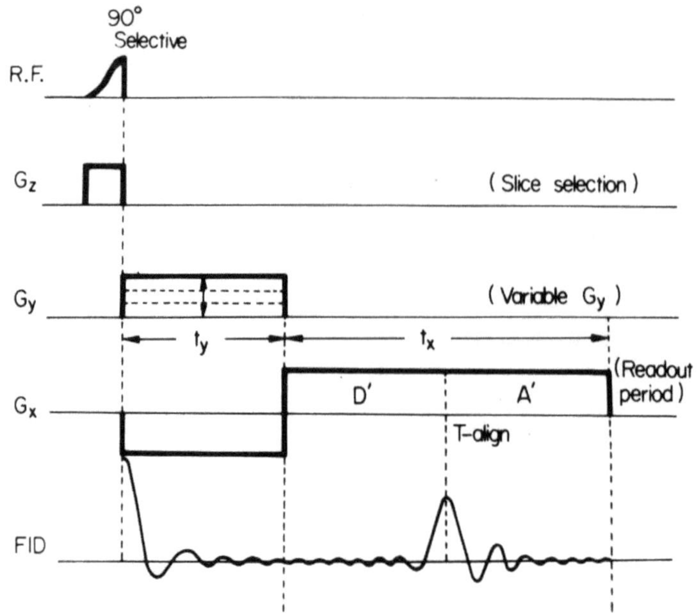

Fig. 7. Pulse sequence diagram of the new Hutchison-KWE method. In this scheme FID's are obtained at a fixed time interval $t_y \leq t \leq t_y + t_x$.

e.g. the original Hutchison type. It is only necessary to vary the amplitude. Second, the position of the FID signal can be fixed and is not not dependent on the gradient pulse transient effect. The Hutchison-KWE method is practical and potentially useful in NMR imaging since it does not require either the high intensity r.f. pulse or the critical gradient pulses.

3. COMPUTER SIMULATION RESULTS

3.1 Simulated Images by the Direct Fourier Reconstruction (DFR) Methods of Mersereau-Oppenheim

Computer simulation studies of the direct Fourier reconstruction method of Mersereau and Oppenheim have been performed with the three different sampling techniques which were discussed in an earlier section, (section ii.a). In these studies, we used the standard Shepp head phantom. Images obtained by different angular sampling schemes are shown in Fig. 8. It is seen that the new optimized concentric square sampling which cor-

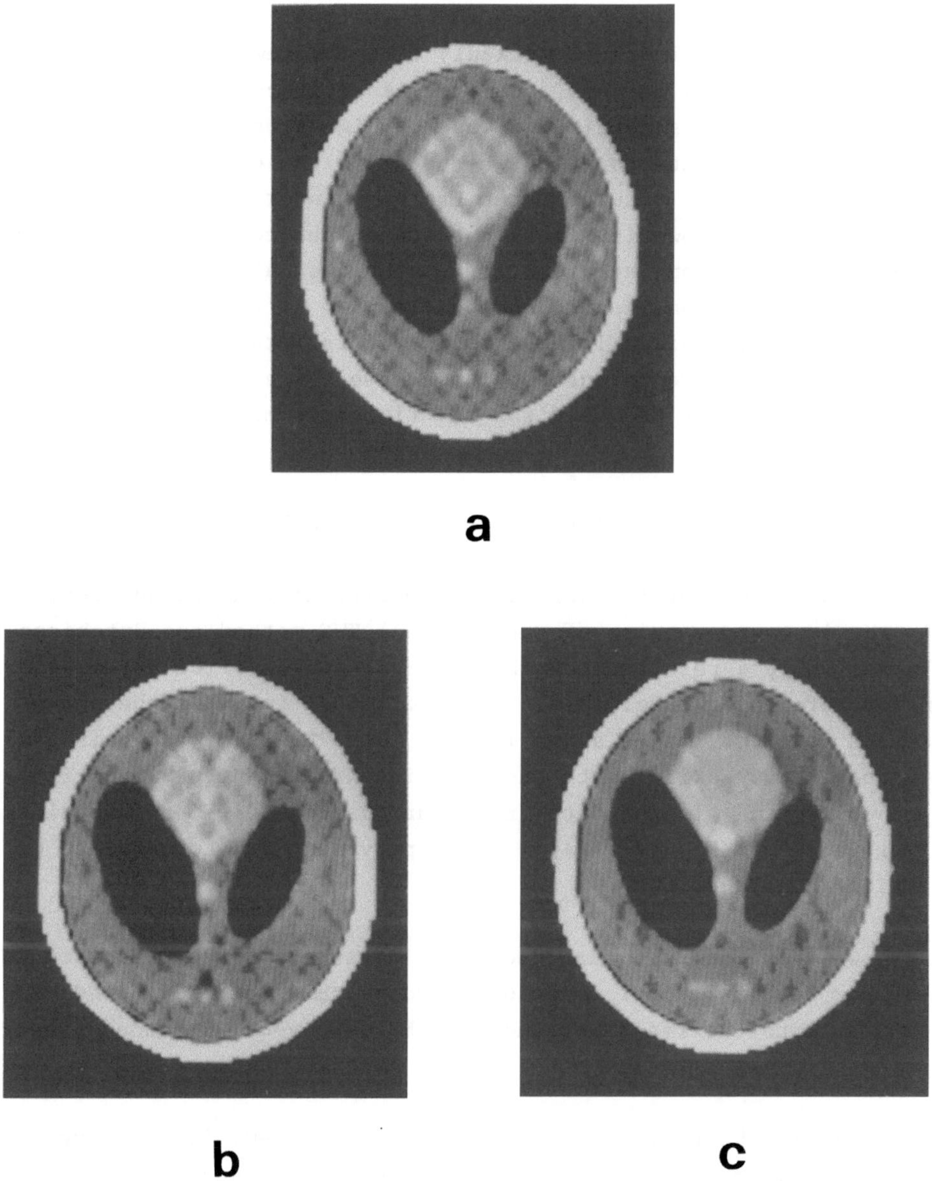

Fig. 8. Images of direct Fourier transform techniques of Mersereau and Oppenheim. (a) Original equal angular interval ($\Delta\theta$ = constant) concentric square raster sampling, (b) equal distance ($\Delta\omega_y$ or $\Delta\omega_y$) concentric square raster sampling or variable angular interval concentric square sampling. (c) new optimized variable angular interval ($\Delta\rho$ = constant) concentric square raster sampling.

responds to the Figure 4(c) appears to be optimum. However, even though the image quality has improved markedly with the new scheme compared with previous sampling methods, some streak artifacts are still visible. Further improvement was obtained by use of higher order interpolation. Four point Lagrange interpolation was used for the elimination of the streak artifacts and a result of the Lagrange interpolation is shown in Fig. 9. The image shown in Fig. 9 appears closer to the image that was obtained by convolution backprojection technique, although a slight degradation of resolution is noticeable. In view of the various advantages of the DFR technique in NMR imaging, this method should be considered as an alternative reconstruction scheme for simple and quick image reconstruction without the need for the time-consuming backprojection operation.

3.2 Simulated Images with the KWE and Hutchison-KWE Methods

We have studied both of the KWE direct Fourier imaging techniques, namely the improved version of the original KWE method and Hutchison-KWE. In Figure 10(b), a result using the original single slice KWE method is shown for reference. Fig. 10(a) is the original image of the phantom. It is seen from the images shown in Fig. 10(b) that the image is symmetric about the y axis. With the modified KWE technique which is capable of collecting all the data from four quadrants, we have obtained the image shown in Figure 11(a). The improvement due to complete data collection is noticable.

Finally, we show the reconstructed image obtained by the Hutchison-KWE technique. The quality of image obtained by this method is good, and is comparable to the image obtained by the improved KWE method (see Fig. 11(b)).

As discussed in the previous section, this new scheme is relatively uneffected by the gradient pulse timing, especially transient rise and fall times of the rapidly switching gradient pulses. This method is, therefore, useful in experimental situations where instrumental uncertainty is expected to be large.

4. CONCLUSIONS

Viewing the increasing interest of the direct Fourier transform imaging, especially in the area of NMR tomography, we have investigated two direct Fourier transform imaging methods in depth and explored their possible

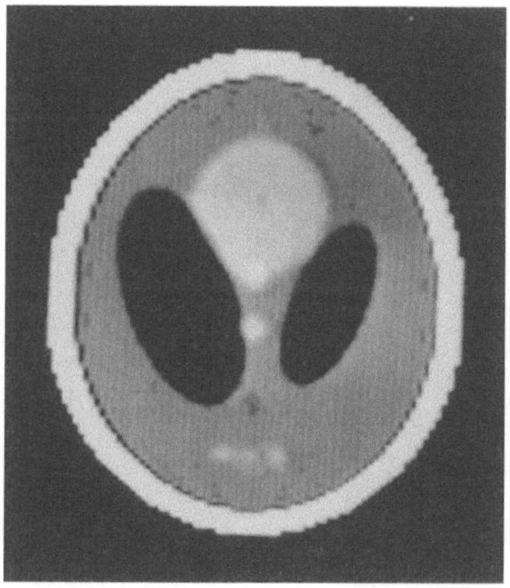

Fig. 9. Effect of interpolation in DFR techniques; Image obtained by 4-point Lagrange interpolation plus optimized concentric square raster sampling technique (compare with image of Fig. 8(c)).

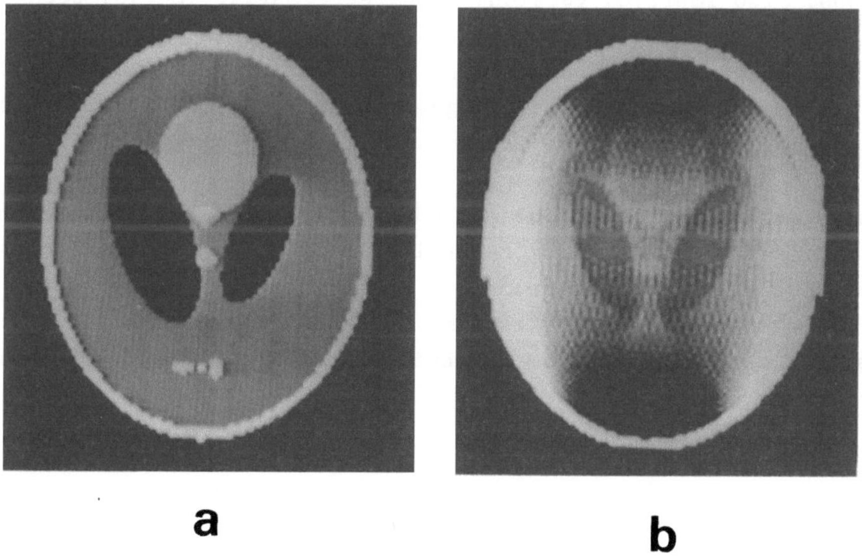

Fig. 10. Original phantom and a simulated image of single slice, original KWE direct Fourier imaging ($G_x = G_y = 0.1$ Gauss/cm with no noise). (a) Original Shepp phantom and (b) The image obtained with original KWE method.

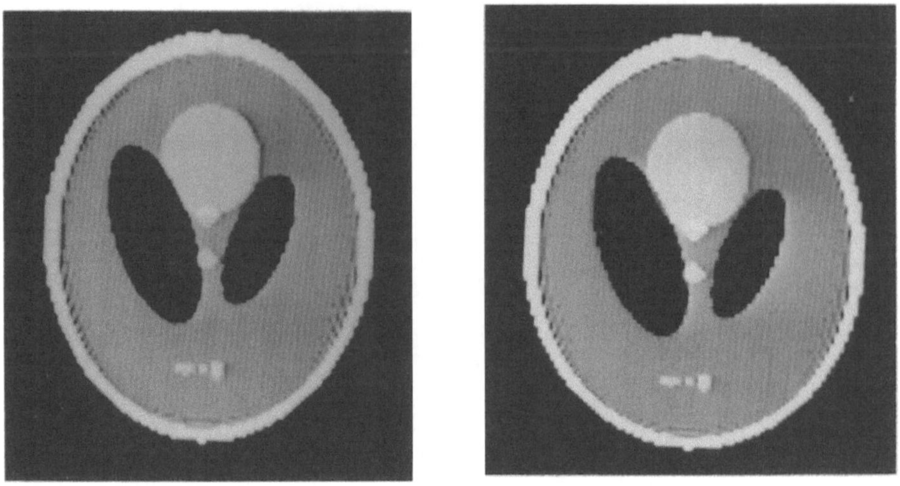

a b

Fig. 11. Simulated images of: (a) Single slice modified KWE direct Fourier imaging and (b) New Hutchison-KWE scheme. Both with $G_x = G_y =$ 0.1 Gauss/cm without noise.

extensions. Two improved versions of direct Fourier imaging techniques, the direct Fourier imaging of Mersereau and Oppenheim with optimized concentric raster sampling scheme and the modified Hutchison-KWE direct Fourier transform technique, are suggested for possible applications to NMR imaging.

Computer simulations of the two methods proved to be superior to the other methods in image quality as well as from a computational consideration. The Hutchison-KWE technique appears to be ideally suited for actual implementation due to its simplicity and absence of the stringent requirements on the r.f. pulse intensity and fast gradient pulse timing.

REFERENCES

1. Lauterbur PC (1973). Image formation by induced local interactions: examples employing nuclear magnetic resonance. Nature 242:190.

2. Mansfield P (1977). Multiplanar image-formation using NMR spin echoes. J Phys C-10, L55.

3. Andrew ER, Bottomley PA, Hinshaw WS, Holland GN, Moore WS and Simaroj C (1977). Phys Med Biol 22:971.

4. Damadian R (1971). Tumor detection by nuclear magnetic resonance. Science 171:1151.

5. Shepp LA (1980). Computerized tomography and nuclear magnetic resonance. J Comput Assist Tomogr 4:94.

6. Marr R, Chen C-N and Lauterbur PC (1980). On two approaches to 3-D reconstruction in NMR zeugmatography. Proc Math Aspec CT, Oberwolfach, West Germany.

7. Cho ZH (1980). Development of methods and algorithms for Fourier transform nuclear magnetic resonance (NMR) tomographic imaging. ISS Lab Report #2, KAIS, Seoul, Korea.

8. Hawkes RC, Holland GN, Moore WS, Roebuck EJ and Worthington BS (1981). Nuclear magnetic resonance (NMR) tomography of the normal abdomen. J Comput Assist Tomogr 5:613.

9. Kumar A, Welti D and Ernst RR (1975). NMR Fourier zeugmatography. J Mag Res 18:69.

10. Mersereau RM and Oppenheim AV (1974). Digital reconstruction of multi-dimensional signals from their projections. IEEE Proc 62: 1319.

11. Brooks RA and diChiro G (1976). Principles of computer assisted tomography (CAT) in radiographic and radioisotopic imaging. Phys Med Biol 21:689.

12. Hutchison JMS, Edelstein WA and Johnson G (1980). A whole-body NMR imaging machine. J Phys E13:947. Also private communication, Lauterbur PC (1980).

13. Song HB (1980). Extended line scanning tomography in spatially resolved Fourier transform NMR imaging. ISS Lab Report #2, KAIS, Seoul, Korea.

14. Shepp LA and Logan B (1974). The Fourier reconstruction of a head section. IEEE Trans Nucl Sci NS-22:344.

15. Song HB, Cho ZH and Hilal SK (1982). Direct Fourier transform NMR tomography with modified KWE method. IEEE Trans Nucl Sci NS-29: 493.

16. Cho ZH, Kim HS, Song HB and Cumming DRJ. Fourier transform nuclear magnetic resonance tomographic imaging. IEEE Proceedings (in press).

RADIATION DETECTORS FOR CT INSTRUMENTATION

M. R. Farukhi*

Rexon Components, Inc.
Willoughby, Ohio

*Previously with the Harshaw Chemical Company, Solon, Ohio

1. INTRODUCTION

The selection of the radiation detector is a basic consideration in the design of instrumentation for transmission and emission computed tomography. Almost every conceivable detector type has been considered at one time or another; however, in general, scintillation crystal detectors have gained prominence in all modalities in medical imaging. Several reviews and references exist in literature that elaborate on the state-of-the-art.

Scintillating Crystals	X-ray CT Positron CT	Farukhi (1,2) Farukhi (1-3)
Gas (Xenon) Ionization Chambers	X-ray CT Positron CT	Boyd et al. (4), Peshman (5) Lim et al. (6), Townsend et al. (7)
Liquid Xenon Multiwire Proportional Chamber	ECT	Zaklad et al. (8)
Liquid Xenon Scintillator	Positron CT	Lavoie (9)
Semiconductors	Positron CT X-ray CT	Kaufman et al. (10) Beinglass et al. (11)
Plastic Scintillators	Emission CT	McIntyre (12)
Gas Scintillators	X-ray	Hoan et al. (13)
Fluoroscopic Screens	X-ray CT	Baily (14)

With the exception of high pressure Xenon ionization detectors in X-ray CT scanners, most instrumentation in x-ray, single photon or positron CT diagnosis utilize scintillators and a study of their properties will be covered in this chapter.

2. HISTORICAL PERSPECTIVE

In diagnostic medical imaging, the advent of CT instrumentation started with:

X-ray Transmission CT: (XCT) 1972	Hounsfield in 1972 with a single NaI(Tl) detector which moved laterally and at each lateral position it made rotational motion around a patient. (15)
Single Photon Emission CT: (SPECT) 1963	Kuhl and Edwards (16) with a set of parallel and opposed NaI(Tl) scintillation detectors extended rectilinear scanning by adding a transverse and rotational motion to their scanner.
Positron Emission CT: (PET) 1953	Brownell and Sweet (17) used two opposing planar array of NaI(Tl) detectors which translated and rotated around a patient.

Tomographic imaging techniques per se in medicine originated much earlier than noted above and have been discussed in various reviews while instrumentation perspectives have been commented upon by others (18-23). In XCT instrumentation, the scanners have undergone four (4) generations of development and are presently in their fifth stage of development, characterized by:

- Over 1,000 detecting channels arranged in a stationary circular array and unquestionably employ scintillation detectors.
- The light sensing element is a solid state diode -- typically silicon photovoltaic diode.
- These DISCO (diode-scintillator combination) are being miniaturized and optimized.

Both SPECT and PCT designs have graduated into a multiple circular array so that the commonality in instrumentation expressed earlier (20) is now a reality and hybrid instrument schemes are tantalizing to consider for the future. The basic element for detection -- scintillator and PMT --

can also be replaced with a disco and the results of a pulse counting combination of CsI(Tl)/Si-PIN diode are discussed in a later section.

The historical perspective in CT detector development is marked with the painful laborings of industrial manufacturers as they vied competitively in the market place. Properties such as afterglow or phosphorescence, decay modes, fast components, etc., were utilized in the selection of scintillators without even quantifying them. Perhaps, it is unique in that new candidates for scintillation detector applications were actually adopted for commercial production and the research on their characteristics followed rather than vice versa.

XCT: The Hounsfield scanner introduced in 1972 utilized a single NaI(Tl) scintillator (15) and was used for transmission imaging of the head. The logical extension for whole body scanning was introduced by Ledley et al (24) in a modified second generation (multi-detector, translate and rotate motion for both source and detector) version of the EMI scanner. The detectors were still NaI(Tl) and were supplied by Harshaw. This effort resulted in the formation of two nucleating companies -- DISCO (a product of the National BioMedical Research Foundation of Georgetown University) which was absorbed by Pfizer Medical Systems, and Neuroscan which was to become the CT/N of General Electric.

In both cases, commercial grade NaI(Tl) was used, resulting in superior signal/noise ratio than the "EMI-blend" and exhibited tremendous difficulty in use for imaging due to afterglow!

The "EMI-blend" of NaI(Tl) is low Tl grown material which typically is 30% the pulse light output of commercially available NaI(Tl) but exhibits 0.3% levels of afterglow compared to 1-5% for the same time-frame (see reference 1). About the same time, Ohio-Nuclear introduced their head- and whole-body scanners, except CaF_2(Eu) was substituted for NaI(Tl) based on better and more controllable afterglow characteristics. Both the Pfizer and the GE NaI(Tl) scanners were eventually converted to CaF_2(Eu) before being eventually phased out of production. CaF_2(Eu) has a similar level of light output and afterglow as the "EMI-based" NaI(Tl). Again, the demand was satisfied, not by doing careful measurements, but by selecting what scintillators were commercially available which were non-hygroscopic, exhibited high light output with little, if any, afterglow under intense x-ray excitation.

$Bi_4Ge_3O_{12}$ (BGO) became available as a new R/D scintillating material in

1975 and was immediately considered by CT instrument manufacturers due to its higher stopping power which allowed for many more detecting channels for the scanning circle. As the need for ever increasing number of channels and the consequent necessity of miniaturizing while maintaining an economic perspective grew, photomultipliers were replaced by silicon photodiodes and the scintillator selection criteria extended to emission spectra and spectral match between scintillator and the light sensor.

Siemens led in this innovation with the well-known combination of CsI (Tl)/Si diode (25) in 1976. Soon, BGO was to be replaced by $CdWO_4$ whose afterglow was comparable to BGO, with a higher light output than BGO, and a comparable spectral match to Si-diode as CsI(Tl). $ZnWO_4$ and other oxide scintillators are being considered as alternates, and we finally observe a more normal pattern where R/D in material science is ahead of commercial adaptation (2).

PCT: Systems for PCT imaging seemed resigned to use NaI(Tl) where the compromise between detection efficiency and spatial resolution caused an impasse, accounting for the image in the developmental cycle compared to XCT instrumentation which, though it was introduced at least 20 years later, it was adopted more universally. Cho and Farukhi's (26) contention in the suitability of BGO in lieu of NaI(Tl) for positron cameras addressed the compromise between detection efficiency and spatial resolution. Based on these results, the Montreal Neurological Institute committed itself to BGO (a bold step but a very monumental one) and commissioned Harshaw to construct the first BGO Positron camera -- the Positome II (27). Since then, most commercial and various research PCT systems use BGO.

The inferior time resolution of BGO compared to NaI(Tl) led to the investigation of other commercially available scintillators, resulting in the discovery of CsF for time-of-flight assisted tomography (28,29). Here again, the Allemand discovery (28) was not based on an apriori study of scintillators, but on a Harshaw supplied CsF crystal commercially grown for fast counting purposes and casually supplied by the author as an interesting candidate for positron studies. Since that event, interest in fast inorganic scintillators was renewed and various candidates have been considered for the time-of-flight tomography (3,30).

SPECT: Systems for studying single photon emission started with Kuhl and Edwards (16) use of NaI(Tl) leading to commercialization by various instrument manufacturers in the form of rotating gamma camera heads. However, the detector of choice has remained NaI(Tl), though attempts to divert attention to BGO (20) have been made. This technique is susceptible to photon statistics, self-attenuation and inefficient photon utilization, and hence growth and the widespread adoption of this imaging technique has been hindered. A scintillator having the light conversion efficiency of NaI(Tl) but all other properties of BGO would cause a major improvement in the pace of development of SPECT instrumentation.

3. BASIC THEORETICAL CONCEPTS

3.A Terminology

The theory of luminescence in solids is beyond the scope of this investigation and several excellent treatises exist on this subject, both from the theoretical and experimental point of view. Some references are listed below:

F.A. Kroger - Reference (31)
D. Curie - Reference (32)
J.B. Birks - Reference (33)
P. Goldberg (eds.) - Reference (34)

It is useful, however, to address certain aspects of luminescence that are germane to the understanding of the scintillation phenomena in the materials considered for this study. We need to define certain terms that are commonly used, but not fully understood.

3.B Luminescence - A two-step process that involves the excitation of the electronic system of a solid and the subsequent relaxation of this system to its original state with a resultant emission of electromagnetic radiation (light photon from ultraviolet to infrared in excess of thermal radiation.

Then, depending upon the mode of excitation, we distinguish:

- <u>Radioluminescence</u> produced by absorption of x-rays, gamma rays, charged particles, fission fragments, etc.
- <u>Cathodoluminescence</u> produced by the absorption of electrons.
- <u>Photoluminescence</u> produced by the absorption of light (usually ultraviolet radiation).

Other modes of excitation include the application of an electric field (electro-), chemical reaction (chemi-), biological (bio-), and mechanical (tribo-).

The resultant emission is categorized into fluorescence (see Fig. 1a) or phosphorescence (Fig. 1b), depending upon the nature of the energy level scheme.

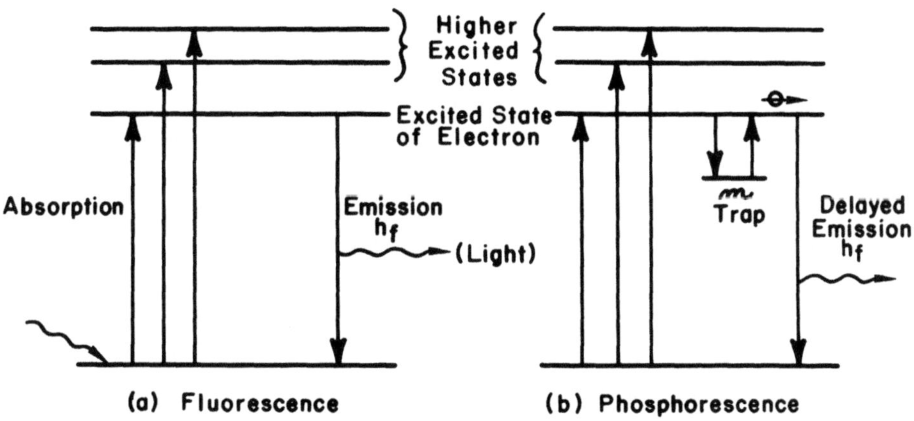

Fig. 1. Energy-level schemes for fluorescence and phosphorescence in solids.

The excitation by high energy radiation such as x-rays and gamma rays excites the electron into various excited states but the net result, in general, is to arrive at the first excited state from which emission occurs by relaxation to the ground state (fluorescence). The lifetime of such a transition is typically in the 10 ns region, about the lifetime of an atomic dipole emission. In phosphorescence, the emission is arrested by a metastable state "m", or a trap, and after a time "t" has elapsed, the electron may acquire sufficient thermal energy from its environment to be raised to the first excited state whereupon emission, albeit delayed, occurs as in the fluorescence case. Time-frame for phosphorescence is in the order of a second or longer.

In the intermediate time range from 10 Ms to 100 ms, it becomes difficult to differentiate between "long fluorescence" (due to forbidden transitions) and unfortunately, one is concerned with the use of phosphors for CT instrumentation.

The distinction can be made by studying the luminescence decay as a function of temperature;

Fluorescence - Is not strongly dependent upon temperature and usually both the decay time and intensity or efficiency decrease with increasing temperature due to the increasing probability of non-radioactive de-exciting modes.

Phosphoresence - Is strongly dependent upon temperature and the duration is a Boltzman function linked to the energy or trap-depth.

In either event, the measurement in the CT mode requires the value of the "persistence of luminescence" at a time "t" greater than 10 Ms after the cessation of excitation. This persistence is termed afterglow. It is of primary importance in the selection of a scintillator for use in XCT, since the afterglow should be negligible or non-existent so as to allow for proper sequential measurements of the luminescent intensity, and thereby the absorption by the patient, without having to calibrate or correct for afterglow.

It is obvious from the discussion that the properties of primary interest to XCT, PCT and ECT imaging are:

1. Afterglow
2. Fluorescence efficiency - Pulse height and integrated light output (current)
3. Luminescence decay modes
4. Temperature dependence of light efficiency and decay

and as we consider the containment of the radiation in the detector, other properties such as stopping power, hygroscopicity, etc. (see references 1,2) can be delineated. In essence, the discussion of these properties is covered in detail (2) and will not be repeated except in those cases where new data and insight is provided.

4. CANDIDATES

The scintillators in use for radiation-based CT instrumentation are categorized as either being (a) halide, or (b) oxide crystal phosphors. The word "crystal" takes on added significance in that a large crystalline section has to be grown which is essentially single and transparent to its scintillation emission throughout its volume. The growth of crystals carries with it a myriad of problems different in contrast to the well-known technology for x-ray intensifying screen phosphors which are more luminescent and are formulated from powders and/or sintered agglomerates of tiny crystals.

In the definition of luminescence terminology, we undertook the case of an essentially free atom such as in a gas and the associated electronic transitions between various states. The crystalline nature of the phosphors with its periodicity, depending upon lattice structure, imposes a crystalline field on the first excited states and thereby modifying the transitions. Essentially the ground and the first excited states stretch into valence and conduction bands and the metastable states become trapping levels or sites in the band gap. Below the conduction band, the electron-hole may form bound combinations known as excitons. For our consideration, it is sufficient to know that an ion such as Tl+ responsible for luminescence may exhibit slightly different characteristics such as intensity, decay, wavelength, etc., depending upon the crystalline lattice (host) it finds itself in. These classes of scintillators are known as impurity activated scintillators.

In other crystal lattices, an impurity-ion is not necessary for scintillation to occur. In BGO, the emission is governed by an excess of one of its constituents, bismuth; and in the case of BaF_2 (35), hole trapping at the halogen site followed by electron-hole recombinations is considered to be the responsible mechanism for scintillation.

Candidates that have been considered for CT application include (1-3):

Halides	Oxides
NaI(Tl)	$Bi_4Ge_3O_{12}$
CsI(Tl)	$CdWO_4$
CaF_2(Eu)	$CaWO_4$
BaF_2	$ZnWO_4$
BaF_2(Cl)	CeP_5O_{14}
BaFCl(Eu)	γ-$(Y,Ce)_2Si_2O_7$
CsF	
CdF_2	

4.A Halides

Among the halide lattices, the alkali halides have been subjected to the most study and provide the most reliable data with correlation to theoretical calculations. In addition to the references (32-35) the contribution edited by Fowler (36) adds to our understanding of the luminescence phenomena in these class of crystals. From the practical users point of view in medical imaging application, the following observations add to our insight into the use of alkali halide scintillators:

- All "pure" alkali <u>iodides</u> scintillate (decay times less than 10 sec) at room temperature and increase in their fluorescence efficiency at lower temperatures.
- Their counterparts in the group II-VII (alkaline earth <u>iodides</u>, e.g. CaI_2, SrI_2, BaI_2, HgI_2) also behave in a similar manner.
- With the exception of the cesium fluoride, the other alkali fluorides are not known to luminesce, or if they do they exhibit an extremely faint luminescence at liquid helium temperatures (4°K). Luminescence efficiency increases as one traveses to the chlorides, to bromides, and finally to the iodides.
- The mechanism causing luminescence in the "pure" halides in hole trapping at halogen sites to form hole centers designated as V_k centers (molecular ion combination comprising of two neighboring anions and the hole).

$$h + 2\,I^- \;---\; I_2^- \qquad 1.0$$

$$h + I^- + Cl^- \;---\; ICl^- \qquad 2.0$$

These centers may be arrested at lattice defects at low temperatures and become mobile as the temperature increases. In either event, a recombination with an electron leads to relaxation to the initial state with emission of light photons. Decay time of the emitted light is usually very fast compared to light emitted by impurity activation of the halide.

- Impurity additions of metallic ions (cations) such as Tl, Mn, Cu, Ag, Hg, etc., cause alkali halides to luminesce over a wide spectrum with varying efficiencies and time constants for decay. By far, the most useful of these ions is thallium (Tl) doping of the alkali iodides. NaI(Tl) has been the mainstay of detectors in medical application for many years.

° Energy transfer in the alkali halide can occur by:

(a) Electron and hole diffusion in their respective bands (37);
(b) Exciton (bound-electron-hole pair) diffusion;
(c) Resonance transfer - absorption of intrinsic luminescence of the host lattice by an impurity center and its characteristic re-emission at a longer wave length;
(d) The existence of hole (V_k) and mixed hole centers (ICl-) which may be mobile at higher temperatures in different lattices. TlCl (I,Be) is a scintillator which scintillates at room temperature only when activated with Iodine (38).

All of the above processes are known to occur in the halides and a strong case could be advanced both pro and con for (a) and (b). The alkali halides are well known to be insulators with very poor photoconductivity indicating limited charge motion while a definitive experiment remains to be performed on exciton mobility in the lattice. The reader is referred to the texts (32-34,36) for further study. However, the following data in Table 1 illustrates why all of these processes augment the emission in NaI(Tl).

The thallous ion has absorption bands (A,B,C) whose location is altered slightly by the lattice surrounding it. The C and A band dominate while the B band is weak. Examination of the data in Table I indicates that the difference between the excition band and the C-absorption band is the least for NaI(Tl), 0.12ev, indicating a propensity for efficient exciton transfer of energy.

The Stokes shift is the measure or indicator of energy loss to non-radiative transitions or the thermal heating of the lattice. Again, the value is the lowest for NaI(Tl), indicating a propensity for efficient excition transfer of energy.

The A-band absorption and the intrinsic emission overlap is the greatest for NaI(Tl) which is again in its favor.

The hole mobility has been calculated by Murray (37) for NaI, KI, RbI ($5x10^{-8}$) and KCl ($4x10^{-10}$). Clearly any charge transport process involving hole motion will proceed the fastest in NaI and coupled with the fast diffusion time (37) will yield a more efficient transfer process.

TABLE I

ABSORPTIONS AND EMISSIONS IN ALKALI HALIDES @ 300°K IN ev

Halide	Exciton Band	C-Band Tl-Abs	Ex-"C"	Intrinsic Emission	Stokes Shift	A-Band Tl-Abs	Ent-"A"	Hole Mobility $cm^2/v\text{-}sec$
NaI	5.42	5.30	0.12	4.00	1.42	4.23	-0.23	$4x10^{-4}$
KI	5.63	5.25	0.38	3.35	2.28	4.32	-0.97	$4x10^{-6}$
CsI	5.62	3.65	0.48	3.65	1.97	4.15	-0.50	
NaBr	6.50	5.74	0.76					
NaCl	7.88	6.23	1.65					
CsBr	6.63	5.79	0.84					
CsCl	7.64	6.33	1.31					
CsF	9.13			3.18	5.95			

4.B Oxides

The oxide lattices have been studied as hosts for laser and maser applications and are utilized primarily for these applications. Luminescence phenomena in oxides has commanded the interest of researchers almost as early as the halides (31,32) and continues to be studied intensely (34, 36). The investigations of Nestor and Huang (39) on BGO and Gillette (40) on $CdWO_4$ stimulated the interest in scintillation applications fueled primarily by the development in CT technology.

Energy transfer and transitions leading to emission in BGO have been recently reviewed by Weber (41) and will not be repeated here. Dopant additions of rare earth ions -- Er, 1/6 and Nd were studied by Dickinson et al (42) and they concluded that BGO may hold some promise as a medium gain laser host. Investigations at Harshaw (2) date back to 1978 and have centered on sensitizing BGO to red emission with BGO doped with Cr, Sn, Nd, Si and (Al, TQ). Pulse height and energy resolution measurements did not indicate any substantial improvement over unacclaimed BGO. Peak shifts in emission spectra were noted. Recent developments in the scintillation aspects in BGO have centered around better crystal growth and in the "purity" of the material. Crystal growth itself is a purification process, and if one used grown crystal sections as the starting material the second re-growth produces a more "pure" crystal. Twice grown crystals have demonstrated energy resolution values between 9.3-9.5% FWHM for 662 keV gamma rays (43). Studies by Farukhi (43) indicate that (2X) grown BGO performs better than "8% ^{137}Cs " NaI(Tl) as a spectroscopic detector for energies 2.6 meV and higher. Even the low energy performance of BGO is suitable for considering it in lieu of NaI(Tl) for many spectroscopic applications.

Later examination of (3X) grown BGO indicated a 8.9% FWHM resolution for 662 keV gamma rays, the best reported value to date (44). This quality material was shown to be better at 2 meV and higher than " 7.5% ^{137}Cs " NaI(Tl). In both of these studies, the response to 122 keV gamma rays significantly differed from he other values, indicating a non-linearity of response at low energies much akin to the behavior in NaI(Tl). This area needs to be investigated. Other areas of investigation include the shape of the decay curve and its components, further regrowth to improve the "purity" of the crystal lattice and possibly controlled impurity activation so that BGO indeed becomes a replacement for NaI(Tl) in a majority of applications.

Cadmium tungstate crystal technology is less understood than BGO and almost every aspect of its being a radiation detector needs to be investigated. Two aspects critical to CT application are:

(1) Emission spectra to gamma excitation;
(2) Temperature coefficient of light output.

A common fallacy in the field lies in the belief that luminescence excited by uv radiation and gamma is identical. In most cases, this is indeed the case and facilitates the experimental study of the mechanism responsible for luminescence. However, the pure halides exhibit differences in emission peaks.

TABLE II

EMISSION SPECTRA MAXIMA FOR KCl, KBr AND KI AT TEMPERATURES BELOW 12°K UNDER uv AND X-RAY EXCITATION (34).

Excitation	KCl	KBr	KI
uv (7.76-8.26 ev)	2.27 ev	2.29, 4.43	3.31, 4.15
X-ray	2.31	2.27, 4.42	3.33, 4.00

A similar situation exists for $CdWO_4$ (2). Excitation in the long wavelength (360 nm) excites both the 480 and the 540 nm emission while the short uv (250-320 nm) excites only the 480 nm. X-ray or gamma excitation results in the preferential emission at 540 nm. The blue emission could be weakly excited and can account for afterglow (20 μs decay constant).

The temperature coefficient of $CdWO_4$ for light output was claimed to be "0" near room temperature (40). Since the material grown today is considerably different than that used in Gillette's investigation (40), the light output was measured in terms of pulse height and current (PMI) as a function of temperature and the data is displayed in Figure 2. The crystal was kept in an oven and was insulated from the PMT with a quartz light pipe of 1" diameter x 6" long. The PMT was kept at $20^{o}C$ and the data was taken for both heating and cooling with dry ice. The temperature was measured with a chromelalunel thermocouple and standard electronics were used.

In Figure 2, the pulse height and resolution remain constant until 40°C before both start to deteriorate as the temperature increases. At 100°C, the pulse height had decreased to 85% of its room temperature value in clear contrast to the behavior in either BGO or NaI(Tl). The better performance at higher temperatures should provide for uses in well-logging applications for natural background measurements. The current is more sensitive to temperature variation than pulse height and can be accounted for by the faint contribution of the long decay (20 microsecond) blue emission at 480 nm.

An interesting feature is the detection of thermoluminescence at or near room temperature. The sample had been subjected to both uv and gamma rays prior to measurement but not in any deliberate fashion. Upon initial heating, the pulse height and resolution varied in a similar manner to that shown in Figure 2, but the current exhibited a peak starting from $20^{o}C$ to $55^{o}C$, reaching a maximum at $25^{o}C$ with a current value three times that at $20^{o}C$ or $60^{o}C$. Further studies along this nature are in order to provide a better understanding of the fluorescence and trap-phenomena.

5. ADVANCED CONCEPTS

5.A Disco or diode/scintillator combinations made their entree as early as the 1960's with the University of Chicago group using combinations of silicon-diodes and CsI(Tl) to detect photons at 50 meV and higher (45, 46). These diodes were used for pulse counting applications and below 50 meV the signal/noise ratio became inadequate to count individual pulses. The application of this genre of diodes to CT instrumentation was introduced by Siemens where one sensed the current increase upon x-ray

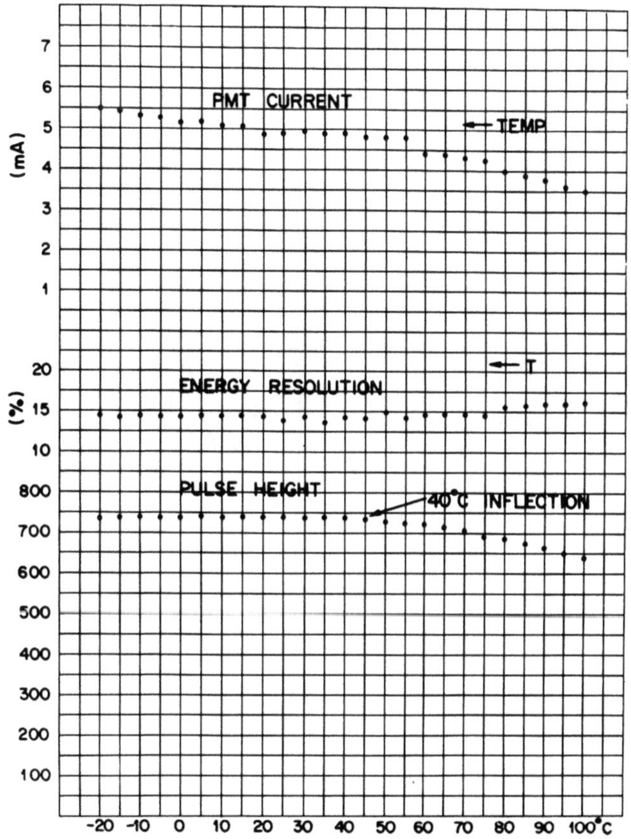

Fig. 2. Current, pulse height and resolution of $CdWO_4$ as a function of temperature.

upon x-ray excitation of the CsI(T1). Since CaI(T1) has considerable afterglow (2) and which varies from sample to sample depending upon the history and trap population, most manufacturers adopted $CdWO_4$ for such applications. The disco combination is utilized either in an array in CT tomography or individually to sense dosage or density in non-CT applications.

The recent introduction of the PIN silicon diode by Hamamatsu is a major improvement in the noise characteristics of these diodes. It is possible by careful selection to choose diodes that, when coupled to selected CsI(T1) crystals, one can form a scintillators/sensor unit which can pulse count to 335 keV ! A device such as this has been constructed and spectra measured at $20^{o}C$ and $0^{o}C$ is illustrated by the series of Figures 3-6.

Fig. 3. Co^{60} and Cs^{137} operation spectra for 20°C operation. CsI(Tl)/Si pin diode. (Crystal size: 20mm x 20mmX20mm).

The question now remains what further improvement can be effected either in the diode or in the scintillator.

5.B New Scintillators

The reawakened interest in oxide scintillators itself has provided a treasure-house to explore. BGO used initially for charge integration application is destined for spectroscopic replacement for NaI(Tl). $CdWO_4$ has replaced BGO but with its temperature behavior advantage (see previous section) could find further use in high temperature spectroscopy creating the need for yet a better and new scintillator replacement from possibly the oxide family.

BaF_2 has gained acceptance by virtue of its fast uv component for time-of-flight tomographic application. The mechanism is most likely linked to hole-trapping or excition decay since the emission is very close to

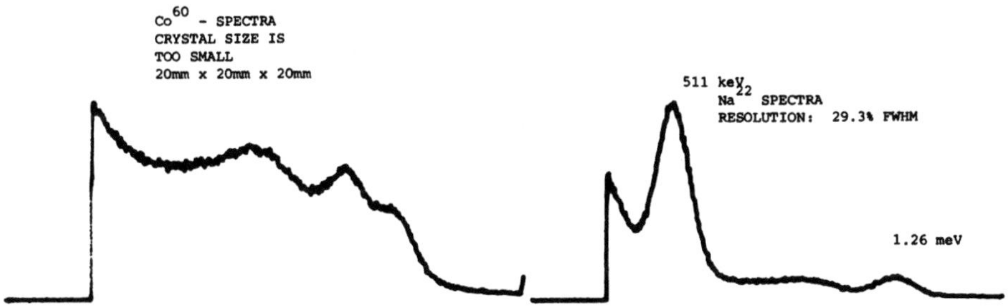

Fig. 4. Co^{60} and Na^{22} spectra for 0°C operation. CsI(Tl)/Si pin diode.

the fundamental absorption edge. Discovery will now lead to examination of the cause for scintillation. Fluorides provide for fast decay scintillators such as CdF_2, CsF_2, BaF_2 (3) and possibly SrF_2. Obviously, one must examine other fluorides for this phenomena and this must be coupled to the study of hole-trapping or hole centers in polar crystals.

6. NEW SENSORS

Considerable interest has been generated in the study of HgI_2 and CdTe, not only as photo diodes for the detection of nuclear radiation but also as photosensors (47) coupled to scintillators. Silicon avalanche diodes have been suggested for providing gain in the detection of light from low energy excitation of the scintillators. However, this investigation presents a low cost silicon AN diode that could be used to detect 511 keV annihilation photons (see figure).

The field of semiconductors and photoconductive compounds is vast and often reviewed. New candidates other than HgI_2 and CdTe should be addressed towards this end.

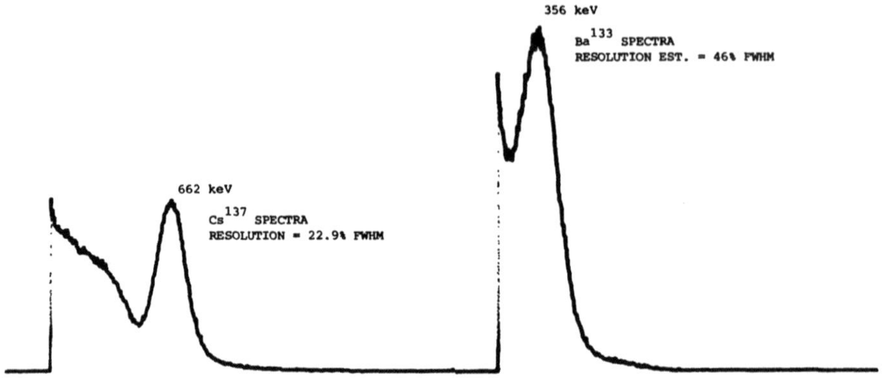

Fig. 5. Cs^{137} and Ba^{133} spectra for 0°C operation. CsI(Tl)/Si pin diode. (Crystal size: 20mm x 20mm x 20mm).

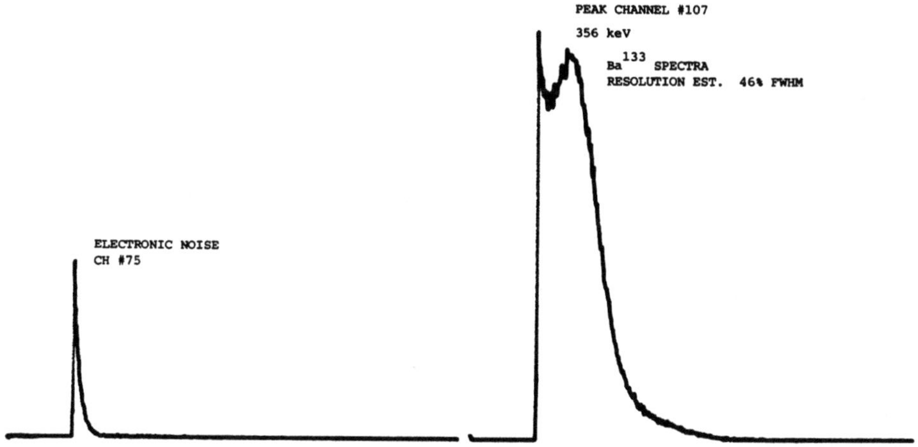

Fig. 6. Electronic noise and Ba^{133} spectra for 20°C operation. CsI(Tl)/Si pin diode. (Crystal size: 20mm x 20mm x 20mm).

REFERENCES

1. Farukhi MR (1978). Scintillation detectors for CT applications - An overview of the history and state-of-the-art. Workshop on Transmission and Emission Computed Tomography, Seoul, Korea.

2. Farukhi MR (1982). Recent developments in scintillation detectors for x-ray CT and positron CT applications. IEEE Trans Nucl Sci NS-29:1237.

3. Farukhi MR (1982). Fast inorganic scintillators for time-of-flight tomography. Workshop on Time-of-Flight Tomography, Washington University, St. Louis, Proc IEEE Comput Soc No. 448.

4. Boyd D, Coonrod J, Dehnert J, et al (1974). A high pressure xenon proportional chamber for x-ray laminographic reconstruction using fan beam geometry. IEEE Trans Nucl Sci NS-21:184-187.

5. Peshman KR (1981). Xenon gas ionization detectors. (In) Radiology of the Skull and Brain, (Newton TH and Potts DG, eds), The C.V. Mosby Company, Vol. 5, pp 4112-4126.

6. Lim CB, Chu D, Kaufman L, et al (1975). Initial characterization of a multiwire proportional chamber positron camera. IEEE Trans Nucl Sci NS-22(1):388-394.

7. Townsend D, Schorr B and Jeavons A (1980). Three-dimensional image reconstruction for a positron camera with limited angular acceptance. IEEE Trans Nucl Sci NS-27(1):463-477.

8. Zaklad H, Derenzo SE, Budinger TF, et al (1974). Liquid xenon multiwire proportional chambers for nuclear medicine applications. Proc First World Congress of Nuclear Medicine, Tokyo, pp 362-367.

9. Lavoie L (1976). Liquid xenon scintillators for imaging of positron emitters. Med Phys 3:283-293.

10. Kaufman L, Williams SH, Hosier K, et al (1978). An evaluation of semiconductor detectors for positron cameras. J Comput Assist Tomogr 2:651.

11. Beinglass I (1980). Med Phys 7:370.

12. McIntyre JA (1980). Plastic scintillation detectors for high resolution emission computed tomography. J Comput Assist Tomogr 4:351-360.

13. Hoan NN, Majewski S, Charpak G and Policarpo AJPL (1979). An efficient, gaseous detector with god low-energy resolution for ($\leq$ 50 keV) imaging. J Nucl Med 20:335-340.

14. Baily NA (1980). Video techniques for x-ray imaging and data extraction from roentgenographic and fluoroscopic presentations. Med Phys 7(5):472-791.

15. Hounsfield GN. Method and apparatus for measuring x- or -radiation absorption or transmission at plural angles and analyzing the data. British Patent No. 1283915 (1972), U.S. Patent No. 3778014 (1973).

16. Kuhl DE and Edwards AQ (1963). Image separation radioisotopic scanning. Radiology 80:653-661.

17. Brownell GL and Sweet WH (1953). Localization of brain tumors with positron emitters. Nucleonics 11:40.

18. Brooks RA and DiChiro G (1976). Principles of computer assisted tomography (CAT) in radiographic and radioisotopic imaging. Phys Med Biol 21:689-732.

19. Phelps ME (1977). Emission computed tomography. Sem Nucl Med 7(4): 337-365.

20. Cho ZH, Nalcioglu O and Farukhi MR (1978). Analysis of a cylindrical hybrid positron camera with bismuth germanate (BGO) scintillation crystals. IEEE Trans Nucl Sci NS-25:952-963.

21. Boyd DP (1979). Status of diagnostic x-ray CT. IEEE Trans Nucl Sci NS-26:2836-2839.

22. Boyd DP (1981). Transmission computed tomography. (In) Radiology of the Skull and Brain, (Newton TH and Potts DG, eds), Vol. 5, Chapter 130, pp 4357, The C.V. Mosby Company.

23. Budinger TF, Derenzo SE, Huesman RH and Cahoon JL (1982). Positron emission tomography: Instrumentation perspectives. Proc International Workshop on Phys and Engr in Medical Imaging, Pacific Grove, IEEE Comput Soc No. 406:3-13.

24. Ledley RS, DiChiro G, Luessenhop AJ and Twigg HL (1974). Science 186:207-212.

25. Fan CY (1974). Detection of scintillation photons with photodiodes. Rev Sci Instr 35(2):158-163.

26. Cho ZH and Farukhi MR (1977). Bismuth germanate as a potential scintillation detector in positron cameras. J Nucl Med 18:840-844.

27. Thompson CT, Yamamoto YL and Mayer E. Positome II (Abstract) J Comput Assist Tomogr 2(5).

28. Allemand R, Gresset C and Vacher J (1980). Potential advantages of a cesium fluoride scintillator for time-of-flight positron camera. J Nucl Med 21:153.

29. Mullani NA, Ficke DC and Ter-Pogossian MM (1979). Cesium fluoride - A new detector for positron emission tomography. IEEE Trans Nucl Sci NS-27:572.

30. Laval M, Moszynski M, Allemand R, et al. Barium fluoride - Inorganic scintillator for subnanosecond timing, (to be published in Nucl Instr and Methods).

31. Kroger FA (1948). Luminescence of solids. Elsevier Publishing Company.

32. Curie D (1963). Luminescence in crystals. Wiley & Sons, Inc.

33. Birks JB (1964). The theory and practice of scintillation counting. The MacMillan Company.

34. Goldberg P (ed) (1966). Luminescence of Inorganic Solids. Academic Press.

35. Farukhi MR and Swinehart CF (1971). Barium fluoride as a gamma ray and charged particle detector. IEEE Trans Nucl Sci NS-18:200.

36. Fowler WB (ed) (1968). Physics of Color Centers. Academic Press.

37. Murray RB. Energy transfer in alkali halide scintillators by electron-hole diffusion and capture.

38. Rosette KH, Farukhi MR, Kramer GR, et al (1970). A high Z scintillator. IEEE Trans Nucl Sci NS-17(3):89-94.

39. Nestor OH and Huang CY (1975). Bismuth germanate: A high Z gamma-ray and charged particle detector. IEEE Trans Nucl Sci 22:68.

40. Gillette RH (1950). Calcium and cadmium tungstate as a scintillation counter crystal for gamma-ray detection. Rev Sci Instr 21: 294.

41. Weber MT (Nov 10, 1982). Discovery of the scintillation properties of BGO; underlying principles. International Workshop on Bismuth Germanate, Princeton University.

42. Dickinson SK, Hilton RM and Lipson HG (1972). Czochralski synthesis and properties of rare earth doped bismuth germanate ($Bi_4Ge_3O_{12}$). Mat Res Bull 7:181-192.

43. Farukhi MR (1982). $Bi_4Ge_3O_{12}$ - A scintillator replacement for NaI(Tl). Proc 1982 Annual Meeting of the Material Research Society, Boston.

44. Farukhi MR (1982). Improvements in the scintillation response of $Bi_4Ge_3O_{12}$. Proc International Workshop on Bismuth Germanate, Princeton University.

45. Tuzzolino AJ, Hubbard EI, Perkins MA and Fan CY (1962). J Appl Phys 33:148.

46. Fan CY (1974). Detection of scintillation photons with photodiodes. Rev Sci Instr 35(2):158-163.

47. Iwanczyk JS, Barton JB, Dabrowski AJ, et al (1983). A novel radiation detector consisting of an HgI_2 photodetector coupled to a scintillator. IEEE Trans Nucl Sci NS-31(1):363-367.

POSITRON EMISSION TOMOGRAPHY - BASIC PRINCIPLES, CORRECTIONS AND CAMERA DESIGN

Christian Bohm, Lars Eriksson and Mats Bergström

From the Department of Physics, University of Stockholm, and Departments of Neurophysiology, Neuroradiology and Radiation Physics, Karolinska Institute, Stockholm.

By using positron emission tomography (PET) it is possible to noninvasively study distributions of subtances labeled with positron emitting isotopes (Fig. 1). A number of positron cameras with widely varying de-

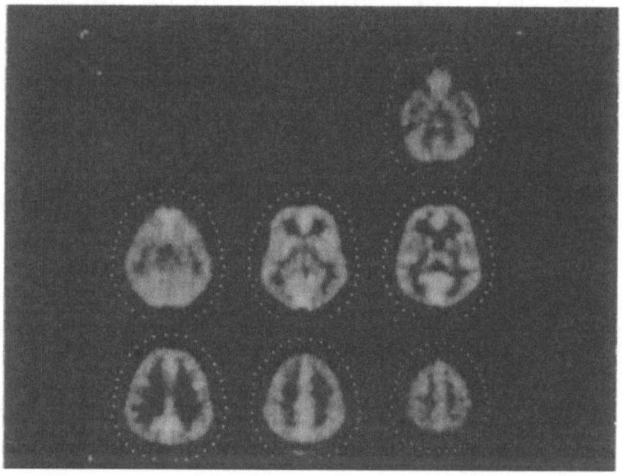

Fig. 1. Seven slices through the patient's head obtained with a 4-ring detector positron camera system (Scanditronix PC-384B). The images show the uptake of C-11-labeled glucose in a normal brain. The separation between the slices is 13.5 mm, with the lowest section in the upper right corner.

sign parameters have been developed for this purpose. Some are designed to study activity distribution in different parts of the body. Others are dedicated to the brain. The data obtained are used to reconstruct images describing the distributions. These images are sometimes further processed to evaluate physiological parameters of special interest within specified regions of the image. In this case it is often necessary to use sequences of images to determine the time dependence of the distribution. Such a quantitative use of the image data demands a high degree of reproducibility and freedom from systematical errors. This, in turn, puts high demands on the positron camera design and on the algorithms and correction schemes used in the data processing. The positron camera should also be designed to have high sensitivity and good spatial resolution. The aim of this chapter is to discuss how this can be done. In order to do this, the chapter is divided into three parts. The first part introduces some basic principles, the second part deals with necessary corrections and the third discusses some actual camera designs.

1. BASIC PRINCIPLES

1.1 Positron Physics

The presence of positron emitting isotopes can be detected by the two 511 keV gamma rays that are created when a positron annihilates with an electron. These gamma rays are emitted in opposite directions. Thus, when two 511 keV gamma rays are detected in a pair of opposed gamma detectors, one in each detector, they may or may not have been emitted from the same positron annihilation event. By requiring that the two gamma detections should coincide or more realistically should occur within a short time interval, it is possible to reduce the probability for an uncorrelated event. The two types of coincidences are known as true coincidences and random coincidences. Now, if the gamma rays were emitted from the same annihilation event (a true coincidence), this event must have occurred in the volume spanned by the two detectors. This volume, which can be regarded as the active volume of a positron detector, is often referred to as the detection channel.

This description is, of course, simplified. In reality, the gamma ray pair is not exactly 180 degrees opposed. There is a small deviation due to the momentum of the positron. Due to this motion, other positron will also travel some distance before the annihilation. This means that the position of the positron annihilation, which is the entity measured by the positron camera, does not exactly coincide with the position of the positron emitting isotope. The kinetic energy of the positron, which is a characteristic of the given positron emitter, will thus impose a lower limit on the possible resolution obtained by a positron camera.

	C-11	Ga-68	Rb-82
Endpoint energy of -spectrum (MeV)	0.961	1.899	3.335
R (50%) in mm	0.54	1.6	4.3

Table 1. The contribution of the positron range to the intrinsic resolution cannot be made on a FWHM and FWTM basis. Derenzo (1) has suggested an alternative parameter, the fraction of annihilations whose projection on the image plane falls within a distance R from the source. The fraction referred to in the Table is 50%.

Even if the annihilation occurred in the detection channel and created gamma rays directed along the detection channel towards the detectors, this does not necessarily lead to a detected coincidence. One gamma ray may pass through the detector without being absorbed (due to detector inefficiency), it may also be absorbed in a photoelectric process before reaching the detector or it may be scattered out of the channel in a Compton process. There is at the same time a distinct possibility that an annihilation which occurred outside the detection channel may be scattered towards the detectors and thus create an undesired scattered coincidence. However, by demanding that the gamma energy must exceed a given value (a threshold value) it is possible to reduce the rate of scattered coincidences, since Compton scattered gamma rays have lower energies. This requirement will also reduce the rate of random coincidences. However, increasing the energy threshold will remove an increasing number of true coincidences.

1.2 Detectors and Detector Materials

The spatial resolution of a positron camera depends on the width of the detection channel. There are two different approaches to minimize this width. One can either use small detectors, or use larger but position-sensitive detectors with high intrinsic spatial resolution.

When the first approach is used, a standard gamma detector design is usually chosen. It consists of a photomultiplier tube (PM-tube), a scintillation crystal and often a light guide to connect the two. The light guide is used when the mechanical design requires a distance between the PM-tube and the crystal. However, since it introduces an additional boundary surface in the detector and thus a potential source of light loss, the light guide is, if possible, eliminated.

Several crystal materials have been used to design gamma detectors.

Their properties (see Table 2) make them suitable in different applications. In early constructions NaI(T1) was used. Since this material, as can be seen in the Table below, is comparatively inefficient, the change to BGO (bismuth germanate) in later constructions leads to a significant efficiency improvement (about a factor 4). Furthermore, the BGO crystal is not hygroscopic and consequently does not need an airtight container. This results in a more compact detector assembly, which also adds to the increased efficiency.

Detector Material	CsF	BGO	NaI(T1)	BaF2	
				Fast	Slow
Density (gm/cm^3)	4.64	7.13	3.67	4.89	
Absorption Coeff. 511 keV Photons (cm-1)	0.44	0.92	0.34	0.47	
Scintillator decay time in ns	5	300	250	0.8	620
Emission wavelength in nm	390	480	410	225	310
Energy resolution in FWHM of 511 keV in percent	23	>10	>7	13	

Table 2. Some relevant properties of different crystal materials used in positron cameras (2).

The second approach using larger but position-sensitive detectors has been tried with proportional wire chambers (3). These detectors possess excellent intrinsic spatial resolution, but the efficiency is low, in the order of 10%. This is a serious limitation, since the relevant efficiency parameter of the instrument is proportional to the square of the gamma detector efficiency. Another way to construct a positron-sensitive detector is to use the light division method (as in Anger cameras) (4,5). Two or more PM-tubes are then mounted onto one crystal. By comparing the light input to the different PM-tubes it is possible to estimate the position of the gamma absorption.

1.3 Detection Elements

The basic information collecting element in a positron camera is the detector pair introduced above connected to a coincidence circuit. This circuit may be a conventional coincidence circuit or a time of flight system. Each detector is also connected to an energy discriminator, which is used to reject events with unacceptable energies. This will reduce the rate of scattered and random coincidences but also the true coincidences.

1.3.1 Coincidence Detection Element

When two detectors are connected to a coincidence circuit, the only events that are accepted are those that occur in both detectors within a time interval T. Since the position within the detection channel is of no consequence, the count rate is, in the ideal case (neglecting corrections), proportional to the total activity in the detection channel, i.e. it is proportional to the projection of the activity distribution on an axis perpendicular to the detection channel.

As was mentioned above, there will also be a contribution from uncorrelated events (randoms) due to the finite time window. The size of the random contribution (R c/s) is given by:

$$R = 2T*S1*S2 \qquad (1)$$

where S1 and S2 are singles count rates in the respective detectors. The choice of the coincidence time window T is made on the basis of the following requirements: it must be wide enough to accept most of the true coincidences, but as small as possible to avoid unnecessary random events. The optimum choice will depend on the singles count rates. At very high count rates T should be reduced. If a fixed time window is chosen for moderate count rates, the FWTM (full width of tenth maximum) of the timing distribution is often a good compromise.

When considering a coincidence detection element, the relevant parameters are: good time resolution, narrow detection channel, high detector efficiency and reasonable energy resolution.

The parameters are, to a large extent, determined by the choice of detector material. The most commonly used materials are in this case NaI(T1) and BGO (bismuth germanate).

1.3.2 Time of Flight

In a time of flight (TOF) design, the time interval between the two events events is also measured. This additional parameter determines the position of the annihilation event within the detection channel. The position information can be used in the reconstruction process to improve the signal/noise ratio of the image. However, the time resolution should be less than 500 ps to obtain any significant improvement. This requirement outrules BGO as a suitable TOF crystal material. Cesiumfluoride (CsF) and bariumfluoride (BaF2) have, on the other hand, been found to be possible (6). Of those, BaF2 gives the best timing resolution, is not hygroscopic and has the additional advantage of allowing a higher packing fraction.

1.4 Reconstruction methods

1.4.1 Tomographic reconstructions

The data obtained from coincidence detection elements determine the projection of the activity distributions along different directions in the object. Assume, for the time being, that there are enough measurements to consider the projections continuous, and that there are enough projections to cover all angles. Let us furthermore assume that the measurements are accurate and that the detection channel is very narrow, giving us a negligible error in the position determination perpendicular to the detection channel. The data are then of the same form as those of CT, which means that the same image reconstruction algorithms apply. As an object to be processed, let us, for the sake of simplicity, choose a point source. This does not limit the applicability of the discussion, since if the image of a point source (the point spread function, the PSF) is known, it is easy to calculate the image of any extended source (by convolution of the source distribution with PSF) due to the linearity of the method. If the projections are measured and backprojected into the image plane (see Fig. 2), the resulting image will show a $1/r$ point spread function, giving a smeared-out image. This image will, however, improve considerably after filtering, provided that the filtering function was chosen as the inverse (in convolution sense) of the $1/r$ distribution. It is easily realized that the order of the backprojection and the filtering process will not affect the result. However, reversing the order will improve the reconstruction speed considerably, since the filtering is then performed on one-dimensional projections instead of on the 2-dimensional image. This method is called the "filtered backprojection" method. The absolute value of the Fourier transform of the corresponding filter (the frequency response function FRF) is proportional to the frequency (Fig. 2d).

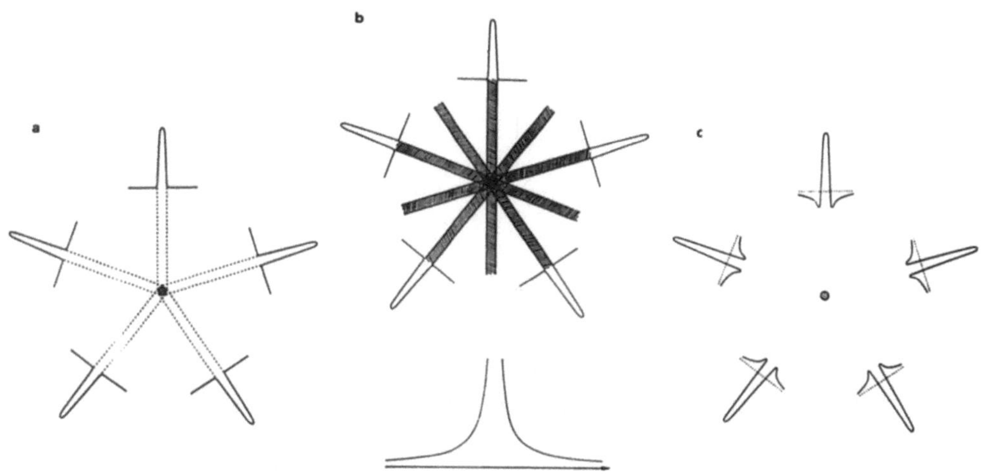

Fig. 2: The filtered backprojection method.
(a) Projections in different angles are obtained from measurement data.
(b) Top: Backprojection of projection data.
Bottom: Corresponding image point spread function.
(c) Backprojection of filtered projection data.

The object distribution is not completely recovered if a realistic width of the detection channel and the positron range are taken into consideration. The point spread function is then no longer ideal. Its projection will then be equal to the detector resolution convolved with the projected positron range distribution.

1.4.2 Time of Flight Reconstructions

In time of flight, the data consists of position estimates where the statistical uncertainty is not isotropic. The position uncertainty perpendicular to the detection channel is much smaller than the uncertainty along the channel. This data can be reconstructed using a variant of the filtered backprojection method (Fig. 3).

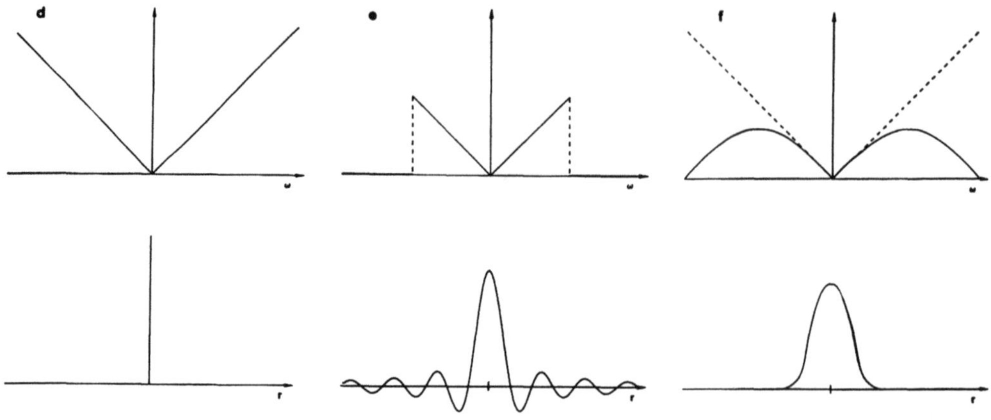

Fig. 2. (d) Top: The modulation transfer function, MTF frequency content) of an ideal reconstruction.
Bottom: Corresponding point spread function, PSF.
(e) Top: The MTF of a frequency range limited reconstruction.
Bottom: Corresponding PSF.
(f) Top: The MTF of a Shepp Logan type filter.
Bottom: Corresponding PSF.

The data is backprojected into the image plane using confidence weighting. This means that when one event is backprojected, each point of the image will be incremented by a value proportional to the probability that a positron annihilation in this point would give rise to the event. After backprojecting all data from a point source, the image will have a PSF:

$$P = \int f(t)f(r-t)dt \qquad (2)$$

where f is the distribution of the TOF information. A deconvolution function corresponding to p can be found, which can be used to restore the image. Even in this case it is possible to permutate the order of the backprojection and the filtering process.

Simulations and some experiments show that a TOF reconstruction using a

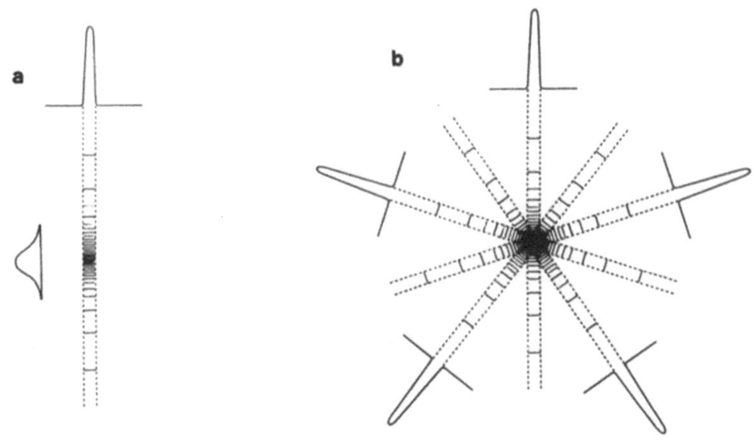

Fig. 3. Backprojection of time of flight information.
(a) Each pixel is incremented by a value proportional to the probability that the pixel was the location of the event.
(b) Backprojection of data from a point source.

time resolution of 300 ps improves the signal/noise ratio by a factor of 4π in an object with a diameter of 30 cm (2). In the extreme limit of very good time resolution the TOF reconstruction will convert into a direct mapping of the activity. In the other limit the TOF reconstruction will be equivalent to a tomographic reconstruction. The TOF reconstruction can also be made using a similar but simplified method where during the backprojection only one point is incremented for each coincidence event. This method is, however, statistically less efficient.

1.5 Noise

So far the projections have been assumed to be exactly determined. This is, of course, not true in a realistic situation. Since the values along the projections are determined by the number of counts in the detection channels and since these numbers are limited and Poisson distributed, there will be Poisson distributed statistical errors in the projections. These errors will appear as uncorrelated noise in the projection data.

In the reconstruction process, however, the noise is coupled through the reconstruction filter during the convolution, making the image noise correlated.

An accurately reconstructed image is usually very noisy. This is due to the shape of the fourier transform of the filter function (Fig. 2d), which suppresses low frequencies (the signal) and enhances the high frequencies (noise). This situation can be improved by setting the filter function to zero outside the frequency range of the signal (Fig. 2e) and thus restricting the frequency content of the signal and the noise simultaneously. However, since such an abrupt frequency cut-off will introduce oscillations in the image, a compromise is usually preferred (Fig. 2f).

1.6 Sampling Considerations

In the discussions above, it has been assumed that the projections were determined at every point and that all projections were present. This is usually not true. In practice, positron camera sampling is rather sparse. According to the Nyquist criterium, the sampling distance must be closer than 1/2 of the wavelength corresponding to the highest spatial frequency present in the projection, or information will be lost. The artefacts created by insufficient sampling density are called aliasing artefacts. When this criterium is applied to projection data, where each detector has a resolution of about half the detector width (d) and a maximum spatial frequency of about 2/d, the desired sampling distance will be greater than d/4.

There are two ways to eliminate aliasing artefacts. One is by increasing the sampling density by introducing other degrees of freedom, such as translations and/or rotations. The other is by limiting the frequency content of the projections by a suitable smoothing procedure (incorporated into the filtering function), which at the same time degrades the resolution. The first method is the common one while the latter requires a sufficiently high intrinsic spatial resolution that can accept the degradation. A comparison of the two methods shows, however, that if the resulting resolution is identical, the latter will give a better signal/noise ratio (7).

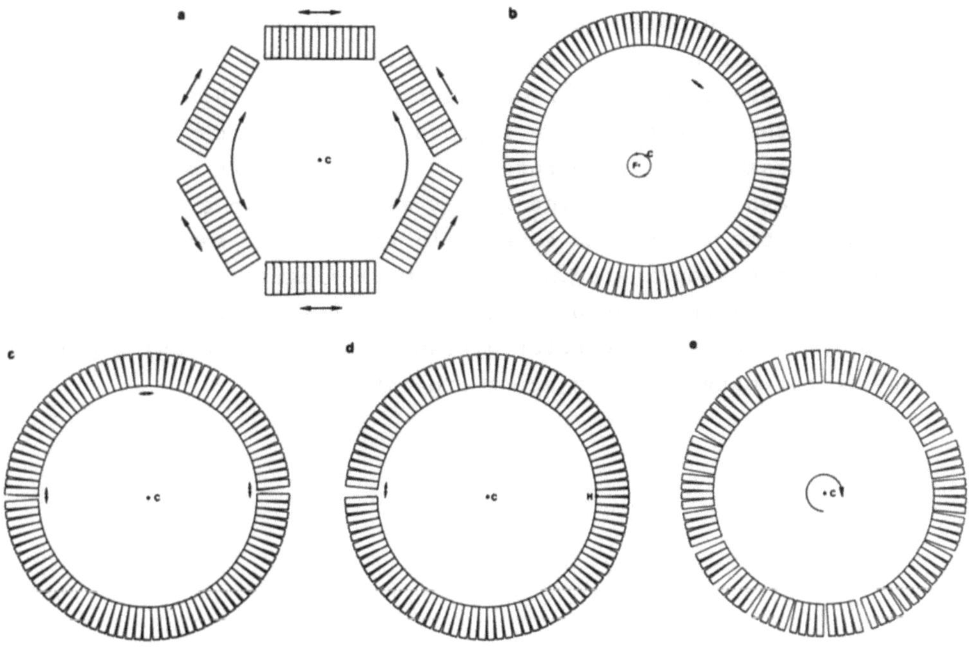

Fig. 4. Sampling density improving motions.

(a) Hexagonal geometry. Two motions are often used: a linear motion in the tangential direction and a rotational motion around the origin.
(b) Circular geometry. In the wobbling motion the center of the detector ring (C) will move along a circle with a fixed origin (F). A rotational motion around the center of the ring (C) is sometimes used, although it is not necessary to fulfill the sampling requirement.
(c) Dichotomic motion. The ring is divided into two parts, one stationary and one movable with the center (c) as a fixed point.
(d) Clamshell motion. The two parts are connected by a hinge (H).
(e) Rotation of a ring with unevenly spaced detectors.

Let us see how these problems are solved in a few different multidetector geometries. In a hexagonal geometry (Fig. 4a) the detectors are mounted next to each other in six detector banks (8). Six projections are measured at the same time. To improve the angular sampling, the whole detector assembly must be rotated. The bank pairs must also be translated sideways to improve the linear sampling. In a ring detector system (Fig. 4b) the linear and especially the angular sampling are better. Regarding the sampling points in a (O,r) diagram, a sinogram (Fig. 5), will lead to the realization that the effective sampling density is actually half the detector-detector distance, since the missing grid points may be estimated from adjacent projections.

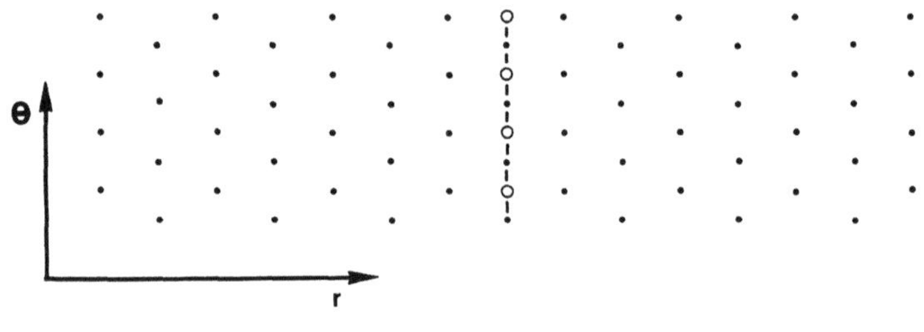

Fig. 5. The location of the sample points in a ring system displayed in a O,r diagram. The sampling density variation along the projection is not taken into consideration.

But in order to eliminate aliasing hazards, the sampling distance must yet be cut in half (or alternatively the frequency content must be reduced to 50%). This can be accomplished in different ways. The whole detector assembly can be moved continuously (wobbled) so that the center traces a circle (9). During this motion the projections can be sampled to any density. The wobbling may also be performed in a discrete mode. However, in both cases the sample point density will not be uniform. This means that some regions will be oversampled in order to satisfy the Nyquist criterium over the whole region. Other possible motions with a more uniform sampling density are the dichotomic motion (10) (Fig. 4c), the clamshell motion (11) (Fig. 4d), or a rotating detector ring with unevenly spaced detectors (12) (Fig. 4e).

Another important aspect of the sampling density improving motion is its repetition rate. A positron camera is often used in a dynamic mode, studying rapidly changing activity distributions. It is then essential that the motion scheme has a high repetition rate so that the activity distribution can be considered stationary during a motion cycle. Sampling density improving motions are difficult to implement when very short measuring times are used. This is, for example, the case when gating techniques are used to freeze heart motions during a positron camera scan. The measurement is then divided into short intervals each of which is stored in the memory according to the corresponding ECG-state.

1.7 Multiple Coincidence Events

When a multidetector system is used, multiple coincidence events will also occur. They are made up by combinations of single and true coincidences that occur within the time window. Although they do contain some information concerning the activity distribution, they are usually disregarded. A reconstruction of this data would be complicated. The rate of these events depends strongly on the number of constituents. Triple events are rare at moderate count rates, but at very high count rates they will contribute to a substantial part of the total number of events, resulting in substantial data losses.

1.8 Space Variant Resolution and Sampling

All positron camera systems are hampered by space variant resolution since the resolution function varies along the detection channel. In true ring systems the shape of the resolution function will also depend on the angle of incidence. If these space variant effects are not corrected for, they will lead to a space variant point spread function. Uneven sampling distances occur in true ring detector systems when they are used in a stationary or fixed point wobble mode. This is solved by interpolation.

2. CORRECTIONS

The properties of a positron camera have, so far, dealt with ideal conditions. In practice, a number of disturbing factors are present which, for artefact free images and adequate quantitation, have to be corrected for. These corrections include random, triple, scatter, attenuation and efficiency corrections.

2.1 Random Coincidences

The first correction is a random subtraction. The importance of this correction depends on the count rate since the number of random coincidences increases by the square of the total count rate, whereas true coincidences increase linearly. Random coincidences are a considerable problem in a BGO or Na(Tl) system when the injected amount of radionuclides exceeds 40-50 mCi (13,14).

A number of different methods for random correction have been described (15). In one method a delayed coincidence window is used. The pulses from the detectors are then individually delayed before entering the coincidence circuit. The events that satisfy the delayed coincidence criterium will, under these circumstances, not contain any true coincidences. They can thus be used to estimate the random count rate within the proper coincidence window. In another method a crude estimate of the fraction of random coincidences is obtained from peripheral detection channels outside the field of view (16). The level obtained there is then subtracted from the projection. A complicating factor in this case is that the scattered coincidences are included in the estimate, and they are also subtracted as a constant level instead of being subtracted according to its true distribution (13,15).

In a third method, the single count rates are monitored during the measurement. The count rates of random coincidences are then calculated using the formula (1) with an appropriate time window. This method determines the random contribution with good statistical accuracy. Thus it will not, contrary to the other two methods, degrade the statistical quality of the measurement (15).

2.2 Triple Coincidences

The occurrence of triple events has already been discussed in section (1.7). The dead time losses it causes as well as other dead time losses in the system can be corrected for by a simple renormalization, using an empirically determined facor (13,14). This correction factor depends exponentially on the coincidence count rate.

2.3 Scattered Radiation

In a typical positron camera study, a major part of the photons emitted in a central detection channel are scattered. The chance that these photons will hit a detector in a different detection channel and thus be erroneously recorded, is not negligible. The fraction of scattered events in a central detection channel is strongly dependent on the features of the PET scanner, for example, its geometry, size and shape of collimators, energy discriminator settings, etc. (17,18,19). Published values of the scatter contributions from different scanners range from about 5 to 25% (13,14,16,18,20,21,22,23,24,25).

The distribution of scattered events along a projection is directly related to the three-dimensional distribution of activity and scattering media. By assuming that the objects to be imaged are distributed along the z-axis, extending outside the region where scattered events can contribute to the count rate, the scattering can be treated as if it was determined by the two-dimensional distribution of activity and scatter media.

Experiments, where line sources were examined at different positions within a cylindrical water phantom (Fig. 6), show that the projection of the scatter distribution change very little as the line source is moved to different positions along the projection lines (26), i.e. the scatter distribution is to a first approximation a function of the activity projection. Moving the line source to different positions perpendicular to the projection lines leads to a definite change in the scatter distribution. Thus, the scatter distribution is not determined by the total amount of activity alone.

To a reasonably good approximation, the scatter distribution can be described by an integral transformation of the unscattered distribution, using an empirically determined scatter distribution function f(x,y) (26,27).

$$S(x) = \int_{-D}^{D} P(t)\ f(t,x-t)\ dt \qquad (3)$$

where $S(x)$ = distribution of scattered events
$P(x)$ = distribution of unscattered events
$2D$ = diameter of image field of view

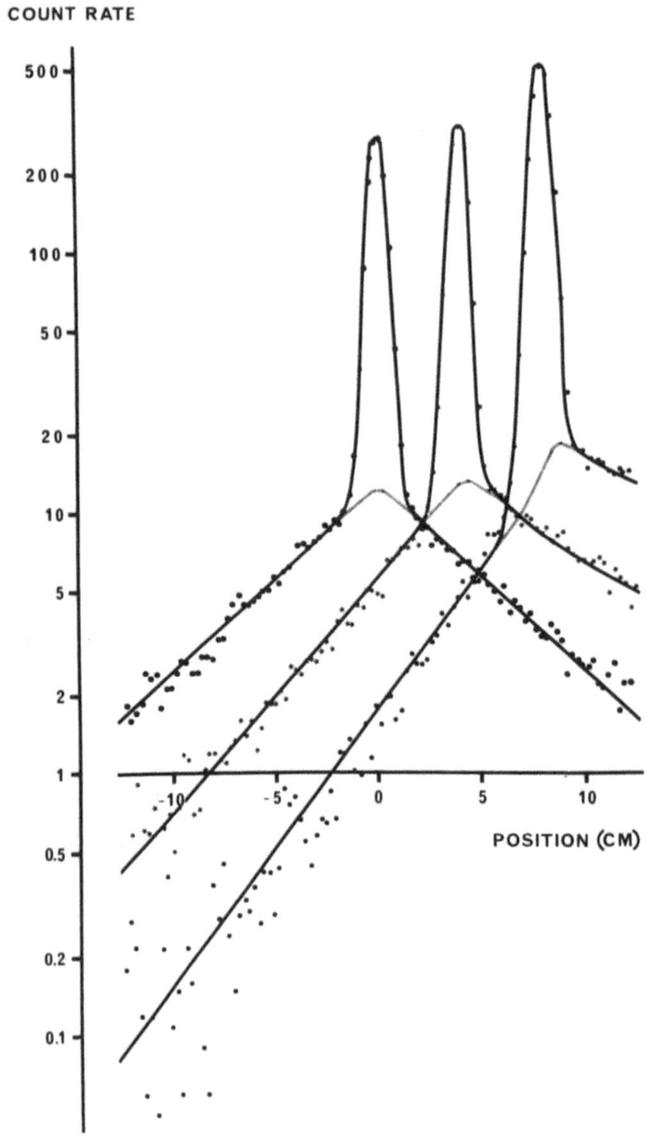

Fig. 6. Projections obtained with a line source at different positions in a 20 cm diameter water phantom. (a) Line source at positions along the projection lines, 0, 4, and 8 cm from the phantom center.

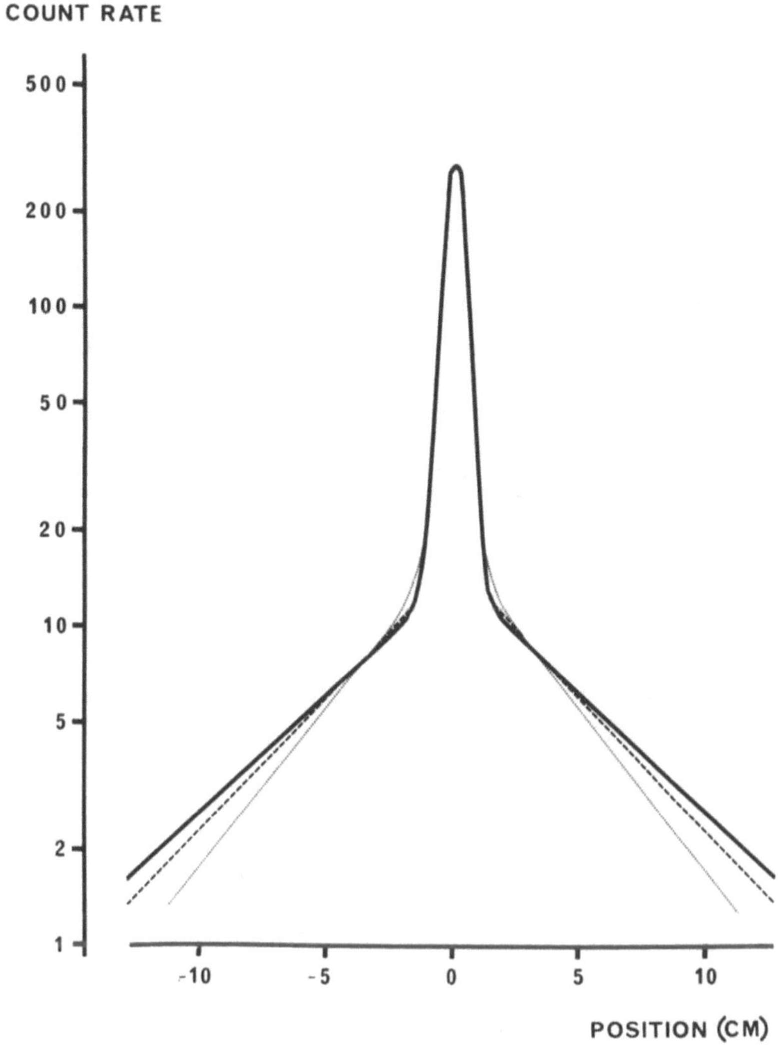

Fig. 6. (b) Line source at positions perpendicularly to the projection lines, 0, 4, and 8 cm from the phantom center.

Since the measured projections after random correction consist of both scattered and unscattered events, a function h(x,y) must be found that satisfies

$$S(x) = \int_{-D}^{D} (P(t)+S(t))\, h(x,x-t)\, dt \qquad (4)$$

A solution to this equation can be found but the most straightforward procedure is to sidestep the function f and determine h directly. The distribution of scattered events from a cylindrical object can thus be described using an integral transformation of the projections. The calculated scatter distribution is then subtracted in each projection. This correction scheme has been shown to work well for different distributions of activity within the object (Fig. 7).

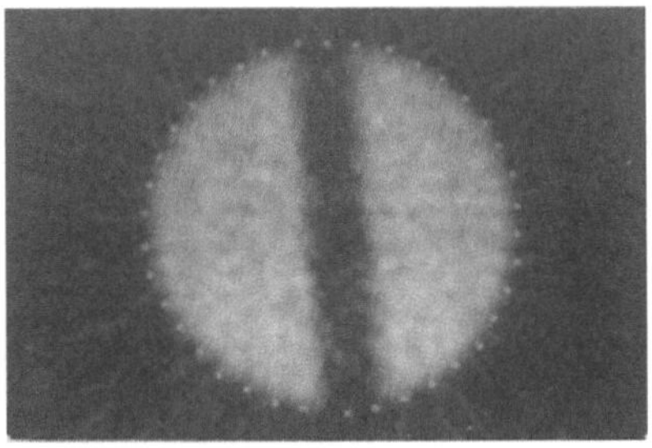

Fig. 7. Phantom with diameter 20 cm filled with activity except for a slit with a width of 3 cm.
(a) Image of a uniform activity except for the slit, reconstructed with scatter correction.

A similar method where the scatter distribution function is obtained analytically from Klein-Nishina cross-sections, has been used by Endo et al (28). A third method which estimates the level of scattered radiation from peripheral detection channels was mentioned in section 2.1 (29). Different contribution curves will, in this case, be fitted to the determined level, depending on the object configuration. For instance, in Ga-68 brain studies with most of the activity in extracranial soft tissues, an overestimation of the scatter level by using a distribution curve relevant to uniform activity within the object, may easily result in negative activity values in the brain tissue (26).

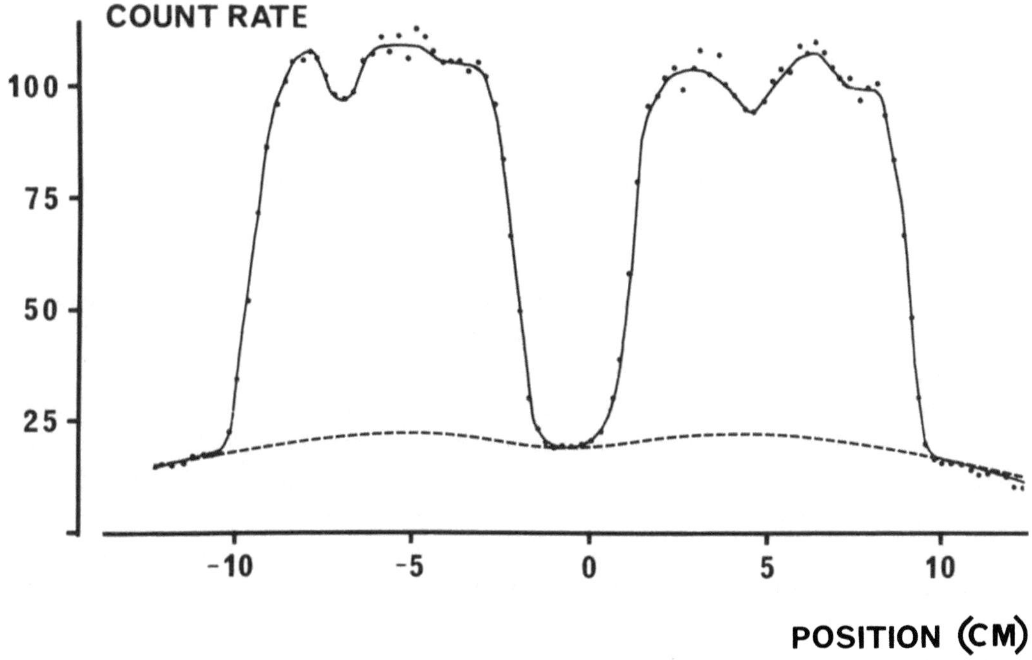

Fig. 7. (b) Projection parallel to the slit plotted without scatter correction. The small ripples of the curve are due to thin walls separating different sections. Dashed line indicates scatter distribution obtained by integral transformation.

2.4 Attenuation Correction

In the image reconstruction the measured data are ideally assumed to be proportional to the total concentration of radionuclides along the detection channels. This is, however, not true in an actual measurement

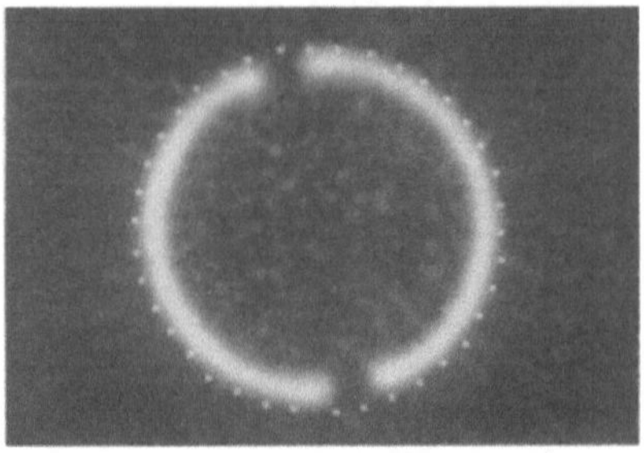

Fig. 7. (c) Image of a phantom with only an annulus of activity except for the slit, reconstructed with scatter correction.

due to the attenuation of photons within the object. Thus in a head examination, central detection channels measure only 1 photon out of 7 emitted. The rest are absorbed or scattered out of the detection channel. Thus the total attenuation along each detection channel has to be known, and the measured values corrected accordingly (29,30). The large correction factors involved make an accurate description of the attenuation crucial for accurate quantitation. Examples of this are given in Fig. 8. In phantoms and in some brain sections the object contours may be described by ellipses, and within the contour a uniform attenuation may be assumed. In most cases, however, such a procedure is not accurate enough.

A contour-finding algorithm may be applied in brain scans that have some activity in the extracranial soft tissue (29,31). This can either operate in the images or, which is faster, in the projections. In such an algorithm the edges of the object are determined in the projections by finding the maximum slope. Edge positions from all projections are then used to "peel off" regions outside the object contour (Fig. 9). A crude correction for the skull bone may be done by extending the contour radially a distance corresponding to the thickness of the bone, approximately 6 mm.

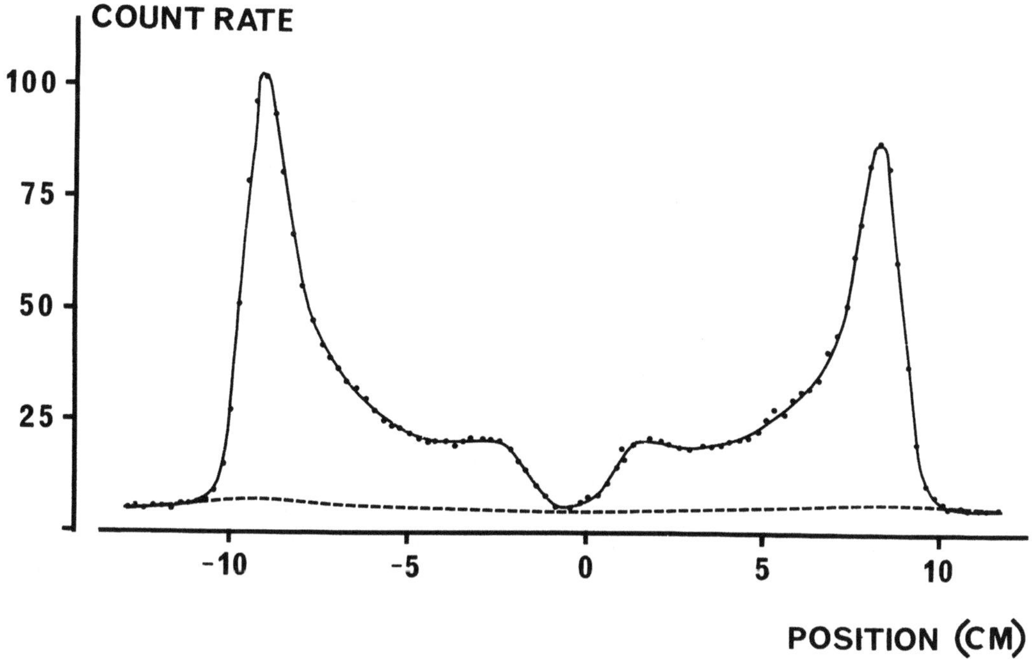

Fig. 7. (d) Projection parallel to the slit obtained with the phantom of (C). Dashed line indicates scatter distribution obtained by integral transformation.

The assumption of uniform attenuation within the object may lead to large errors in interior structures with greatly differing attenuation, like structures with bone or air (13,29). Air structures without activity may thus falsely be represented as areas of high activity, whereas bone structures may be represented within large negative values of activity.

A way to overcome problems of attenuation that will also handle internal structures of different attenuation, is by using transmission scan information. A transmission scan can be made in the positron camera by

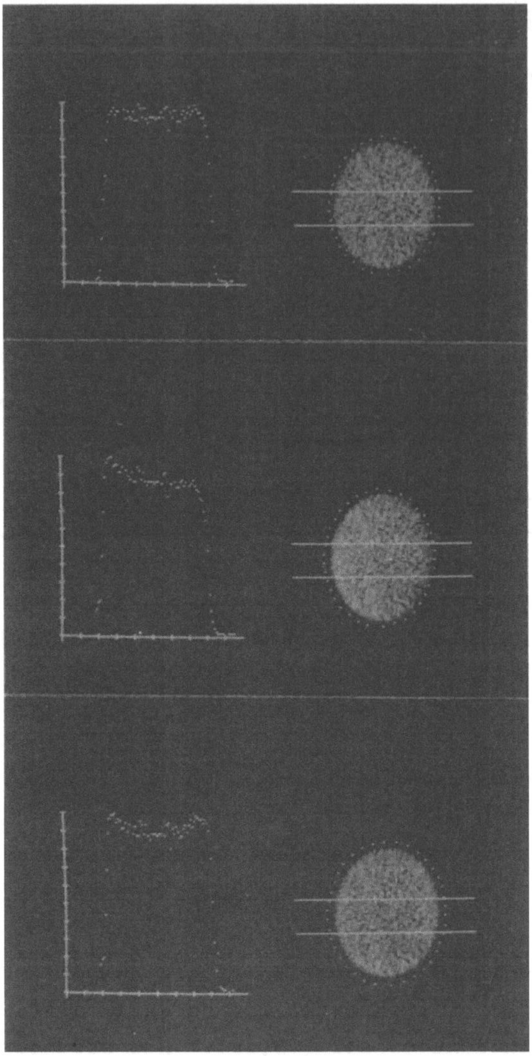

a

b

c

Fig. 8. Phantom with uniform activity in an elliptic container with full axes 16-20 cm reconstructed with attenuation described by ellipses of different dimensions.
(a) Attenuation contour properly selected. A profile across the image shows a flat distribution.
(b) The same phantom reconstructed with the attenuation contour displaced 5 mm to the left. A large asymmetry is introduced with peripheral values deviating with about 13%.
(c) The same phantom reconstructed with an attenuation contour with full axes too large by 10 mm in both directions. Centrally the values are increased by 6% compared to (A) and peripherally the increase is 16%.

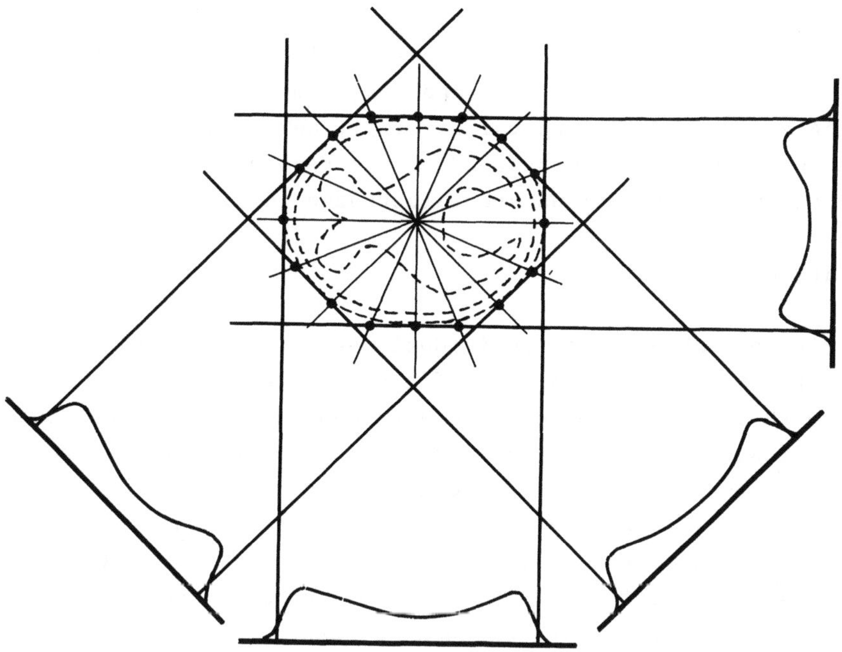

Fig. 9. Principle for the contour-finding algorithm. The borders of the activity region are determined by finding the maximum slopes. These points are backprojected in the image. A number of vectors are used to define points showing the periphery. The lengths of the vectors are given by the smallest distance from the origin to the intersection with the backprojected border points.

using an external source surrounding the patient. One disadvantage when using this method is that the absorption data will be noisy and this noise will, in turn, propagate into the emission tomography data (30,32). The noise can be reduced if an edge detecting algorithm is applied on the reconstructed image to find the borderlines between different media and if the absorption is assumed constant in the intermediate regions

(33). The correction can also be made using X-ray CT, which is statistically better determined. In this case projections are calculated from CT images. The CT values of bone and tissue must be transformed to corresponding attenuation values valid for 511 keV photons. The resolution must also be degraded to correspond to the resolution of the PET camera. Examples of these corrections are illustrated in Fig. 10.

2.5 Efficiency Variations

The counting efficiencies of individual detection channels will show relatively large variations. A part of these variations will change with time, like those due to temperature dependent light coupling, coincidence channel drifts, etc. Other causes of efficiency variations are constant in time and are directly related to geometrical factors. In peripheral detection channels in a ring detector system, the coincident photons have a skew angle of incidence to the detector surface. This may lead to a number of different effects. The detector surface shows a smaller solid angle. Photons may penetrate through adjacent detectors or intradetector shieldings. The photons may experience a shorter length of detector material, etc. These different factors may lead to efficiency variations along a projection like the example shown in Fig. 11 (13,21,29,34).

Unless corrected for, the efficiency variation will result in streaks, noise-like patterns and sometimes circular artefacts (29).

The usual method for correcting efficiency variations is by calculating the efficiencies of each detection channel from a scan of a known object, usually a cylindrical annulus or a cylinder. The recorded number of counts in a patient scan is then divided by the efficiency for each detection channel prior to image reconstruction.

2.6 Artefacts

Over- or undercorrection for the different factors mentioned above will lead to quantitative errors in the images. In some cases these errors may contain high frequency components which will degrade the quality of the image, whereas in some cases the visual interpretation will not be complicated while the quantitative values are affected. These erroneous patterns are usually denoted as artefacts.

The artefacts can, in most cases, relatively easily be traced back to

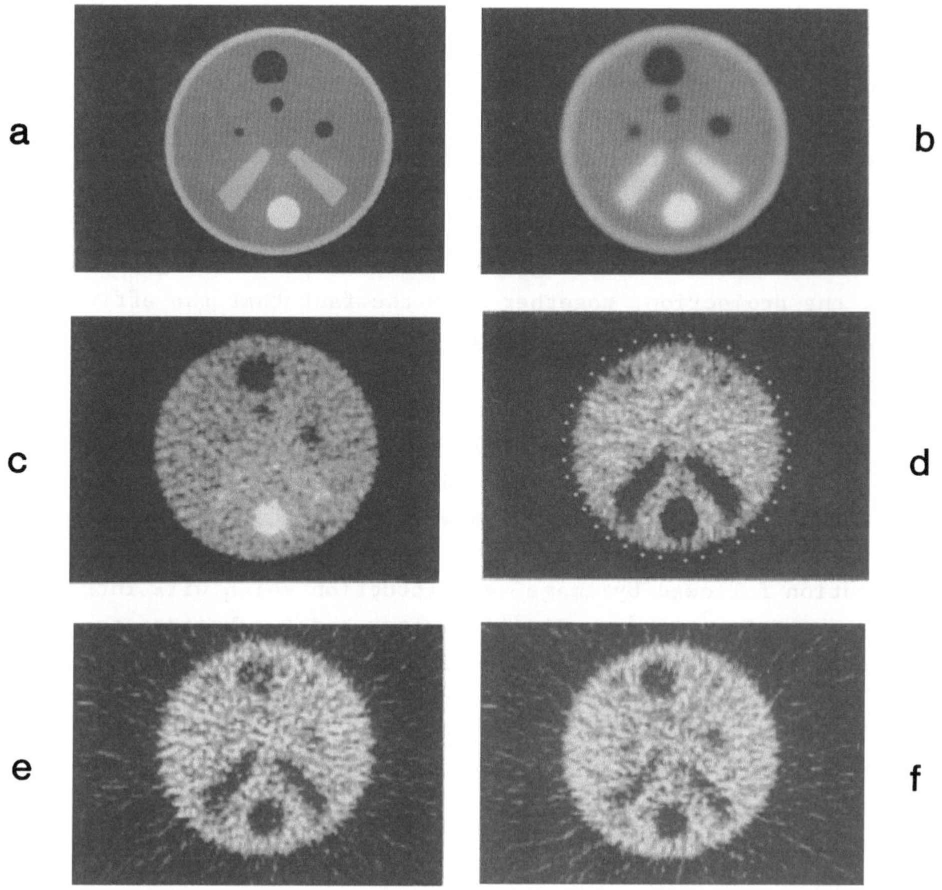

Fig. 10. Emission and transmission images of a water-filled phantom of 20 cm diameter containing air cavities of diameters 8.5, 12, 15 and 30 mm, and structures of PVC (density 1.45 g/cm3) and aluminum (density 2.7 g/cm3).
(a) CT scan of the phantom.
(b) CT scan modified to have the same resolution as a PET scan.
(c) Transmission scan taken with the positron camera.
(d) Emission scan corrected for attenuation assuming a cylinder with uniform attenuation.
(e) Emission scan corrected for attenuation using the transmission scan.
(f) Emission scan corrected for attenuation using CT scan.

the fundamental corrections described above. Examples of this have been mentioned above, like the effect of inaccurate attenuation correction leading to major nonuniformities, or the circular symmetric nonuniformity caused by efficiency variations along the projection. Most of the high frequency artefacts can be related either to sampling errors which give star artefacts, efficiency errors in one detection channel resulting in a streak, or efficiency errors of one detector in all its combinations, resulting in a fork artefact (Fig. 12). The stepwise change in efficiency along the projection, together with the fact that the efficiency is different for true and scattered coincidences, may easily lead to ring artefacts (29,34).

For some corrections like, for example, the random correction, an error will give more complicated patterns. Since the distribution of randoms is uniform, an error in the time window will lead to a base level shift in the projection. This base level is, however, subjected to an absorption correction followed by image reconstruction which will introduce a centrally symmetric pattern. Finally, after being submitted to "efficiency correction", the nonuniformities in the detector efficiencies will introduce traces of streak- and fork-artefacts (Fig. 12e) (29).

3. POSITRON CAMERA DESIGN

3.1 Historical Remarks

Positron imaging has been known since the early 1950's when it was suggested independently by Wrenn et al (35) and Brownell (according to Sweet (36)). Brownell and Sweet (37) described the first positron scanner in 1953, a two detector system with NaI detectors used for brain tumor detection. The early systems all utilized one single detection channel. It was not until the beginning of the sixties that the development of multi-detector systems started. Rankowitz et al (38) presented, as early as in 1962, a ring detector system using 32 discrete NaI (Tl) detectors (Brookhaven). The main obstacle was then the data at the time. Transverse section imaging with single photons was pioneered by Kuhl and Edwards in 1963 (39), using a scan technique similar to modern emission tomography scanning. They also relied on the simple backprojection technique for reconstruction of the images. In 1972 another multi-detector system was reported, the PC I of the Massachusetts General Hospital (MGH) (40) with two banks of detectors and 127 detectors per bank.

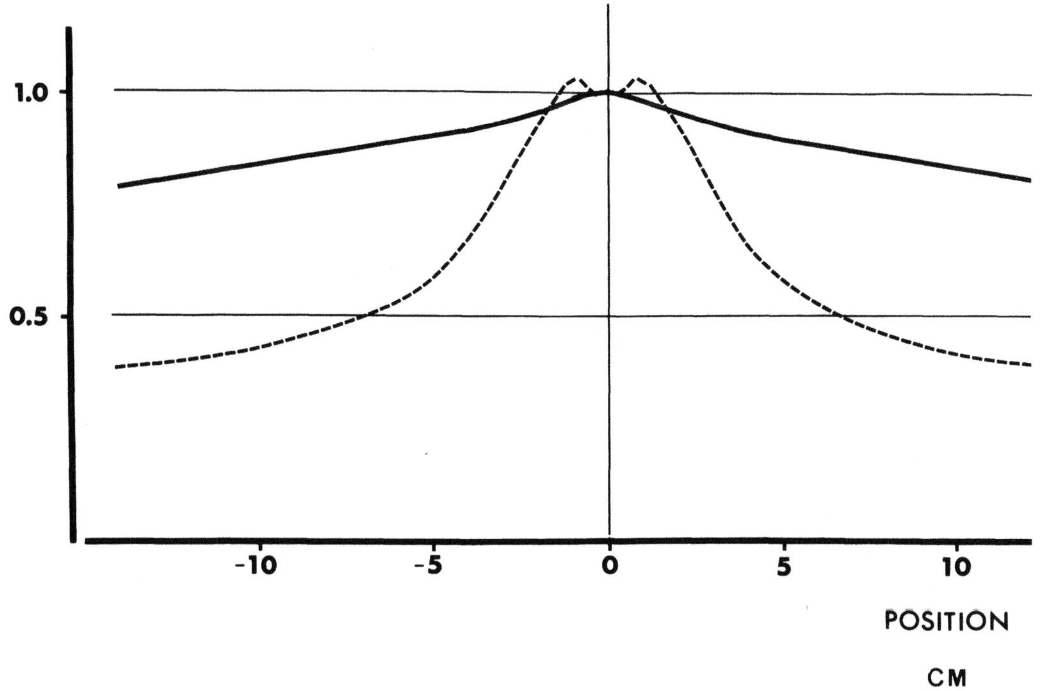

Fig. 11. Variation of the efficiency along a projection for two different positron cameras. Full line represents a 96-detector ring BGO camera, and dashed line represents a 95-detector ring NaI camera.

The breakthrough for positron imaging, however, came from another field, transmission CT. With Godfrey Hounsfield's EMI scanner a method for reconstructing images was available (41). The first systems to utilize these new ideas were the PETT machines from the Ter-Pogossian group in St. Louis, ranging from PETT I (42) to SUPER PETT I (43).

In 1976 the UCLA 64 detector system was completed (44,45). The system performed half a crystal rotation to improve sampling. It is now in op-

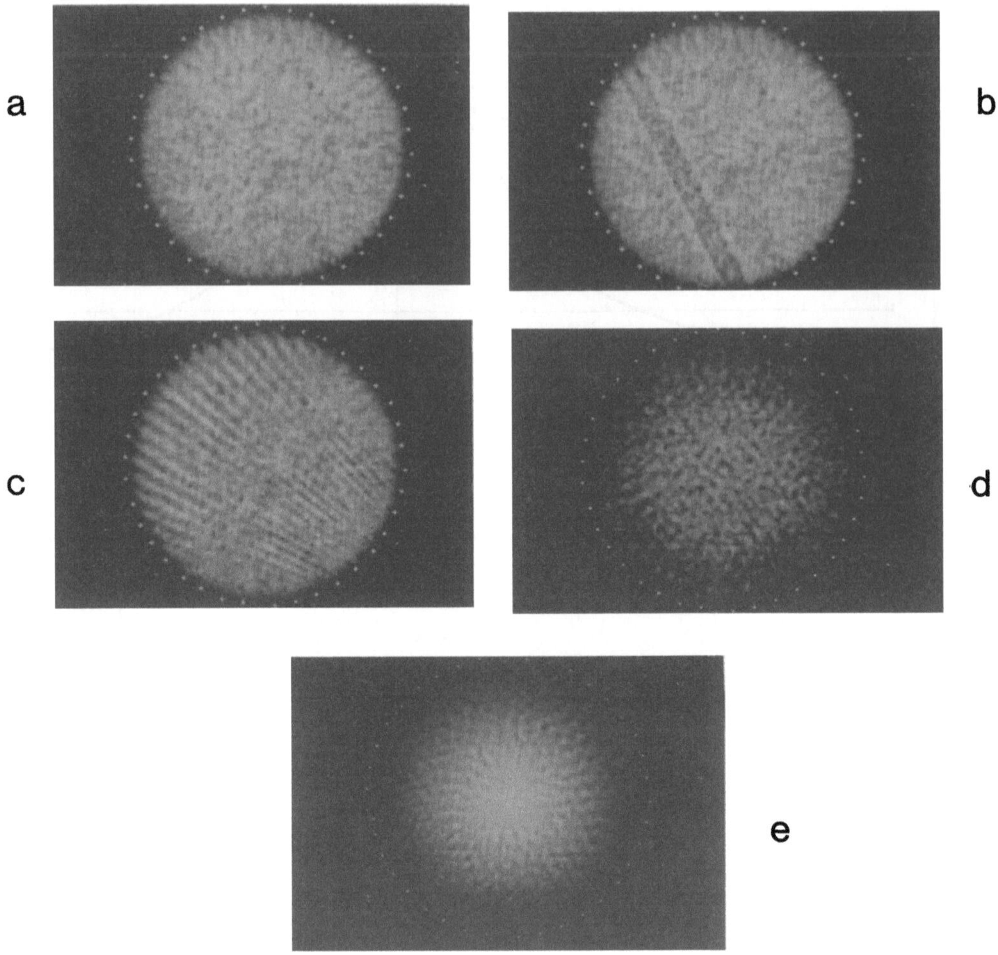

Fig. 12. Uniform container of activity with different artefacts introduced.

(a) Original image.

(b) A factor of 4 reduced number of counts in one detection channel.

(c) A factor of 2 reduced number of counts in all combinations including one specific detector.

(d) An image reconstructed without random or scatter correction, the efficiency difference in true relative to random and scattered coincidences introduces ring artefacts.

(e) Only random coincidences, reconstructed with attenuation correction as if they were true coincidences.

eration at Columbia University, New York, performing a different sampling scheme, dichotomic sampling (1982). The old Brookhaven scanner came into routine clinical use in 1975 when it was taken over to Montreal and was equipped with a modern image reconstruction algorithm. It was in 1978 replaced by Positome I (46) and later on by Positome II, the first BGO detector PET. Cesiumfluoride crystals were for the first time used in PETT VI, followed later on by SUPER PETT I, the first system operating with TOF techniques, both systems from Ter-Pogossian's group in St. Louis.

At the end of the seventies, several commercial companies started to pay attention to PET systems. Cyclotron Corporation made, in 1975, a commercial version of the MGH positron camera, called TCC 4200. One of these systems was, for a long time, operated at the Sloan-Kettering Institute but is now replaced by a later model TCC 4600, a ring detector system with 5 detector rings. ORTEC started also in 1975-76 to develop a modified PETT III version called ECAT. The latest version, NeuroECAT (47), is a dedicated brain scanner. The Swedish cyclotron manufacturer, Scanditronix, has constructed a BGO ring system called PC384-B (48). There are two Japanese systems currently under development, Headtome III (49) (Schimadzu) and Positologica II (50) (Hitachi), also BGO ring systems.

3.2 Geometry

Ideally, all the annihilation photons originating from the investigated patient should be detected. However, in a practical situation less efficient devices must be accepted. Current technology necessitates the use of discrete detector elements. A near 4π arrangement with small discrete detectors is not realistic due to its complexity and cost. The conventional approaches to positron camera systems are either dual plane systems (MGH-system, Dual Angercamera systems, MWPC-systems) or ring systems. In a multi-ring system, several rings are stacked upon each other. Multi-ring systems have the advantage of high sensitivity and provide 2N-1 tomographic slices simultaneously, N being the number of rings, since intermediate slices can be reconstructed from cross coincidences between adjacent rings. The thickness of a cross slice will, however, strongly depend on the distance from the center. An advantage with ring or multi-ring systems, apart from providing the highest packing fraction and the most uniform coverage of the imaged volume, is the fact that most uniform coverage of the imaged volume, is the fact that the angular sampling requirement generally is fulfilled. A camera rotation is usually not necessary. The high sensitivity and the highly symmetric geometry make very short scan times, in the order of a few seconds, feasible. If a reconstructed spatial resolution of the order of the crystal width is accepted, then also the linear sampling requirement is fulfilled. The ring(s) can then be operated completely stationary giving scan times in the subsecond range, making gated studies possible. Normally a reconstructed resolution in the order of half the crystal width size is attempted. In this case a sampling enhancement motion must be added, as discussed previously.

The current PET scanners can in a broad sense be divided into two groups, planar systems and ring- or quasiring(s)- systems. In the following three paragraphs, current positron camera systems belonging to these two groups will be described.

3.3 Planar Positron Camera System

Two different types of planar positron camera systems exist. To the first category belong the MGH systems and TCC 4200 with discrete detector elements distributed on the two planes. The surface-sensitive devices belong to the second category, for example, two Anger cameras or two multi-wire chamber cameras coupled in coincidence.

PC II consists of two planes with 144 NaI(T1) detectors per plane (40). The planes are displaced linearly in 6 steps for each of the 15-30 angles being measured, giving a total of 90 to 180 data sets. The detector banks thus rotate and translate around the patient being imaged. Typically 160 3-second images are recorded. The crystal size is 20 mm dia * 40 mm crystals, giving a reconstructed resolution of roughly 10 mm. The maximum count rate is around 100,000 counts/sec.

The former Searle positron camera system (4) is an example of surface-sensitive detectors. This device is based on two Anger cameras placed on a rotating stand. The standard electronics for the Anger cameras was modified to accept higher count rates without pile-up phenomena (pulses will occur on the tail of the previous pulse causing level shifts). The system had problems with random coincidences and scatter coincidences due to the open geometry. For head studies the count rate was limited to roughly 4000 c/s. The resolution was of the order of 9 mm.

The other group of surface-sensitive cameras, the wire chambers (3), are still in an early stage of development with low detection efficiencies, limited energy discrimination possibilities and with the same problem as above, scattering and randoms due to the open geometry. The maximum count rate has so far been of the same order as above, i.e. around 4-5,000 c/s. However, a recent report (51) shows that it is possible to obtain significantly higher detector efficiencies. The wire chambers can achieve the highest spatial resolution reported so far of the order of 2-3 mm FWHM.

3.4 Sampling Schemes of Ring Detector Systems

Let us first repeat what was said about the sampling requirement. Generally, the activity has to be sampled in steps smaller than one-quarter of the crystal width. For the quasiring systems with hexagonal or octagonal geometry, such as ECAT and NeuroECAT, the detector banks are displaced in a desired number of steps. Since only a small number of projections can be recorded simultaneously in such a system, the detector assembly must be rotated (Fig. 3a).

However, for the ring systems, the sampling requirement is more tricky to implement because of the ring geometry. In a stationary ring, only a sampling which is equal to half the crystal-crystal distance can be obtained. That is far from the one-quarter requirement. The wobble technique was the first solution to increase the linear sampling of ring detector systems (9). During continuous wobbling the whole ring is moved along a circle with maintained orientation (Fig. 3b). The wobble diameter is often chosen in the order of the center-to-center distance of the crystals. This means that every detection channel will sweep over a region covered by the wobble diameter (often set to d, the cyrstal-to-crystal distance). When an event is recorded it will be stored in a memory array corresponding to the projection angle. The position in the memory array will depend on the instantaneous position of the detection channel in a fixed coordinate system. For Scanditronix PC384-B, the resulting sampling distance is less than 1.5 mm, i.e. well below the required 3 mm of the 12 mm crystals used in Scanditronix machines.

Other sampling schemes employed for circular systems are: (a) the dichotomic motion (10), (b) the clamshell motion (11), and finally (c) the continuous rotation with unevenly spaced detectors (positology ref. 12). For performing the dichotomic sampling motion the ring is cut into two halves and these two parts perform an oscillating motion relative to each other (Fig. 3c). To obtain a d/4 sample distance, a four-point sampling is considered. This sampling scheme is implemented in the old UCLA ring now moved to Columbia University, NYC.

The clamshell motion is a special case of the dichotomic ring motion with only one end opened, the other being hinged together (Fig. 3d). This sampling scheme is presently being implemented at the Donner ring at Berkeley. Only two positions are used for fulfilling the sampling requirement.

The technique to increase linear sampling by using rotation and displaced detector positions was to our knowledge originally suggested by Kuhl and Edwards (52) for single photon emission scanners, and the same scheme was also used for the early PETT machines. A generalized version of this technique is now implemented in a Japanese ring detector system, Positologica, which performs a continuous rotation motion. The scintillators are arranged around the circumference of the ring in a non-uniform way (Fig. 3e).

The actual positions have been estimated by computer simulations. Since every detection channel will give a contribution to every projection twice, the number of sample points per projection will be high, of the order of 2-3,000. The sampling density is very uniform. The complications, however, are many. Slip rings have to be used in order to provide high voltage and power to the rotating electronic system. Data transfer between the moving and stationary part is performed with light diodes. The ring can be moved with 30-60 rpm thus making fast dynamic studies of the order of 1 second feasible. This data collection speed is common for all ring systems.

3.5 Different Ring Detector Systems

In Table 3 the figures of merit of several current multi-ring systems are shown. However, the figures are difficult to compare, and due caution is necessary. Most of the positron camera systems shown are dedicated to the brain. The size of the ring diameters of these brain scanners vary, from 38 cm (Neuropet) to 75 cm (Headtome III). The ring diameter is one of the most debated issues in ring detector constructions and the wide span of selected radii illustrate the different opinions of the different groups. For a given scintillator size, an increasing diameter will result in the following positive features: the number of view angles increases, the number of randoms decreases, scatter rejec-

	SCX Univ. of Sthlm, Karolinska PC 384-B	TCC TCC 4600	ORTEC UCLA NeuroECAT	NIH NeuroPet	Nucletronix Washington University PETT VI	Nucletronix Washington University SUPER PETT I	AECL MNI Montreal PositomeII	Hitachi NIRS Tokyo PositologicaII	Shimadzu Research Inst RBV Akita Headtome III	LETI Grenoble TOF-PET
Detector configuration diameter (cm)	Ring 48	Ring 60	Octagon 60	Ring 38	Ring 57	Ring 90	Ring 42	Ring 85	Ring 75	Ring 92
Number of crystals/ring	96	96	88	128	72	96	64	160	160	96
Number of rings	4	5	3	4	4	4	2	3	3	4
Number of slices	7	9	5	7	7	7	3	5	5	7
Crystal material	BGO	BGO	BGO	BGO	CsF	CsF	BGO	BGO	BGO	CsF or BaF
Crystal size (mm)	12x20x30	20(dia)x40	17x28x30	8.25x20x35	20x24x65	25(dia)x45	20x30x30	15x24x24	13.4x25x40	24(dia)x40
Axial extension (cm)	10	11	10	9	11	12-13	7	12	10	12-13
Sampling scheme	continuous wobble	discrete wobble+ rotation	translation + rotation	discrete wobble	continuous wobble+ rotation	continuous wobble+ rotation	discrete wobble	continuous rotation	discrete wobble+ rotation	discrete wobble+ rotation
Size of image field (cm)	26	25	25	23	27		30	50	30	50
Highest reconstructed resolution (FWHM) Transaxial (mm) Axial (mm)	 8 12	 10 10	 8 15	 6 12	 8 15	 8-9 12	 11 16-17	 9-10 12-13	 7 12-13	 7-8
Sensitivity (c/s/uCi/ml) direct plane cross plane total	 22000 (5500) 29000 (7300) 175000 (44000)	 24000 (6000) 36000 (9000) 264000 (66000)	 15000 (1900) 22000 (2800) 	 42000 (10500) 	 15000 (2600) 36000 (9000) 111000 (19300)	 5500 (1900) 22000 (2800) 44500 (7000)	 75000 (8400) 	 33000 (5700) 39000 (6800) 180000 (31200)	 15000 (2400) 	 35000 (6100)
Minimum scan time (sec)	1	1		1	1	1	1	1	1	1
Computer configuration Type Memory (primary/disc MB)	 VAX 11/780 512kB 67MB	PDP 11/60 128kB or VAX 11/780 512kB	 PDP 11/60 128kB 10MB		 Interdata	 Interdata	 PDP 11/34			 SOLAR
Status (dec 1982)	in operation	in operation	in operation	In partial operation	In operation	In partial operation	In operation	In partial operation	Completed in march-april 83	With CsF in partial operation
Comment	Brainscanner	Brainscanner	Brainscanner	Brainscanner	Brainscanner	Whole body scanner	Brain scanner	Whole body scanner	Brain scanner	Whole body scanner

*) Values normalized to 1 cm of crystal height a) Estimated values

Table 3. Some commercially available positron cameras.

tion and resolution uniformity is improved. On the negative side, a decreased sensitivity and increased costs are found. Derenzo finds an optimum choice of diameter for brain scanners of 55 cm, considering only sensitivity, randoms and scatter (17). However, the optimum region is flat and a region between 45 to 65 cm diameter is acceptable.

In order to compare sensitivities, the best resolution available with the different instruments has been selected. The sensitivity is strongly dependent on the resolution, especially when beam blockers are used to improve the resolution, as is the case for most of the shown systems. Beam blockers usually reduce the sensitivity with a factor of 2-4, and furthermore, if a spatial resolution gain is attempted due to the increased intrinsic resolution, then the sampling requirement has to be adequately fulfilled in the new situation. The selected sampling conditions must always be adequate for the image reconstruction filter in order to obtain artefact-free reconstructions. The existence of beam blockers will also change the scatter geometry for the positron camera system and must be considered in the scatter correction scheme. The sensitivity values must also be random and scatter corrected for a fair comparison. The axial extension of the crystals determines the axial resolution or the slice thickness. There is usually a mismatch between the two modes of resolution, in the axial and the transaxial direction. The same situation is met in CT where the ratio between these two resolution modes can be a factor of 5-10. In PET the difference is much less, perhaps only 50%. The differences in axial extension must, however, be considered when comparing the sensitivities. In order to make the sensitivities independent of the slice thickness, the sensitivity values for 1 cm axial extension have been estimated. These values are given in parenthesis. Not included in Table 3 is the "quality" of the spatial resolution, i.e. the ability of the instrument to detect contrast differences in different sized objects. This type of information would constitute a one-dimensional figure of merit. The detectability can be presented in a graph as a function of the object size. Such a figure of merit already exists for CT and should be used also for positron camera systems (53).

In the following a series of ring detector systems will be described. The discussion will be limited to second generation machines where BGO or CsF scintillation crystals are used and which have multi-slice capabilities. As a reference, the Scanditronix brain scanner PC384-B is used (13,48). This system utilizes 4 rings with a ring diameter of 48 cm. Each ring has 96 detectors organized into modules with eight detectors per module. BGO scintillation crystals with the size 12*20*30 mm are being used. This implies an intrinsic resolution of about 6.5 mm FWHM. The reconstructed resolution is 8 mm. The four rings are interconnected electronically so that they provide coincidence data for 7 slices. The sensitivity, when measured for a 20 cm phantom, is around 22,000 c/s for direct slices and 32,000 c/s for cross-slices per μCi per ml. The imaging properties of the system are reliable up to system count rates of 1 MHz. The total system sensitivity is around 200,000 c/s per μCi per ml.

The PC384-B provides a good spatial resolution and imaging uniformity as is shown from the image obtained from the so-called Derenzo phantom (22) (Fig. 13). The smallest sector consists of 10 mm tubes placed 10 mm from each other and they are all clearly resolved. The sytem at the Karolinska Hospital in Stockholm has been used in clinical routines for Ga-68 blood-brain barrier studies, C-11 for metabolic studies (glucose) (Fig. 1), and for blood volume (carbonmonoxide) studies.

The NeuroPet (54) is a development of NIH, Bethesda, USA. It is a dedicated brain scanner with four rings and a ring diameter of 38 cm. The crystals used are BGO's with the size of 8.25*20*35 and every ring has 128 crystals. The coverage of the brain is 9 cm. The detector ring performs a discrete wobbling motion, i.e. the detector system measures the activity from different positions along the wobble circle. The image field is 23 cm in diameter and the reconstructed resolution is 6 mm. Seven slices are simultaneously measured. The sensitivity is quite high, 42,000 c/s per μCi per ml for direct slices. The compact geometry, however, gives problems with scatter and random coincidences.

Positome II is a two-ring system developed at the Montreal Neurological Institute in Montreal, now available as Therascan 3128 (46). The crystal is around 18*30*30 mm giving a spatial resolution of 11 mm FWHM and an efficiency of around 75,000 c/s per ring. The camera performs a discrete wobble and rotation to achieve an adequate sampling density. Positome II has been in use since 1978 for metabolic studies, blood flow studies and blood-brain barrier studies.

The Cyclotron Corporation now offers a ring detector positron camera system (55) with 5 rings, 9 slices, with 96 BGO crystals per ring, each with the size of 20 mm (dia) *40 mm which provide an 11 mm resolution

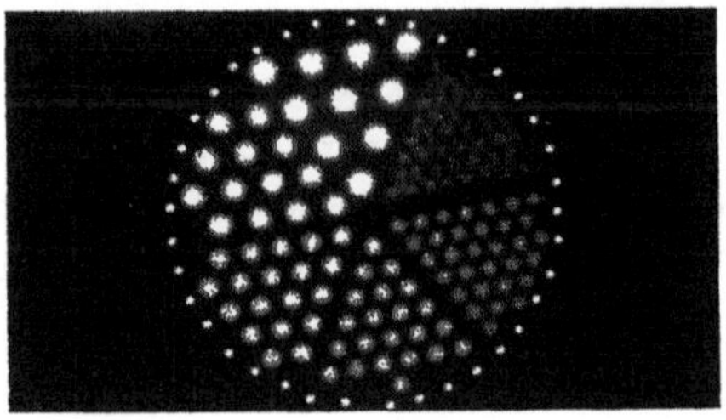

Fig. 13. An image of a Derenzo phantom from a Scanditronix PC384-B camera. This phantom is used for image resolution and uniformity studies.

and a sensitivity of 24,000 c/s for direct and 36,000 c/s for cross-slices. This camera performs a wobble and rotation to achieve the d/4 sampling.

The PETT VI machine from the Ter-Pogossian group in St. Louis (56) is a four-ring system with 72 20*24*65 mm CsF scintillators per ring. The ring diameter is 57 cm and 7 slices are simultaneously recorded. The axial field of view is 11 cm. The CsF provides high time resolution and there are plans to incorporate them into TOF positron cameras. PETT VI has a 6 ns time window. The detection efficiency relative to a comparable BGO system is, however, low. The detector system performs a continuous wobble plus rotation to achieve the necessary d/4 sampling condition. The wobble accuracy is 16 points. The best reconstructed resolution is 8 mm with a sensitivity of 13,000 c/s in direct slices and 17,000 c/s for cross-slices.

The Research Institute for Brain and Blood Vessels in Akita, Japan is constructing a positron camera system called HEADTOME III together with the commercial company, Shimadzu. Unlike earlier HEADTOME's which were hybrid systems (49) with capabilities for single photon emission tomography, HEADTOME III is a dedicated positron camera. The system will have three detector rings with 160 BGO crystals per ring. The crystal size is 13.4*25*40 mm and the ring diameter is 75 cm. Estimated resolution in the high resolution mode is approximately 7 mm. The sensitivity is estimated to be 15,000 c/s in the high resolution mode for direct slices. To improve sampling, the system will perform a discrete wobble plus a rotation.

Another collaboration between scientific institutions and industry for positron camera construction is found in Japan, involving the National Institute of Radiological Sciences, Chiba, Hitachi and Hamamatsu. Their design is based on the prototype POSITOLOGICA and is therefore called POSITOLOGICA II (50). It is a whole body machine with 3 detector rings and a ring diameter of 85 cm. Each ring has 160 detectors organized in groups of 4 (quad-detector). Every quad is mounted on two PM-tubes and with a spatial coding technique the appropriate scintillator can be selected. Five slices can simultaneously be recorded, with a resolution of 9-10 mm (crystal size is 15*24*24 mm). The sensitivity is 33,000 c/s in direct slices and 39,000 c/s for cross-slices. The time resolution is quoted to be 3.6 ns FWHM and 7.1 ns FWTM. The different quads are placed on the ring circumference in a non-uniform way and with a continuous rotation (30-60 rpm) a better than d/4 sampling is achieved.

Ortec's and UCLA's development is NeuroEcat (14) which is somewhat changed compared to earlier versions. The detector ring has octagonal geometry with 11 BGO detectors per bank, giving 88 detectors per ring. The crystal size is 17*27*29 mm. The best resolution is 8 mm at a sensitivity of 14,300 c/s for direct slices and 22,000 c/s for cross-slices. Sampling is, as in earlier ECAT versions, achieved through a translate-rotate motion.

Several groups are currently working with the development of TOF systems for positron emission tomography. Ter-Pogossian's group in St. Louis has already developed a prototype with TOF capabilities (43), SUPER PETT I, with four detector rings and 96 CsF crystals per ring. The ring diameter is 90 cm. Crystal size is 25 mm (dia)*45 mm giving a 67% packing fraction. With beam blockers, a high resolution of 8-9 mm FWHM can be obtained. The sensitivity is, at a high resolution, ca 5,500 c/s for direct slices and 7,500 c/s for cross-slices. With TOF a factor of four is gained, thus corresponding to a sensitivity of 22,000 c/s (30,000 c/s) in a "classical" positron camera.

A group in France (LETI Grenoble) is working with a similar system (6), a whole body positron camera with TOF possibilities. The camera has four detector rings, each ring having 96 detectors, either CsF or BaF2. Investigations have shown that time resolutions of 300 ps is possible. The ring diameter of the system is 92 cm. The CsF crystals mentioned in the ref (6) had the size 24 mm(dia)*40 mm. A high resolution of 7-8 mm is expected with a total sensitivity of 35,000 c/s for the seven slices. The sampling enhancement motion will be similar to the one in SUPER PETT I, a discrete point wobble plus rotation.

3.6 Future Developments

As has already been discussed, an ideal PET system requires a nearly 4π geometry and a spatial resolution close to the limits of the positron imaging, i.e. ~ 3 mm. Data should be recorded with a sensitivity high enough to permit regional uptake studies in volumes ≤ 1 cc on a time scale of 1-5 seconds.

With current technology, individual detectors have to be used. A 3 mm reconstructed resolution requires a crystal width of ≤ 5 mm and a linear sampling distance of ~ 1 mm.

To be able to build up a nearly 4π geometry with < 5 mm wide crystals, several thousands of detectors are needed, mounted on a large number of PM-tubes.

How can these design goals be realized? Different approaches have been tried. The S-PET project (57) is an ambitious project which will use 17 rings with together almost 8,000 small BGO crystals with a planned resolution of = 3 mm. The rings will perform a dichotomic motion to improve sampling. The design concepts are currently being evaluated from both theoretical and experimental points of view.

Other high resolution instruments have been suggested at UCLA, Massachusetts General Hospital, Donner Laboratory (UCB) and by the PET collaboration in Stockholm.

All these developmental efforts are based on the ring concept utilizing very small BGO crystals. The MGH system is actually near completion and consists of 360 4 mm wide BGO crystals mounted on a ring-shaped light guide (5). The number of PM-tubes is 90 and the system relies on an analogue readout technique, realizing a maximum likelihood estimate of the position. This ring can thus be characterized as being pseudocontinuous compared to a ring with discrete detector elements. The expected spatial resolution is 5-6 mm FWHM. The ring will be operated stationary.

The ring systems suggested by the UCLA group and the Stockholm PET collaboration are based on digital readout techniques. The new UCLA ring being planned is a whole body system consisting of 640 small BGO crystals less than 5 mm wide (58). The crystals will probably be organized into groups, mounted on a common fast photomultiplier tube. The individual crystals of the group are subsequently being identified by means of light sensitive semiconductor elements. Mercuric iodide is one of the candidates. The Donner group has discussed a similar solution (59).

All these future PET ring systems with small discrete crystal elements with widths < 5 mm are intended to operate stationary. With close packed crystals, the inherent linear sampling will be ~ 2.5 mm, i.e. the same as the intrinsic detector channel resolution. By designing a reconstruction filter of say, 6-7 mm, the reconstructed images will be free from aliasing artefacts and furthermore, the high intrinsic resolution yields a high signal-to-noise ratio (7).

Another way to increase the sensitivity is to use the TOF technique. Then other crystal materials such as CsF or BaF2 rather than BGO or NaI (Tl) crystals have to be relied on. The French LETI group reports timing resolutions of 500 ps for CsF and 300 ps for BaF2, consistent with a position uncertainty along a detection channel of 7.5 cm and 4.5 cm, respectively. With a time resolution of 300 ps as obtained with BaF2 crystals, the sensitivity of the system is virtually increased with a factor of four relative to the situation when the same system is operated conventionally. However, if the BaF2 crystals are replaced with BGO crystals, the higher detection efficiency of the latter compensates for this factor of four. The TOF operated small crystal systems (width < 5 mm) will give more problems than conventionally operated BGO systems with the same crystal width due to higher amounts of intercrystal spillage resulting in loss of efficiency and degraded spatial resolution.

The added timing requirement also poses difficult problems with the PM-tube coupling.

When compared to PET systems with small crystals giving intrinsic resolutions in the 3-4 mm range, the TOF techniques will not yet compete with

with conventional BGO systems but will become important for systems with intrinsic resolutions > 5 mm, i.e. for crystal widths exceeding 10 mm. The TOF concept will probably, with present technology, be more useful in whole body systems than in brain systems.

One question not addressed so far concerns random noise in TOF operated PET systems. These systems have a superior true-to-random coincidence ratio due to time windows which are more than one order of magnitude lower than conventional systems. Also, the deadtime losses due to extracted triple coincidences will be significantly lower in a TOF system. These two noise factors will be of importance for high activity imaging, e.g., imaging with O-15. However, current commercial systems such as NeuroECAT and PC384-B can handle count rate situations corresponding to 40-50 mCi injected O-15 activities in patients, by applying simple deadtime correction schemes and accurate determinations of random coincidences.

REFERENCES

1. Derenzo SE (1979). Precision measurement of annihilation point spread distributions for medically important positron emitters. Proc 5th Int Conf Positron Annihilation (Japan):819-823.

2. Laval M, Allemand R, Campagnolo R, Garderet P, Gariod R, et al (1982). Contribution of the time-of-flight information to the positron tomographic imaging. Proc Third World Conf in Nucl Med Biol Paris, Pergamon Press, ed C. Raynaud, pp 2315-2323.

3. Jeavons AP, Schoir B, Kull K, Townsend D, Ficy P and Donath A (1981). A large area stationary positron camera using wire chambers. (In) Med Radionuclide Imaging 1980, Vol 1, pp 49-77.

4. Muehllehner G, Buschin MP and Dudek JH (1976). Performance parameters of positron imaging camera. IEEE Trans Nucl Sci NS-23:528-537.

5. Burnham CA, et al (1982). A positron tomograph employing a one dimension BGO scintillation camera (to be published).

6. Laval M, Aallemand R, Bouvier A, Campagnolo R, et al (1982). Design and performance of a new positron computed tomograph (PCT) using the time-of-flight (TOF) information. Proc Third World Conference in Nuc Med Biol Paris, Pergamon Press, ed C. Raynaud, pp 554-557.

7. Phelps ME, Huang SC, Hoffman EJ, Plummer D and Carson R (1982). An analysis of signal amplification using small detectors in positron emission tomography. J Comput Assist Tomogr 6(3):551-565.

8. Phelps ME, Hoffman EJ, Mullani N, et al (1976). Design considerations for positron emission transaxial tomograph. IEEE Trans Nucl Sci NS-23:516-522.

9. Bohm C, Eriksson L, Bergstrom M, et al (1978). A computer assisted ring detector positron camera system for reconstruction tomography of the brain. IEEE Trans Nucl Sci NS-25:624-637.

10. Cho ZH, Hong KS, Ra JB, Lee SY, Hilal SK and Corell JW (1981). A new sampling scheme for the ring positron camera. Dichotomic ring sampling. IEEE Trans Nucl Sci NS-28:94-98.

11. Derenzo SE, Budinger TF, Huesman RH, et al (1981). Imaging properties of a positron tomograph with 280 BGO crystals. IEEE Trans Nucl Sci NS-28:81-89.

12. Tanaka E, Nohara N, Yamamoto M, et al (1979). Positology - the search for a suitable detector arrangement for a positron ECT with continuous rotation. IEEE Trans Nucl Sci NS-26:2728-2731.

13. Bergstrom M (1982). Performance evaluation and improvements of quantitation accuracy in transmission and positron emission computer assisted tomography. Thesis, University of Stockholm, Stockholm.

14. Hoffman EJ, Phelps ME, Huang SC, Plummer D and Kuhl DE (1982). Evaluating the performance of multiplane positron tomographs designed for brain imaging. IEEE Trans Nucl Sci NS-29:469-473.

15. Hoffman EJ, Huang SC, Phelps ME and Kuhl DE (1981). Quantitation in positron emission computed tomography: 4. Effect of accidental coincidences. J Comput Assist Tomogr 5:391-400.

16. Yamamoto Y, Ficke DC and Ter-Pogossian MM (1982). Performance study of PETT VI, a positron computed tomograph with 288 cesium fluoride detectors. IEEE Trans Nucl Sci NS-29(1):529-533.

17. Derenzo SE (1980). Method for optimizing side shielding in positron-emission tomographs and for comparing detector materials. J Nucl Med 21:971-977.

18. Tanaka E, Nohara N, Tomitani T and Endo M (1982). Analytical study of the performance of a multilayer positron computed tomography scanner. J Comput Assist Tomogr 6:350-364.

19. Ueda K, Tanaka E, Takami K, Kawaguchi F, Okajami K and Tshimatsu K (1982). Evaluation of shield collimators for multilayer positron emission computed tomographs. IEEE Trans Nucl Sci NS-29(1):563-566.

20. Brooks RA, Sank VS, Friauf WS, Leighton SB, Cascio HE and DiChiro G (1981). Design considerations for positron emission tomography. IEEE Trans Biomed Eng BME-28:158-177.

21. Derenzo SE, Budinger TE, Cahoon JL, Greenberg WL, Huesman RH and Vuletich T (1979). The Donner 280-crystal high resolution positron tomograph. IEEE Trans Nucl Sci NS-26:2790-2793.

22. Derenzo SE, Budinger TF, Cahoon JL, Huesman RLH and Jackson HG (1977). High resolution computed tomography of positron emitters. IEEE Trans Nucl Sci NS-24(1):544-558.

23. Derenzo SE, Zaklad H and Budinger T (1975). Analytical study of a high resolution positron ring detector system for transaxial reconstruction tomography. J Nucl Med 16:1166-1173.

24. Kanno I, Uemura K, Miura S and Miura Y (1981). Headtome: A hybrid emission tomograph for single photon and positron emission imaging of the brain. J Comput Assist Tomogr 5:216-226.

25. Phelps ME, Hoffman EJ, Huang SH and Kuhl DE (1978). ECAT: A new computerized tomographic imaging system for positron emitting radiopharmaceuticals. J Nucl Med 19:635-647.

26. Bergstrom M, Eriksson L, Bohm C, Blomqvist G and Litton J (1983). Corrections for scattered radiation in a ring detector positron camera by integral transformation of the projections. Accepted for publication in J Comput Assist Tomogr.

27. King PH (1981). Noise identification and removal in positron imaging systems. IEEE Trans Nucl Sci NS-28(1):148-150.

28. Endo M, Iinuma TA, Nohara N, Tomitani T and Tanaka E (1982). Software correction of scatter coincidence in Positologica: A positron CT device for head. Proc Third World Conference in Nucl Med Biol Paris, Pergamon Press, ed C. Raynaud, pp 3629-3632.

29. Bergstrom M, Bohm C, Eriksson L and Litton J (1980). Corrections for attenuation, scattered radiation and random coincidences in a ring detector positron emission transaxial tomograph. IEEE Trans Nucl Sci NS-27:549-554.

30. Huang S-C, Hoffman EJ, Phelps ME and Kuhl DE (1979). Quantitation in positron emission computed tomography: 2. Effects of inaccurate attenuation correction. J Comput Assist Tomogr 3:804-814.

31. Bergstrom M, Litton J, Eriksson L, Bohm C and Blomqvist G (1982). Determination of object contour from projections for attenuation correction in cranial positron emission tomography. J Comput Assist Tomogr 6:365-372.

32. Gullberg GT and Huesman RH (1979). Emission and transmission noise propagation in positron emission computed tomography. Presented at the Society of Nuclear Medicine 26th Annual Meeting, Atlanta, Georgia, June 26-29.

33. Huang S-C, Carson RE, Phelps ME, Hoffman EJ, Schelbert HR and Kuhl DE (1981). A boundary method for attenuation correction in positron computed tomography. J Nucl Med 22:627-637.

34. Bergstrom M, Eriksson L, Bohm C, Blomqvist G, Greitz T, Litton J and Widen L (1982). A procedure for calibrating and correcting data to achieve accurate quantitative values in positron emission tomography. IEEE Trans Nucl Sci NS-29(1):555-557.

35. Wrenn ER, Good ML and Handle P (1951). The use of positron emitting isotopes for the localization of brain tumors. Science 113: 525-527.

36. Sweet WH (1951). Uses of nuclear disintegrations in diagnosis and treatment of brain tumors. N Eng J Med 245-875.

37. Brownell GL and Sweet WH (1952). Localizations of brain tumors with positron emitters. Nucleonics 11:40.

38. Rankowitz S, Robertson JS, Higinbotham WA and Niell AM (1962). Positron scanner for localizing brain tumors. IRE Int Conv Rec 9: 49-56.

39. Kuhl DE and Edwards RO (1963). Image separation radioisotope scanning. Radiology 30:653-661.

40. Brownell GL and Burnham CA (1973). MGH positron camera. (In) Tomographic Imaging in Nuclear Medicine (Freedman GS, ed) Soc Nuclear Medicine, New York, pp 154-164.

41. Hounsfield GN (1973). Computerized transverse axial scanning tomography) Part I. Description of system. Brit J Radiol 46:1016-1022.

42. Phelps ME, Hoffman EJ, Mullani NA and Ter-Pogossian MM (1975). Application of annihilaton coincidence detection to transaxial reconstruction tomography. J Nucl Med 16:210-223.

43. Ter-Pogossian MM, Ficke DC, Yamamoto M and Hood JT (1982). Design characteristics and preliminary testing of SUPER PETT I, A positron emission tomograph utilizing photon time-of-flight information (TOFPET). Proc Int Workshop on Time-of-Flight PET, Washington University (to be published).

44. Cho ZH, Chan JK and Eriksson L (1976). Circular ring transverse axial positron camera for 3-D reconstruction of radionuclide distribution. IEEE Trans Nucl Sci NS-23:613-622.

45. Cho ZH, Hilal SK, Ra JB, Hong KS, Bigler RE, Yoshizumi T, Wolf AD and Fowler JS (1982). High resolution circular ring positron tomograph with dichotomic sampling Dichotom I (to be published).

46. Thompson CJ, Yamamoto YL and Meyer E (1979). Positome II: A high efficiency positron imaging sytem for dynamic brain studies. IEEE Trans Nucl Sci NS-26:582-589.

47. Hoffman EJ, Phelps ME, Huang S-C, Kuhl DE, Crabtree M, Burke M, Burgiss S, Keyser R, Highfill R and Williams C (1981). A new tomograph for quantitative positron emission tomography of the brain. IEEE Trans Nucl Sci NS-28:99-103.

48. Eriksson L, Bohm C, Kesselberg M, et al (1982). A four-ring positron camera for emission tomography of the brain. IEEE Trans Nucl Sci NS-29:539-543.

49. Kanno I, Uemura K, Miura Y and Miura S (1982). Design concepts and performances of Headtome, a multi-ring hybrid emission tomograph of the brain. Conference on Positron Emission Tomography of the Brain, Cologne, May 1982. Eds. WD Heiss and ME Phelps (Springer).

50. Tanaka E, Nohara N, Tomitani T, Yamamoto M, Murayame H, Tinuma T, Tateno Y, Ishimatsu K, Takemi K and Hayashi T (1982). A whole body positron tomograph, Positrologica II. Design and performance evaluation. Proc Third World Conf in Nucl Med Biol Paris, Pergamon Press, ed C. Raynaud, pp 535-541.

51. Anderson DF (1982). Extraction of electrons from a liquid photocathode into a low pressure wire chamber. Phys Letter 118B:230-232.

52. Kuhl DE (1975). Private communication.

Kuhl DE, Edwards RQ, Ricci AR, Yacob RJ, et al (1976). The Mark IV system for radionuclide computed tomography of the brain. Radiology 121:405-413.

53. Cohen G and DiBianca FA (1979). The use of contrast-detail-dose evaluation of image quality in a computed tomographic scanner. J Comput Assist Tomogr 3:189-195.

54. Brooks RA, Sank VM, DiChiro G, Friauf WS and Leighton SB (1980). Design of a high resolution positron emission tomograph: The Neuro-Pet. J Comput Assist Tomogr 4:5-13.

55. The Cyclotron Corporation. TCC 4600 Specifications.

56. Ter-Pogossian MM, Ficke DC, Hood JT, Yamamoto M and Mullani NA (1982). PETT VI: A positron emission tomograph utilizing cesium fluoride scintillation detectors. J Comput Assist Tomogr 6(1):125-133.

57. Cho ZH, Hong KS, Ra JB, Hilal SK and Correll J (1981). High resolution spherical positron emission tomograph (S-PET) and its design analysis. Presented at 1981 IEEE Nucl Sci Symposium.

58. Ricci AR, Hoffman EJ, Phelps ME, Huang S-C, Plummer D and Carson R (1982). Investigations of a technique for providing a pseudocontinuous detector ring for positron tomography. IEEE Trans Nucl Sci NS-29:452-456.

59. Budinger TF, Derenzo SE, Huesman RH and Cahoon JL (1982). Positron emission tomography: Instrumentation perspectives (to be published).

SINGLE PHOTON EMISSION COMPUTED TOMOGRAPHY: POTENTIALS AND LIMITATIONS

Chun Bin Lim and Steve Gottschalk

Technicare Corporation,
Solon, Ohio

1. INTRODUCTION

Single Photon Emission Computed Tomography (SPECT) is a diagnostic 3-D imaging procedure of isotope distribution inside an object following internal administration of the isotopes that emit single photons such as Tc-99m (140 KeV) (1,2). The physical operation of SPECT stands on the basic 2-D computed tomography (CT) concept of projection data collection from an object slice over 2π angle, and subsequent reconstruction process of filtered backprojection. A final 3-D volume image is formed by a stack of contiguous 2-D slice images reconstructed. Projection data of isotope distribution can be collected by either a stack of coplanar arrays of small discrete detectors or two-dimensional position-sensitive detectors such as Anger Scintillation gamma cameras.

Two major promises of SPECT are improved lesion detectability and quantitation potential of local activities (3,4). The objective of this chapter is to assess the intrinsic imaging potential of SPECT from physical parameters standpoint. The promise of better lesion detectability over conventional projection imaging stems from the elimination of superposition background in the latter, and the quantitation potential of local activity distribution is due to the intrinsic 3-D imaging nature of SPECT. The improved information quality can be invaluable in the diagnostic process, for example, in monitoring the progress of disease state such lesion shape and size changes over the time course of therapeutic interventions.

A common important requirement in the clinical situations is the statistical reliability of detected lesions or the measured values of local activities. This is true not only within each individual image, but also in the images taken at different times. In this chapter we will try to assess a reasonable limit of SPECT capability in this regard in the presence of major physical parameters affecting SPECT image quality. For this purpose we will take the signal-to-noise approach by dealing with the deterministic and statistical aspects in SPECT imaging. By the word "deterministic" we mean the signals or the mean values of SPECT image in noise-free situation. The major physical processes affecting the deterministic aspect in SPECT are spatial resolution, photon attenuation in an object, and tissue scatter contribution due to imperfect nature of detectors used. The deterministic aspects will be treated in terms of lesion contrast change in the process of SPECT image formation as a function of lesion size. In quantitation of local activities the deterministic components affect the accuracy of measured values by introducing distortions between the local activity level in the object and the SPECT image intensities. The statistical aspect is mainly determined by the amount of isotopes injected into an object, the sensitivity of photon detection system, and the imaging time, eventually translating into the total photon counts collected per SPECT image. The statistical aspect deals with the reliability or reproducibility of detected lesions or quantitated values, either within an image or the images taken at different times. This is similar to the concept of precision in a normal measurement process. In this paper, for physical imaging model, we will follow the approach of lesion detectability in SPECT along the line of Rose model in terms of lesion contrast and size (contrast-detail) in a simplified situation (5,6). As to the quantitation aspect, relevant comments will be made as needed at the appropriate places.

2. REVIEW OF SPECT SYSTEM CONFIGURATION

Before we go into the main body of analysis we will have a brief review of existing SPECT system configurations in order to have a proper perspective of SPECT technology evolution at this time. Photon detection system configuration is particularly important due to the intrinsic scarcity of available photons in gamma imaging. Historically the first section imaging SPECT system was started by D. Kuhl during early 1960's, initially based on simple backprojection reconstruction (7,8). Since then it has undergone steady evolution with increasing research effort, especially during the last five years. At present, the photon detection system configurations for projection data collection are broadly divided into two major groups: a discrete detector-based multi-slice approach and a gamma camera-based contiguous slice approach. The former group can be subdivided into traditional square geometry (9-13), circular geometry with either sweeping vane collimator (14) or its rotating version (15), a set of multiple moving detectors with short focal length focused collimators (16-18), and recently a new circular geometry with pinhole fan-beam collimation based on fan-beam reconstruction (19). All these systems have been designed for brain imaging only so far.

In general the discrete detector-based systems provide single or several slices simultaneously with discontiguity between them. These systems have been optimized mostly for high individual slice sensitivity by both complete 2π angle coverage and axial collimation at the expense of discontinuity between slices. They also have the potential advantage of high count rate capability due to individual detector channel electronic processing, which is good for dynamic studies. However, during the course of recent research applications it has been apparent that the limited number of slices in axial direction turns out to be a serious drawback in clinical utility (20). Rigorously speaking they fall short of true 3-D imaging device, not easily rendering sagittal, coronal, or oblique angle image displays, in spite of the above advantage.

On the other hand, the rotating camera-based systems (21-26) have the natural advantage of slice contiguity with equal resolution, both in transverse and axial directions. This is due to the fact that the projection data are collected in a 2-D detector plane, which is later decomposed as a contiguous stack of projection profiles for reconstruction. Currently the camera-based systems are available mostly in single head systems, and some in dual head systems. Thus the detector periphery does not fully cover 2π angle for each slice even though the overall volume sensitivity may be higher than the discrete detector-based systems. With the advancing SPECT technology and inherent necessity for maximum photon utilization, the multiple head systems that cover full 2π angle are bound to appear eventually. Two notable trends in this technology area are the advent of rectangular detectors and non-circular detector rotation orbit for projection data collection (27,28). The former is to meet the natural projection data requirement of equal length profile stack at each angle, and the latter to improve the spatial resolution in the projection data by bringing the detectors closer to the object whenever possible.

3. PHOTON IMAGING PROCESS FOR LESION DETECTION

In diagnostic photon imaging for lesion detection based on gray scale display, we are usually interested in the local activity variations, particularly the local activity differences in some region relative to the local background. This region is often called a lesion. Since the visual response to the gray scale display depends on the ratio of this activity difference relative to the background level, the activity difference in the lesion over the background level is usually defined as a signal in lesion detection. The ratio of this signal over the background is in turn defined as a lesion contrast, originated from the ratio dependence of visual response (30). In absolute activity quantitation the local activity level would be defined as signal. The lesion contrasts in an object distribution and its image are called the object and the image contrasts, respectively (see Fig. 1).

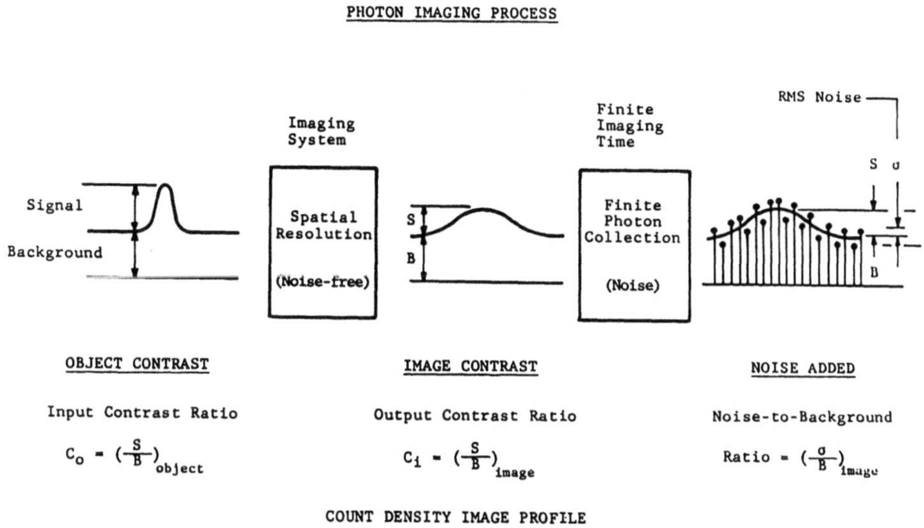

Fig. 1. Photon imaging process for lesion detection, illustrating the éffect of spatial resolution and finite photon collection.

The imaging transformation from an object distribution to an image involves a measurement process, and the measurement process brings in finite spatial resolution of the instrument used as well as the intervening physical processes occurring both within the object imaged and the instrument used during measurement process. The spatial resolution is represented by a point spread function (psf) or a modulation transfer function (MTF), which is the Fourier transform of the psf. The imaging process from an object is mathematically a convolution process of the given object distribution by the local point spread function. The impact of the finite spatial resolution is thus the distortion of local signal level and subsequent degradation of image detail from a true object distribution. In lesion detection it causes reduction of image contrast and edge response. In addition to the finite spatial resolution, the major intervening processes in SPECT that affect local activity signal level during measurement process are photon attenuation within an object and the scatter photons contribution, both due to tissue scattering in the object. The latter is collected in the projection data due to the imperfect energy resolution of the detectors used. Thus, the major factors that affect the lesion contrast and distort the quantitation accuracy in SPECT imaging are the spatial resolution, tissue photon attenuation, and scatter photon contribution (6). Note that all the concepts above have been defined for the noise-free case that unlimited photons are collected for imaging.

In reality, a finite amount of photons are collected for image formation. The discrete and random nature of emitted photons introduce statistical fluctuation of the photons detected according to Poisson statistics. The deviation of this fluctuation from its mean is called a noise, and it is usually represented by a root-mean-square (rms) value. In SPECT the fluctuation of detected photons propagates through numerical reconstruction process, resulting in image value fluctuations at each pixel from image to image even under identical situation (2,4,6,31). This intrinsic statistical variation of image values cause false alarms in lesion detection in an image, and affect the reliability or precision of the measured values in local activity quantitation. For lesions of given size and contrast, the level of false alarm probability and the reliability of the quantitated values are directly affected by the total amount of photons collected and subsequent noise propagation process. The SPECT image noise can be analytically predicted from the detected photon noise by the propagation of the Poisson statistics through the image reconstruction process. From the image rms noise we can in turn predict what the noise averaged over a given region or lesion is (6). Then the statistical reliability of a measured signal or detected lesion is determined statistically by the ratio of the signal level relative to the statistical noise level. The level of false alarm or the reliability of the quantitated values can be assessed in terms of this signal-to-noise ratio, and thereby we can define the region of reliable lesion detection or quantitation in terms of lesion contrast and size (contrast-detail) for a given amount of photons detected. The signal level in a SPECT image can be analytically obtained by the parameters defined above in the noise-free situation.

Once the level of lesion signal and the rms noise corresponding to the equivalent lesion areas are known, the signal-to-noise (S/N) ratio should be set sufficiently high to allow for the many false alarm possibilities at different locations within an image, which increase with the image area (see Fig. 2). In order to have a reasonable reliability for a les-

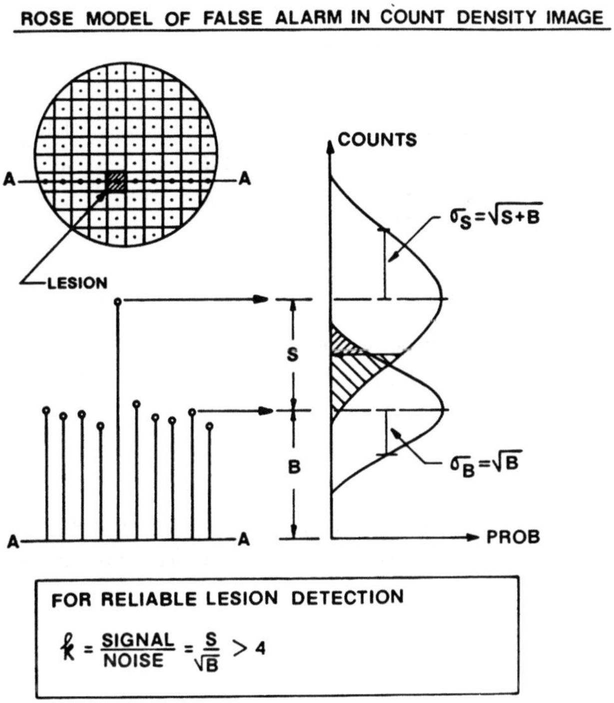

Fig. 2. Statistical process of false alarm generation in photon count density images.

ion detected within an image, the chance of having any single statistical false alarm within an entire image in a set of images should be kept below 5-10% level. This means that, for example, among the 20 images generated under identical situation, only one or two images will have a statistical false alarm in it. In order to ensure this level of statistical reliability per image, the probability of a false alarm rate within a lesion multiplied by the number of equivalent lesion areas within the image should be 5 to 10%. Table I shows this false alarm probability per image as a function of signal-to-noise figure k, for the two numbers 400 and 1000 on the number of lesion equivalent areas per image.

TABLE I

STATISTICAL FALSE POSITIVE PROBABILITY VS. S/N RATIO k

k	Cumulative Prob. Exceeding $k\sigma$	False Positive Rate Picture Elements 400	1000
1	0.15		
2	0.023		
3	1.35×10^{-3}	54%	135%
3.5	2.33×10^{-4}	9.3%	23%
4.0	3.17×10^{-5}	1.3%	3.2%
4.5	3.4×10^{-6}	.14%	.34%
5	3×10^{-7}		
6	2×10^{-9}		

Thus in order to maintain the above level of reliability the S/N ratio k should be approximately 3.5 to 4.0. This concept has been established by Rose previously (5).

Since we have established a conceptual model for physical lesion detectability in terms of signal-to-noise ratio, we can now express the requirement for reliable lesion detection as follows for a lesion of given signal S_L

$$S_L > k \cdot \sigma_L \tag{1}$$

where S_L is a lesion signal level in SPECT image defined as the activity difference over the background as shown in Fig. 1, the rms lesion noise figure averaged over lesion-equivalent areas in the background, and k is the threshold S/N ratio to guard against statistical false alarms. If we divide both sides of the above equation by the background level B we get the following equation expressed in terms of SPECT lesion contrast and the relative lesion noise figure:

$$C_i \equiv \left(\frac{S_L}{B}\right)_{image} > k \cdot \left(\frac{\sigma_L}{B}\right)_{image} \tag{2}$$

From the above equation and previous arguments we can state that the major factors that affect SPECT lesion detectability and quantitation are:

1. Factors reducing object contrast.

2. Factors affecting image noise.
3. Lesion size as averaging area.
4. Threshold S/N ratio k.
5. Availability of lesion-equivalent areas per image.

we will deal with each item in more detail in the following sections.

4. FACTORS AFFECTING SIGNAL LEVEL AND LESION CONTRAST IN SPECT

The major factors that affect the local image amplitude and lesion contrast in SPECT imaging are:

i detector spatial resolution in projection data collection
ii reconstruction spatial resolution
iii photon attenuation due to body tissue
iv scatter photon contribution

(a) SPECT spatial resolution: The spatial resolution in the reconstructed SPECT image is determined by the combined effect of detector system resolution and reconstruction resolution (6). The overall SPECT image formation process from an isotope distribution to an image is shown in Fig. 3 in terms of a transfer function in a frequency domain. During the collection of projection data, individual isotope profile distributions are convolved by the detector psf, and these data are in turn convolved by the reconstruction psf that corresponds to the Fourier transform of the window function for the reconstruction filter. The ramp part of the filter cancels out the (1/r) blurring function of backprojection process. The window function affects both the spatial resolution and image noise in the reconstructed image (32,33). Thus the system resolution response of an image can be represented by

$$\text{System MTF} = (\text{Detector MTF}) \times (\text{Reconstruction MTF}) \tag{3}$$

or

$$\text{System PSF} = (\text{Detector PSF}) ** (\text{Reconstruction PSF}) \tag{4}$$

and the following approximation holds for the Gaussian psf shape assumption

$$R_S^2 = R_d^2 + R_r^2 \tag{5}$$

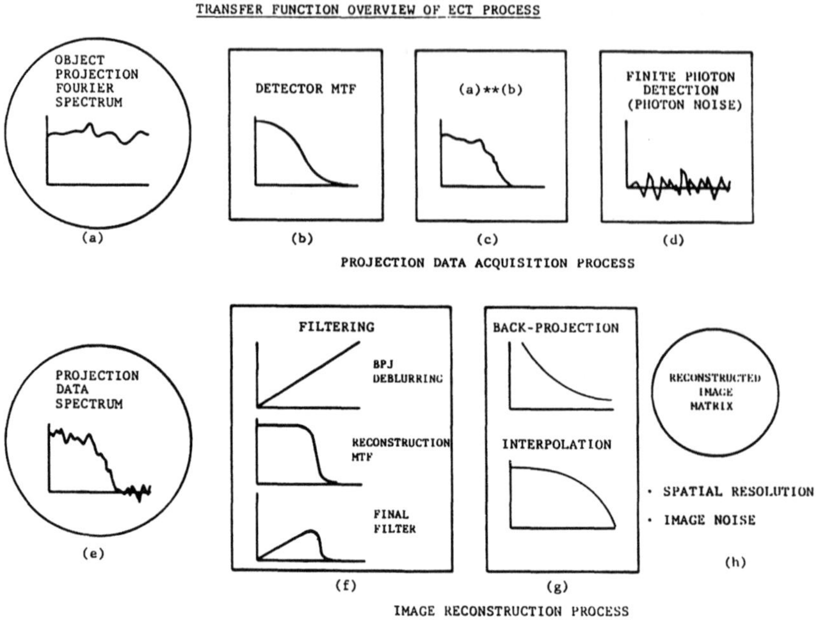

Fig. 3. Transfer function overview of emission CT reconstruction process in frequency domain. Upper row illustrates convolution of projection profile spectrum by detector MTF and addition of photon noise effect. Lower row illustrates multiplication of backprojection deblurring ramp filter and reconstruction window function. System MTF is generated by convolution of detector and reconstruction MTF's.

where R_s, R_d, R_r are the system, detector, and reconstruction resolution distances respectively, and ** represents two-dimensional convolution.

Once the system resolution response for SPECT is established, the contrast reduction of a lesion in imaging from an object distribution to an image depends on the lesion size relative to the system resolution distance. For example, Figure 4 shows the signal amplitude reduction factor for a uniform activity spherical lesion as a function of lesion diameter, represented in FWHM system resolution distance unit, when the lesion is convolved by a Gaussian psf of given FWHM in 3-D space. In SPECT imaging the FWHM system resolution distance is in the order of 8 to 20 mm. Figure 5 shows the lesion contrast reduction factor for the lesion size to 50 mm for the above range of FWHM system resolution distance. Note that the lesion image contrast decreases rapidly from its object contrast with the decreasing lesion size, starting from approximately twice the system FWHM. This is one major factor that degrades the quantitation accuracy for small lesions.

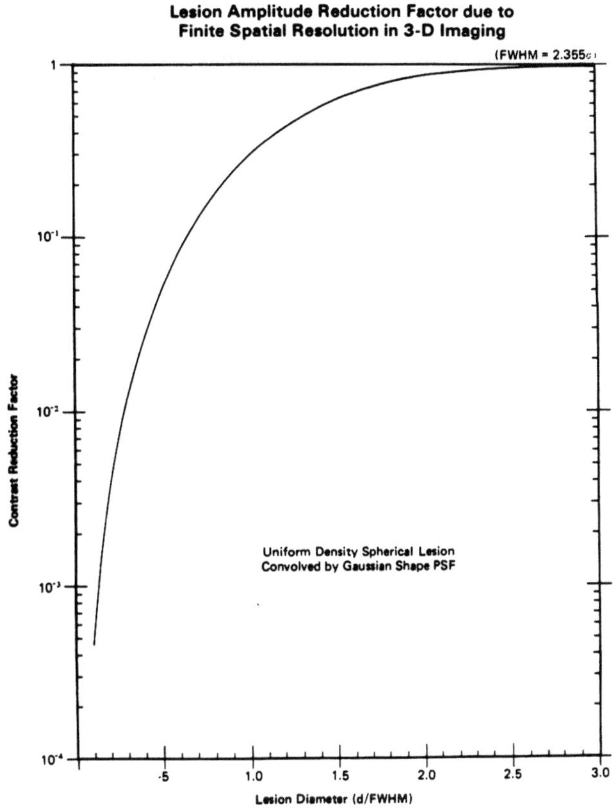

Fig. 4. Lesion amplitude reduction factor due to finite spatial resolution in 3-D imaging.

(b) Photon attenuation: The attenuation of emitted photons due to tissue scattering has two major impacts: (i) distortion of projection profile amplitude in the collected data, and (ii) reduced number of photons (unscattered) detected. The former causes SPECT image amplitude distortion propagated through the image reconstruction process, usually resulting into cupping of image inner values. The latter degrades the SPECT image signal-to-noise ratio due to the increased Poisson noise spread of the detected photons relative to its mean value. Because of the above image distortion, SPECT requires compensation for photon attenuation due to tissue scattering (34,35). This compensation is important in the local activity quantitation. Photon attenuation, however, does not change the SPECT lesion contrast since the lesion signal and the local background levels are reduced by the same ratio.

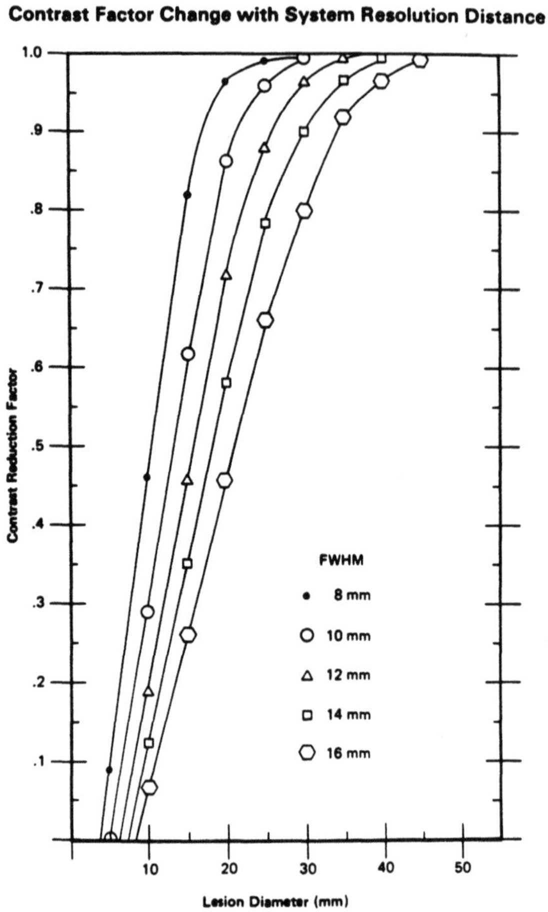

Fig. 5. Contrast reduction factor change with system resolution distance in 3-D imaging.

(c) Scatter photon contribution: The scatter photons in the object are contributed to the detected projection data due to the imperfect energy resolution of the detectors used in their collection. The amount and spatial shape of the scatter photons contained in the projection data in general vary with the depth of a lesion within the object and the object and the energy window used (36-38). Table II shows the scattered photon fraction for a point source located at different depth as a function of energy window width for three object shapes of a rectangle, an elliptical cylinder, and a sphere, when a detector of perfect energy resolution is used (38). These values were obtained by a Monte Carlo simulation of scatter photon transport process for the above shape objects when their sizes shown below in Table II were used. A sphere is meant to be representative of a human brain, and the other two shapes of the torso.

TABLE II

SCATTER FRACTIONS* FOR THE THREE PHANTOMS

Window Width (%)	Phantom	Point Source Depth (cm)				
		0	5	10	15	20
100%	R[1]	0.752	2.77	4.58	6.17	6.99
	C[2]	0.688	2.67	4.60	6.35	7.23
	S[3]	0.462	1.99	3.36	4.39	4.54
25%	R	0.085	0.837	1.37	1.87	2.29
	C	0.061	0.800	1.37	1.93	2.42
	S	0.025	0.687	1.21	1.69	2.01
20%	R	0.0033	0.652	1.12	1.56	1.91
	C	0.0016	0.615	1.10	1.60	2.01
	S	0.0022	0.537	0.981	1.43	1.74
15%	R	0.00	0.466	0.861	1.21	1.52
	C	0.00	0.444	0.833	1.22	1.60
	S	0.00	0.402	0.756	1.12	1.44
10%	R	0.00	0.298	0.598	0.855	1.09
	C	0.00	0.290	0.582	0.843	1.12
	S	0.00	0.275	0.550	0.799	1.06
5%	R	0.00	0.146	0.293	0.440	0.588
	C	0.00	0.144	0.290	0.435	0.585
	S	0.00	0.141	0.280	0.429	0.583
2%	R	0.00	0.058	0.120	0.177	0.241
	C	0.00	0.058	0.119	0.176	0.242
	S	0.00	0.058	0.117	0.174	0.240

[1]Rectangular phantom: L = 50 cm, W = 30 cm, D = 22.5 cm
[2]Elliptical cylinder: L = 50 cm, W = 30 cm, D = 22.5 cm
[3]Sphere: Diameter = 22.5 cm

*Take half of the above values for approximation in actual imaging with gamma cameras because of symmetrical placement of energy window around the photopeak.

Scatter photon profiles are distributed mostly in the region surrounding the point source image of unscattered photons like halos, with its intensity slope decreasing as the source depth increases. In general, scatter photons introduce a low frequency tail to the psf. Thus, when they are convolved with the object source distribution, scatter photons affect the lesion contrast mostly by raising the background level in the image. Therefore the lesion contrast reduction factor due to scatter, C_s, can be approximated (39) by

$$C_s = \frac{1}{1 + f_s} \qquad (6)$$

where f_s denotes a scatter fraction depending on the energy window and the lesion depth. In SPECT, imaging reconstruction process involves angular averaging of detected projection data in the backprojection and thus f_s should be replaced by the averaged scatter fraction $\langle f_s \rangle$.

(d) Lesion contrast in SPECT image: The contrast of a lesion in SPECT image, C_1, can then be stated as follows

$$\begin{aligned} C_i &\equiv \left(\frac{S_L}{B}\right)_{\text{image}} \\ &= \left(\frac{S_L}{B}\right)_{\text{object}} \cdot C_r \cdot C_s \qquad (7) \\ &= \left(\frac{S_L}{B}\right)_{\text{object}} \cdot \frac{C_r}{1 + \langle f_s \rangle} \end{aligned}$$

where S_L and B are the lesion signal and background level both in the object and image as denoted, C_r and C_s are the contrast reduction factors due to spatial resolution and scatter contribution respectively, and $\langle f_s \rangle$ is the averaged scatter fraction in SPECT image.

5. FACTORS AFFECTING SPECT IMAGE NOISE

The image noise in SPECT is generated by the detected photon fluctuation in the projection data set propagated through the reconstruction process (6,31-33). The amount of detected photons is affected by the isotope activity in the object, the SPECT system sensitivity, counting time, and photon attenuation inside an object depending on the object size. The final image noise figure relative its background level is determined by the total number of photons detected, the object size, the sampling interval in the projection data collection, and the filter used in the reconstruction (6). Our objective here is to obtain a reasonable estimate of the relative noise figure averaged over a given lesion size, so that it can be compared with the image contrast of the lesion obtained in the previous section. To simplify, this will be done for the case of a spherical lesion embedded in a uniform background activity of cylindrical shape.

The relative rms noise in SPECT image and an approximate estimate of the relative noise figure averaged over circular lesion area have been derived previously for the circular cylinder of uniform activity (6). The latter has been derived from the former by adopting an effective sampling interval equivalent to the given lesion size. We will describe the essence of these derivations.

(a) Relative rms noise in SPECT image: Recall that the SPECT image values in a slice are computed over the square lattice points of image matrix in reconstruction from a set of projection profiles of detected photons (29). Each profile consists of contiguous strip integral values integrated over a given sampling interval determined by a camera ADC or a discrete detector size. The image noise represents the fluctuation of the image values in a set of experiments conducted under identical situation, originated from the Poisson fluctuation of detected photons. See Fig. 6 for the overall propagation process of detected photon noise through reconstruction.

Suppose we have a uniform disk source of diameter D, and there are N sampling intervals of width w within D such that w N = D. Where w represents the integration interval for the photon count projection sample and N represents the size of the object for given w. If we represent the total number of photons in the projection data set by N_T, then the relative rms noise, defined by the rms noise σ (I) divided by the mean activity level $\bar{I}$, is approximately given by (6).

$$\frac{\sigma(I)}{\bar{I}} \approx b \cdot \frac{N^{3/2} \cdot \langle FF \rangle^{1/2}}{\sqrt{N_T}} \cdot f(r), \tag{8}$$

where N is the number of linear samples covering the object, $\langle FF \rangle$ the normalized filter factor relative to the ramp fliter in noise power, f(r) a noise amplitude shape factor relative to the image center, and b a constant approximately equal to .656. The relative rms noise amplitude di-

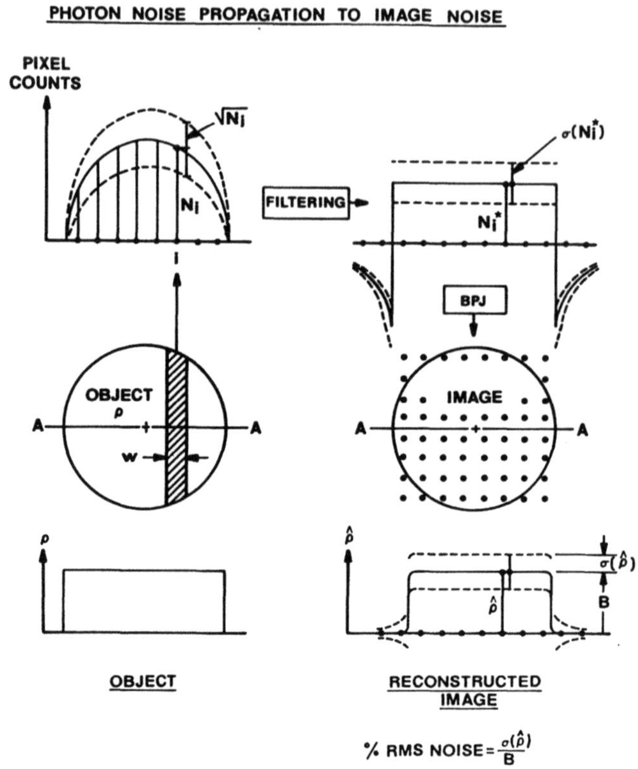

Fig. 6. Schematic illustration of detected photon noise propagation to image noise in SPECT for uniform activity circular disk.

minishes slowly from the center to the edge with a more gradual change in the central area than in the peripheral area. This tendency is represented by f(r). Note that the Nyquist frequency effect is contained in the parameter N in connection with w and D, and the effect of frequency roll-off in the window function is contained in the filter factor <FF> . In an approximate sense, setting the roll-off frequency below Nyquist frequency in the window function has the similar effect the factor filter can disappear in the above equation with the equivalent modification in the value N in terms of an effective sampling interval.

(b) Relative rms noise averaged over lesion area: The above derivation of SPECT noise is based on the projection bin size w. Since, for the ramp filter with cut-off at the Nyquist frequency, over 90% of the power of the filter is contained within one sampling distance of the origin (33) the relative rms noise derived above represents the image value fluctuations averaged over the area element that correspond to approximately twice the sampling interval. The averaging area is also changed by using filters that lower the roll-off frequency below the Nyquist frequency as explained above. In lesion detection or quantitation the signal-to-noise analysis should be applied to the area elements of lesion size to be compatible. Therefore the relative rms noise figure used in this context must represent the image fluctuation over the lesion-equivalent area. This can be approximated by taking the effective sampling interval in the above equation as one-half the lesion size, namely, $N = D/(d/2) = 2D/d$ with d representing lesion diameter. This is equivalent to using photon count samples in the projection data integrated over half the lesion diameter. Then we can represent the relative rms noise over the lesion area of diameter d by

$$\begin{bmatrix} \text{Relative rms noise} \\ \text{over lesion size d} \end{bmatrix} \equiv \sigma_L = b \cdot \frac{(2D/d)^{3/2}}{\sqrt{N_T}} \cdot f(r) \qquad (9)$$

6. SPECT LESION DETECTABILITY EQUATION

Now we can express the relation of physical lesion detectability requirement explicitly using the equations (7) and (9), which were derived above for the SPECT lesion contrast and the relative rms noise averaged over the lesion-equivalent areas in the background, respectively. For a lesion of diameter d and object uptake contrast, $C_o = (S/B)$, embedded in a uniform activity disk background of diameter D, the equation (2) can be given in terms of the total photon count, N_T, the lesion contrast reduction factor, C_r, and the average scatter fraction $\langle f_s \rangle$ as follows:

$$C_o \cdot \frac{C_r}{1 + \langle f_s \rangle} > k \cdot \frac{b \cdot (2D/d)^{3/2}}{\sqrt{N_T}} \cdot f(r) \qquad (10)$$

where k is the threshold S/N ratio, b is approximately .656 from equation (9), f(r) a radial noise shape function normalized to the image center. Note that C_r is a function of lesion size relative to a resolution distance, and its values can be obtained from Fig. 4 for a spherical lesion of uniform activity. An approximate value for scatter fraction can be obtained from Table II. Finally, f(r) is approximately equal to 1.

The threshold S/N ratio k can vary with the lesion size and the number of lesion-equivalent areas in the region of interest in SPECT image, for example in the liver imaging the lesion-equivalent areas should be counted only within the liver area rather than within the entire SPECT image, to be more precise. The region of interest would be smaller in brain imaging because the brain size is smaller than the liver size. Lesion size affects the false alarm probability in two ways. As we discussed above, lesion size is a factor in determining the effective values of image noise because it determines the area for averaging the SPECT rms noise. Also the lesion size directly affect the number in image locations available for the occurrence of false alarms within the given region of interest. Thus the k factor should decrease with increasing lesion size.

The above equation can also predict the total photon counts, N_T, in a SPECT slice necessary for lesion detection, once given the size and uptake contrast of the lesion, the object size, the region size of interest, and the desired false alarm level determined by the k factor. The total counts necessary for detection of a lesion within a uniform activity SPECT slice can be given by

$$\begin{aligned} N_T &= b^2 \cdot \frac{k^2 \cdot (2D)^3 \cdot f^2(r)}{C_i^2 \cdot d^3} \\ &= b \cdot \frac{k^2 \cdot (2D)^3 \cdot (1 + \langle f_s \rangle) \cdot f^2(r)}{C_o^2 \cdot C_r^2 \cdot d^3} \end{aligned} \qquad (11)$$

where C_i represents the lesion contrast in a SPECT image.

Note the difference in the dependency of N_T on C_i and d in SPECT compared to the conventional projection imaging case. In an ideal photon-limited conventional image N_T is inversely proportional to $(C_i \cdot d)^2$ (5) rather than $C_i^2 \cdot d^3$. The above SPECT lesion detectability requirement has been tested by reconstruction simulations with lesions of varying size and contrast contained in a circular uniform activity disk arranged according to a Rose pattern (6).

When the lesion size is sufficiently larger than the SPECT resolution distance and thus the contrast reduction factor C_i becomes unity, the total photon counts for reliable lesion detection is determined mainly by the lesion size and contrast, dictated by the image statistics due to photon count limitation. However, when the lesions become smaller than the system resolution distance, the contrast reduction factor C_i plays a significant role in affecting the photon count limit for their detection, depending on the degree of difference between the lesion size and the resolution distance.

In this case the photon count limit increases further by a factor inversely proportional to the square of the contrast reduction factor C_r due to the geometrical spatial resolution. If one examines carefully the contrast reduction factor in Fig. 4, C_r decreases sharply with the decreasing lesion size for the lesions smaller than approximately twice the resolution distance. Note that its slope of decrease increases with the decreasing lesion size in the logarithmic scale. The impact of this tendency can be better appreciated in Fig. 5. For example, for the three spherical lesion sizes of 30, 20 and 10 mm in diameters, the improvement of the FWHM resolution distance from 16 mm to 8 mm will increase the image lesion contrast by the factors of 1.0/.8 = 1.25, .96/.46 = 2.1, and .46/.07 = 6.6 respectively. According to the equation (11) the above resolution improvement will bring down the necessary photon counts for detecting the above lesions by the factors $1/(1.25)^2$ = .64, $1/(2.1)^2$ = .23, and $2/(6.6)^2$ = .023 respectively by the $1/C_r^2$ factor. This means that in detection of small lesions the improvement of system spatial resolution is critically important in reducing the necessary photon counts for their detectability (40). This tendency predicts that for the detection of extemely small lesions relative to the SPECT resolution distance the pay-off from the resolution distance improvement is far greater than the system sensitivity improvement of SPECT systems.

7. SPECT LESION DETECTABILITY ESTIMATE

Since we have established a numerical expression for the SPECT lesion detectability through statistical signal-to-noise consideration, now we can estimate the SPECT lesion detectability potential in terms of contrast-detail curve for given photon counts and object size. For this estimate we will use two circular uniform activity disks of 20 and 30 cm diameters for reference objects, to be representative of brain and body imaging. For the total photon counts we will use 1 million counts per SPECT slice. This is a reasonable photon count range for camera-based

SPECT systems in static SPECT imaging. For the above two object sizes we can use the SPECT system resolution of 10 mm and 12.5 mm FWHM respectively, based on their estimates at the image center shown in Table III. The slight differences between these numbers and the numbers shown in Table II are to allow for some tolerances in projection data collection and backprojection alignment in reconstruction.

TABLE III

SPECT RESOLUTION DISTANCE ESTIMATE

Parameters	Head	Body
Object Diameter (D)	20 cm	30 cm
Detector Rotation Radius	11 cm	16 cm
Profile Sampling Distance (Δx)	2 mm	3 mm
Profile Sampling Matrix	128	128
Profile Length	25.6 cm (10.1")	38.4 cm (15.1")
Reconstruction Resolution Distance ($R_r = 1.4 \cdot \Delta x$)	2.8 mm	4.2 mm
Detector Intrinsic Resolution (R_d)	4 mm	4 mm
Collimator Resolution (R_c)	8.2 mm	10.1 mm
Detector System Resolution ($R_s^2 = R_d^2 + R_c^2$)	9.1 mm	10.9 mm
SPECT System Resolution (FWHM)	9.5 mm	11.7 mm

For the SPECT lesion detectability estimate we have used the parameters shown in Table IV for SPECT contrast-detail curve in conjunction with the equation (11). For the threshold S/N ratio k we have used k=4 to keep the false alarm probability per SPECT slice within the entire object area below 5% level for both size objects. As for the scatter fractions we have selected 40 and 60% respectively, for brain and body imaging from Table II to allow for scatter photon contribution at 10 and 15 cm tissue thickness respectively, for their image center. The SPECT contrast-detail curves calculated based on the parameters in Table IV are shown in Figs. 7 and 8 for brain and body imaging, respectively. In order to indicate the ideal photon-limited and spatial resolution-limited regions they also show the contrast reduction factors due to spatial resolution, translated from Fig. 4. The contrast-detail curves for brain and body imaging are shown together in Fig. 9 for comparison. Note the

TABLE IV

PARAMETERS USED IN SPECT LESION DETECTABILITY ESTIMATE

Parameters	Head	Body
Object Diameter (D)	20 cm	30 cm
Detector Rotation Distance	11 cm	16 cm
Detector System Resolution (FWHM)	9.6 mm	11.8 mm
Projection Sampling Distance (Δx)	2 mm	3 mm
Reconstruction Resolution Distance (FWHM)	2.8 mm	4.2 mm
SPECT System Resolution Distance (FWHM)	10.0 mm	12.5 mm
Scatter Fraction (15% EW)	40%	60%
Threshold S/N Ratio (k)	4	4
RMS Noise Coefficient (b)	.7	.7
Total Photon Counts (N_T)	1M	1M

steeper photon count increase in the resolution-limited regions to compensate for the corresponding lesion contrast reduction due to a finite spatial resolution.

At this point we have to warn that the SPECT lesion detectability limit shown in these figures are rather conservative estimates for several reasons. First, in nuclear medicine imaging, the false alarm rate is often tolerated to 10 to 15% level in current imaging practice. Second, the region where the false lesion appears due to statistical photon fluctuation applies only to the area where the organ is, not to the entire image area. This is particularly true in body organ imaging. If we consider this practical aspect, the threshold S/N ratio can be lowered to 3.5 level. Thirdly, in actual SPECT images dealing with disturbed source, the activity is mostly concentrated in the area occupied by the organ being imaged and thereby the local background relative rms noise level is reduced compared to the above case of uniform activity distribution due to the particular photon noise propagation characteristics in SPECT imaging (31). This means that the local signal-to-noise ratio in actual clinical SPECT images could be better than the uniform activity distribution case. With these understandings in the clinical SPECT images, we can reasonably assume that the SPECT contrast-detail curves shown in Fig. 9 could be the worst case estimate for the SPECT lesion detectability for the SPECT slice images formed from approximately 1 million photon counts.

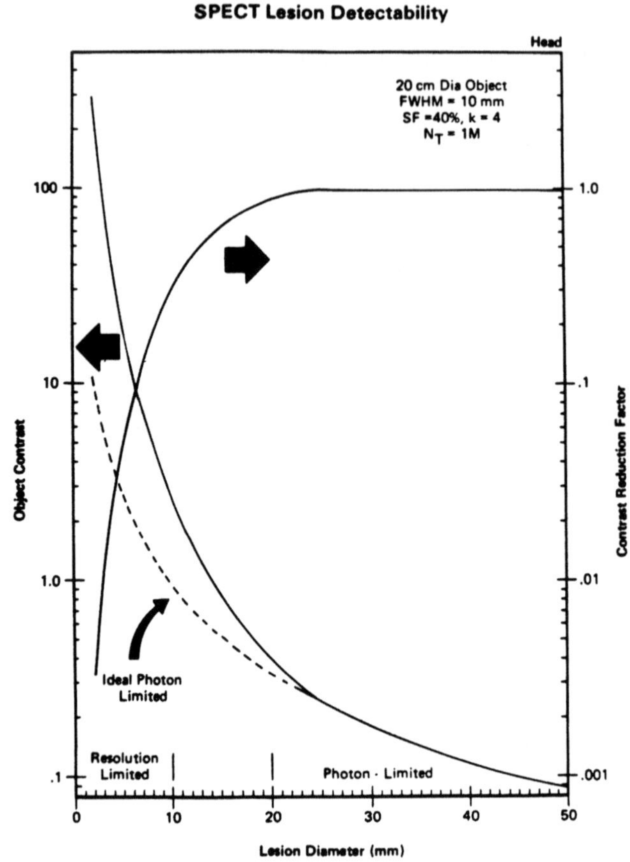

Fig. 7. SPECT lesion detectability contrast-detail curve for brain imaging for 1 million photon count image. It also shows contrast reduction factor due to spatial resolution.

When the photon count per SPECT slice is different from 1 million counts, the contrast-detail curve shown in Fig. 9 should be adjusted by a factor proportional to $1/\sqrt{N_T}$ relative to 1 million photon count. For the same photon counts the contrast-detail curve can be changed with the object size by a factor proportional to $(\text{object size})^{3/2}$. Also, as we use high energy resolution detectors for projection data collection such as high purity Germanium (HPGe) camera, the SPECT lesion detectability can be improved by a factor proportional to $(1+f_s)$ (41,42).

With the above understandings we can safely state the following conclu-

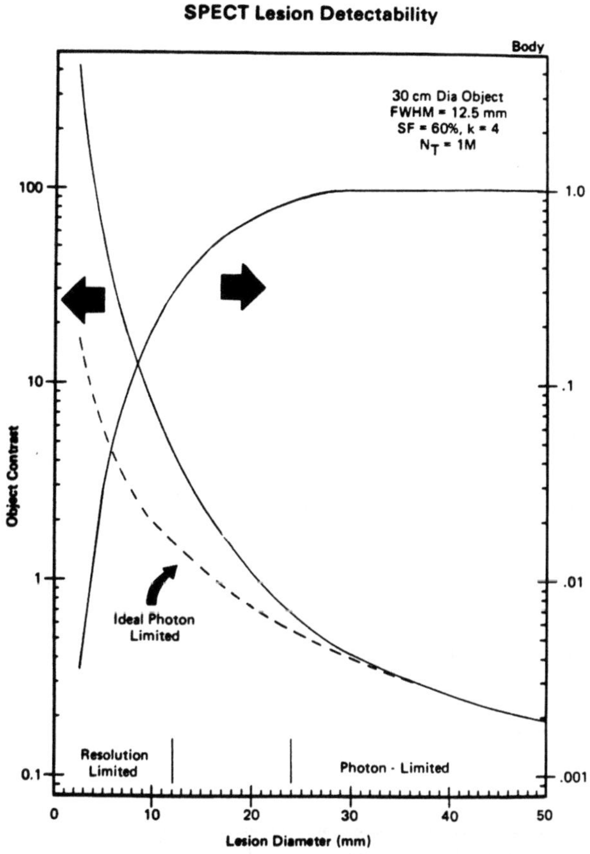

Fig. 8. SPECT lesion detectability contrast-detail curve for body organ imaging for 1 million photon count image. It also shows contrast reduction factor due to spatial resolution.

sions on SPECT lesion detectability near the image center in a 1 million photon count image by examination of Fig. 9. In brain imaging based on the present NaI(T1) scintillation cameras of 10-15% energy resolution, 1 cm lesions of object contrast somewhere between 2 and 3 should be detectable with good reliability. For the 2 and 3 cm lesions the detectable object lesion contrasts can be lowered to approximately .4 and .2, respectively. In body organ imaging the detectable object lesion contrasts are approximately 9, 1, .45 and .3 for the respective lesion sizes of 1 cm, 2 cm, 3 cm and 4 cm. Note that the detectability of the above lesions should be better when they are located in the peripheral region

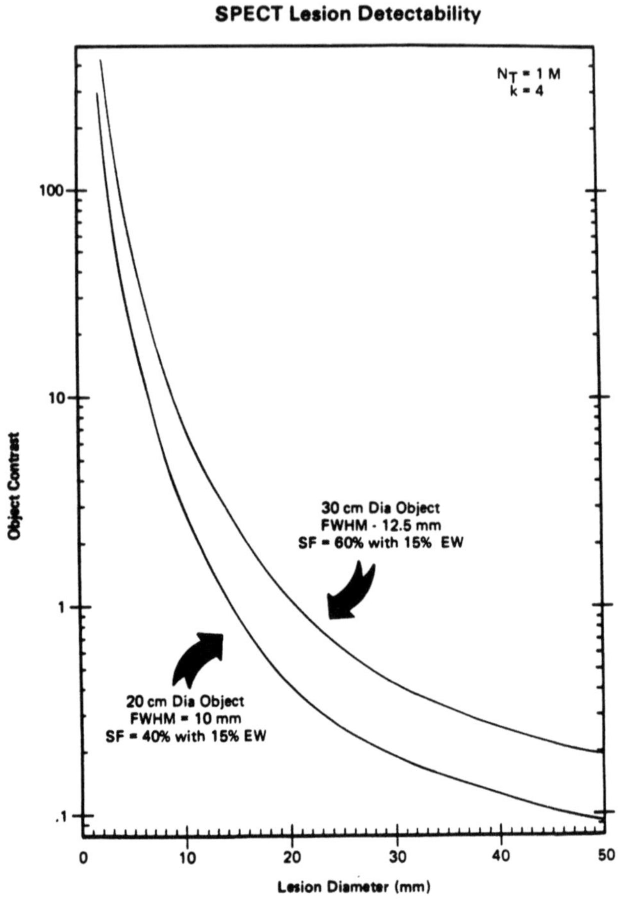

Fig. 9. SPECT lesion detectability contrast-detail curves for brain and body organ imaging for 1 million photon count images.

of the images off the image center, because the SPECT spatial resolution improves outward along the image radius due to the depth-dependent response of collimator resolution in camera-based systems. SPECT image noise also decreases slightly in radial direction.

8. SUMMARY AND DISCUSSIONS

We have established a conceptual model of SPECT lesion detectability from a statistical signal-to-noise consideration. The reliability of a detected lesion can be assessed from the relative ratio between the lesion signal strength and the local relative rms noise averaged over lesion-equivalent background areas, both in the SPECT image. In order to keep the reliability of detected lesions in a SPECT image below 5% false alarm probability level, the threshold S/N ratio should be approximately 3.5 to 4 for the lesion sizes of up to 5 cm. For the larger lesion size the S/N ratio should decrease.

A specific analytical model of SPECT lesion detectability has been derived by an examination of the parameters affecting the signal level and image noise in SPECT image formation process. The major factors which affect SPECT image signal level are spatial resolution, photon attenuation, and scatter contribution, both due to tissue scattering. The spatial resolution reduces the lesion signal and thus degrade SPECT lesion contrast when the lesion size is smaller than twice the SPECT resolution distance. Photon attenuation distorts the amplitude of detected projection data from true projection profile and reduces their signal-to-noise ratio by decreased amount of collected photons. Although photon attenuation introduces SPECT image distortion, resulting into a cupping shape in general, it does not affect the SPECT lesion contrast because both the signal and local background levels are reduced by the same ratio. Scatter contribution reduces the SPECT lesion contrast generally by raising the local background level. The major factors which affect SPECT rms noise are the amount of isotopes injected, the sensitivity of SPECT system, data collection time, object size, and the reconstruction parameters such as a sampling interval for projection data collection and the filter used in reconstruction. For lesion detection analysis, the noise figure averaged over lesion-equivalent areas in the background should be used in comparison with the lesion signal. This lesion noise figure can be approximated by using an effective sampling interval of half the lesion size in the SPECT rms noise expression. A final analytic expression on reliable SPECT lesion detectability limit is obtained by requiring the SPECT lesion signal to be larger than the lesion noise figure by a threshold S/N ratio k, in order to keep the statistical false alarm rate below a certain level. Based on this physical model, the contrast-detail curves corresponding to below 5% level false alarm rate have been obtained for brain and body organ imaging. These curves are shown in Fig. 9. These curves have been generated in reference to the SPECT image center. The lesion detectability in the peripheral region of SPECT should be better mainly because of improved spatial resolution in connection with the depth-dependent response of collimator resolution.

In local activity quantitation the quality of the measured data are affected both by the signal component change and the statistical random fluctuation noise component. The former is related to the accuracy aspect of the measured data and the latter to the precision of reproducibility aspect of the measured data. In absolute activity quantitation correct compensation for tissue photon attenuation and scatter contribution is important. In small lesion activity quantitation the signal distortion due to finite spatial resolution would be the dominant factor. In light of the formidable nature of true compensation for photon attenuation and scatter contribution, the reliable reproducibility aspect may probably have more practical importance in the near future rather than the absolute activity quantitation aspect. With the steady advance and evolution of the technology, SPECT will stimulate the development of new radiopharmaceuticals, which is already in motion (43,44), and will open a new dimension in clinical nuclear medicine in the near future.

REFERENCES

1. Jaszczak RJ, Coleman RE, Lim CB (1980). SPECT: Single photon emission computed tomography. IEEE Trans Nucl Sci NS-27:1137-1153.

2. Budinger TF (1980). Physical attributes of single-photon tomography. J Nucl Med 21:579-592.

3. Jaszczak RJ, Coleman RE, Lim CB, et al (1982). Lesion detection with single photon emission computed tomography (SPECT) and conventional imaging. J Nucl Med 23:97-102.

4. Jaszczak RJ, Coleman RE and Whitehead FR (1981). Physical factors affecting quantitative measurements using camera-based single photon emission computed tomography (SPECT). IEEE Trans Nucl Sci NS-28:69-80.

5. Rose A (1974). Vision, Human and Electronic. Plenum Press, New York.

6. Lim CB, Han KS, Hawman EG, et al (1982). Image noise, resolution, and lesion detectability in single photon emission CT. IEEE Trans Nucl Sci NS-29:500-505.

7. Kuhl DE and Edwards RQ (1963). Image separation radioisotope scanning. Radiology 80:622-653.

8. Kuhl DE and Edwards RQ (1964). Cylindrical and section radioisotope scanning of the liver and brain. Radiology 83:926-935.

9. Kuhl DE, Edwards RQ, Ricci AR, et al (1976). The Mark IV system of radionuclide computed tomography of the brain. Radiology 121: 405-413.

10. Lassen NA, Sveinsdottir E, Kanno I, et al (1978). A fast moving single photon emission tomograph for regional cerebral blood flow studies in man. J Comput Assist Tomogr 2:661-662.

11. Stokely EM, Sveinsdottir E, Lassen NA, et al (1980). A single photon dynamic computer assisted tomography (DCAT) for imaging brain function in multiple cross-section. J Comput Assist Tomogr 4:230-240.

12. Bonte FJ and Stokely EM (1981). Single-photon tomographic study of regional cerebral blood flow after stroke: Concise communication. J Nucl Med 22:1049-1053.

13. Cho ZH, Yi W, Jung KJ, et al (1982). Performance of single-photon tomographic system - GAMMATOM-1. IEEE Trans Nucl Sci NS-29:484-487.

14. Kanno T, Uemura K, Miura S, et al (1981). A hybrid emission tomograph for single photon and positron emission imaging of the brain. J Comput Assist Tomogr 5:216-226.

15. Hirose Y, Ikeda Y, Higashi Y, et al (1982). A hybrid emission CT-headtomer II. IEEE Trans Nucl Sci NS-29:520-523.

16. Zimmerman RE, Kirsch C-M, Lovett R, et al (1981). Single photon emission computed tomography with short focal length collimator: (In) Single Photon Emission Computed Tomography and Other Selected Compute Topics, (Sorenson JA ed), Society of Nuclear Medicine, New York, pp 147-157.

17. Jarritt PH, Ell PJ, Myers MJ, et al (1979). A new transverse-section brain imager for single-gamma emitters. J Nucl Med 20:319-237.

18. Kirsch CM, Moore SC, Zimmerman RE, et al (1981). Characteristics of a scanning multidetector, single-photon ECT body imager. J Nucl Med 22:631-731.

19. Rogers WL, Clinthorne NJ, Stamos J, et al. SPRINT: A stationary detector single-photon ring tomograph for brain imaging. J Nucl Med (to be published).

20. Stokely EM (1982). A contiguous-slice design for single-photon emission tomography (SPECT). J Nucl Med 23:355-356.

21. Budinger TF and Gullberg GT (1974). Three-dimensional reconstruction in nuclear medicine emission imaging. IEEE Trans Nucl Sci NS-21:2-20.

22. Jaszczak RJ, Murphy PH, Huard D, et al (1977). Radionuclide emission computed tomography of the head with 99m-Tc and a scintillation camera. J Nucl Med 18:373-380.

23. Keyes JW Jr, Orlandea N, Heetderks WJ, Leonard PF and Rogers WL (1977). The humongotron -- a scintillation camera transaxial tomograph. J Nucl Med 18:381-387.

24. Jaszczak RJ, Chang LT, Stein NA, et al (1979). Whole-body single-photon emission computed tomography using dual, large-field-of-view scintillation cameras. Phys Med Biol 24:1123-1143.

25. Soussaline FP, Todd-Pokropek AE, et al (1981). A rotating conventional camera single-photon tomographic system: Physical characterization. J Comp Assist Tomogr 5:551-556.

26. Jaszczak RJ and Coleman RE (1980). Selected processing techniques for scintillation camera-based SPECT system. (In) Single Photon Emission Computed Tomography, (Sorenson JA, ed), Society of Nuclear Medicine, New York, pp 44-59.

27. Gottschalk S and Salem D (1982). Effect of an elliptical orbit on SPECT resolution and image uniformity. Proc Third World Congress Nucl Med Biol Vol I:1026-1029.

28. Gottschalk S, Salem D, Lim CB, et al. SPECT resolution and uniformity improvements by non-circular orbit. J Nucl Med (to be published).

29. Lim CB, Chang LT and Jaszczak RJ (1980). Performance analysis of three camera configurations for single-photon emission computed tomography. IEEE Trans Nucl Sci NS-27:559-568.

30. Whitehead FR (1977). Quantitative analysis of minimum detectable lesion-to-background uptake ratios for nuclear medicine imaging systems. (In) Medical Radionuclide Imaging, Vol. I, International Atomic Energy Agency, Vienna.

31. Budinger TF, Derenzo SE, Greenberg WL, et al (1978). Quantitative potentials of dynamic emission computed tomography. J Nucl Med 19: 309-315.

32. Tanaka E and Iinuma TA (1975). Correction functions for optimizing the reconstructed image in transverse section scan. Phys Med Biol 20:789-798.

33. Chesler DA, Riederer SJ and Pelc NJ (1977). Noise due to photon counting statistics in computed x-rays tomography. J Comp Assist Tomogr 1:64-74.

34. Chang LT (1978). A method for attenuation correction in radionuclide computed tomography. IEEE Trans Nucl Sci NS-25:638-643.

35. Tanaka E. Quantitative image reconstruction with weighted backprojection for single-photon emission computed tomography. J Comp Assist Tomogr (to be published).

36. Dresser MM and Knoll GF (1973). Results of scattering in radioisotope imaging. IEEE Trans Nucl Sci NS-20:266-272.

37. Beck JW, Jaszczak RJ, Coleman RE, et al (1982). Analysis of SPECT including scatter and attenuation using sophisticated Monte Carlo modeling. IEEE Trans Nucl Sci NS-29:506-511.

38. Bieszk JA and Lim CB (1982). Point source scatter contributions from finite size objects in radioisotope imaging. Proc Int'l Workshop on Physics and Engineering in Medical Imaging, IEEE Catalog No. 82CH1751-7:20-31.

39. Beck RN (1968). The scanning system as a whole: General considerations. (In) Fundamental Problems in Scanning, (Gottschalk A and Beck RN, eds), Charles C. Thomas, Publishers, Springfield, Illinois.

40. Phelps ME, Huang SC, Hoffman EJ, et al (1982). An analysis of signal amplification using small detectors in positron emission tomography. JCAT 6(3):551-565.

41. Beiszk JA, Lim CB and Hawman EG (1983). A method of improving photon detection efficiency of high purity germanium cameras by including double interaction photopeak events. IEEE Trans Nucl Sci NS-30 (in press).

42. Kaufman L, Lorenz V, Hosier K, et al (1978). Two-detector 512 element high purity germanium camera prototype. IEEE Trans Nucl Sci NS-25:189.

43. Kuhl DE, Barrio JR, Huang SC, et al (1982). Quantifying local cerebral blood flow by N-isopropyl-p-(I-123) iodoamphetamine (IMP) tomography. J Nucl Med 23:196-203.

44. Hill TC, Holman BL, Lovett R, et al (1982). Initial experience with SPECT (single-photon computerized tomography) of the brain using N-isopropyl I-123 p-iodoamphetamine: Concise communication. J Nucl Med 23:191-195.

ENERGY SELECTIVE DIGITAL RADIOGRAPHY

Robert E. Alvarez

DigiRad Corporation
Palo Alto, California

1. INTRODUCTION

The attenuation of body materials changes dramatically with x-ray energy. Table 1 shows the attenuation coefficients of three types of body tissues at energies which span the diagnostic spectrum. Two facts are evident. First, there is a dramatic overall change from low to high energy for all the tissue types. Second, the amount of variation is different for each type of tissue.

TABLE 1

ATTENUATION COEFFICIENTS OF BODY MATERIALS
FOR X-RAY ENERGIES IN THE DIAGNOSTIC REGION
(data from Phelps, Hoffman, Ter-Pogossian) (10)

ENERGY (KeV)	FAT	MUSCLE	ENERGY (KeV)	DENSE BONE
31.1	0.2760	0.3750	30	.958
41.4	0.2189	0.2743	40	.510
52.0	0.1950	0.2333	50	.347
59.6	0.1850	0.2160	60	.272
84.3	0.1660	0.1889	80	.208
103.2	0.1618	0.1773	100	.180
136.3	0.1465	0.1636	150	.149

The information found in these changes is virtually ignored by conventional x-ray imaging systems. The conventional system responds only to the total flux of x-rays irrespective of their energy much as a "black and white" photograph shows the total visible light flux irrespective of its color. Just as color adds extra dimensions to visible light imaging, the use of energy dependent information adds important diagnostically useful information to x-ray images. This added information can be used to remove the effect of intervening tissues from an x-ray image and thus greatly reduce the "clutter" which can severely reduce the probability of detection of a lesion. The information can also be used to characterize the average atomic number of a lesion once it has been detected, a parameter which is obviously useful in diagnosis.

In this article we discuss the type of informaton present in the energy dependent changes in x-ray attenuation. Next we give examples of how this information can be extracted and used in digital radiography systems.

2. VECTOR SPACE DESCRIPTIONS OF THE X-RAY ATTENUATION COEFFICIENT

Conceptually, complete energy dependent information could be extracted from a set of conventional images using monoenergetic x-ray sources with many different energies. Although complete, this sort of information would not be very useful. It would consist of a table of numbers for every point in the image such as those in Table 1. This would not only be difficult to display, but also, the interpretation of the results would be almost impossible.

An alternative to this technique is to use a vector space expansion of the attenuation coefficient as a function of energy. That is, express the attenuation coefficient $\mu(E)$ as the sum of functions of energy multiplied by constant coefficients:

$$\mu(E) = \sum_{i=1}^{\infty} a_i f_i(E) \tag{1}$$

Since the functions of energy $f_i(E)$ are known beforehand, this approach transforms the problem to one of estimating the coefficients a_i. Clearly, this technique will only be practical if the number of functions required to accurately represent the attenuation coefficient (the "dimensionality" of the space) is small.

Application of the known physical properties of the interactions of x-rays and matter lead to important reductions in the required number of functions. Because x-ray energies are so much higher than chemical binding energies, the mixture rule may be applied. This states that the total attenuation for a mixture containing N_i atoms of element i with cross-section σ_i is

$$\mu(E) = \sum_{i=1}^{m} N_i \sigma_i(E) \tag{2}$$

Applying this rule shows that the dimensionality of the space is at most the total number of elements found in the object.

Another important fact about x-ray interactions is that they consist of a small number of distinct physical processes. These processes are mutually exclusive so that the total attenuation of a particular element is the sum of the attenuation due to each process.

$$\mu(E) = \mu_{Compton}(E) + \mu_{Photoelectric}(E) + \mu_{Coherent}(E) \quad (3)$$

Now, if the attenuation for each process were separable into factors which depended on atomic number only and energy only, then equation (3) could be used with equation (2) to show that the dimensionality of the space is only the number of distinct interaction types.

Unfortunately, the interaction attenuations are not exactly separable (1). However, they are very close to it. Techniques based on numerical analysis can be used to show that with an accuracy suitable for even the most demanding medical application, the attenuation coefficient can be represented with a two-function basis set (2).

$$\mu(E) = a_1 f_1 (E) + a_2 f_2 (E) \quad (4)$$

The numerical analysis techniques give optimal basis set functions which consist of carefully chosen linear combinatons of the attenuation coefficients of elements found in the body. However, for practical purposes, the attenuation coefficients of two real materials give sufficient accuracy. This is shown in Table 2 where the errors with a basis set consisting of the attenuation coefficients of two elements are shown. As will be discussed later, the basis set consisting of the attenuation coefficients of two calibration materials has important advantages in the procedure used to solve for the energy dependent information from measurements.

3. ATTENUATION COEFFICIENTS AND LINE INTEGRALS

The x-ray attenuation coefficient summarize all the information which can be extracted by an imaging system using attenuation measurements. X-ray imaging systems cannot measure an attenuation coefficient directly. Rather, they measure quantities which depend on the total x-ray flux transmitted through the patient. In order to understand the relationship between the desired quantity and the measurement, consider the following. The simplest situation consists of a uniform object with attenuation coefficient $\mu(E)$ and a monoenergetic x-ray source with energy E. The transmitted flux is related to the attenuation by

$$I_{transmitted} = I_{incident} \exp[-\mu(E)\ell] \tag{5}$$

TABLE 2

ACCURACY OF CALIBRATION MATERIAL BASIS SET
Calibration Materials Z = 7 & Z = 18
Least squares fit, 30-150 KeV
Tabulated cross-sections are from Veigele (11)

Z	a1	a2	rms fractional error (%)
1	.16218	-.01128	.34
2	.32343	-.02168	.28
3	.48262	-.03104	.37
4	.63909	-.03765	.08
5	.78244	-.03775	.12
6	.90711	-.02748	.09
7	1.00000	.00000	.00
8	1.04809	.05137	.05
9	1.04164	.13397	.10
10	.94337	.26298	.12
11	.77059	.43747	.24
12	.48116	.66930	.11
13	.00000	1.00000	.00
14	-.51247	1.36197	.17
15	-1.37348	1.90736	.26
16	-2.36867	2.52720	.30
17	-3.58552	3.27936	.43
18	-5.12029	4.19594	.48
19	-6.96346	5.29614	.54
20	-9.07623	6.53662	.67

If the object attenuation is not uniform, the product of attenuation coefficient and object thickness is replaced by an integral.

$$I_{transmitted} = I_{incident} \exp[-\int\mu(x,y,z;E)d\ell] \tag{6}$$

Finally, if the source is not monoenergetic but has a spectrum S(E), the transmitted flux is the sum (integral) of the monoenergetic flux at each energy.

$$I_{transmitted} = \int S(E) \exp[-\int\mu(x,y,z;E)d\ell]dE \tag{7}$$

With the two-function basis set decomposition of the attenuation coefficient, the line integral becomes

$$\int\mu(x,y,z;E)d\ell = f_1(E)\int a_1(x,y,z)d\ell + f_2(E)\int a_2(x,y,z)d\ell \tag{8}$$

For a fixed source spectrum, the total flux is therefore a function of only two variables, the line integrals of the basis set coefficients A_1 and A_2 .

$$A_1 = \int a_1 (x,y,z)d\ell \qquad A_2 = \int a_2(x,y,z)d\ell \tag{9}$$

Because of the linear properties of attenuations implied by the mixture rule, the relationship of the attenuation coefficients and the line integrals can be given the simple geometric interpretation shown in Fig. 1. Part (a) shows a two-dimensional plot of the basis set coefficients of several materials. Each point represents the pair of coefficients of a single material. Using a similar two-dimensional presentation where each axis represents the value of each basis set coefficient line integral, the representation of a single material, as shown in part (b), is a straight line through the origin with length proportional to the object thickness. The slope of the line is the ratio of the basis set coefficients of the material.

If there are several types of materials in the object, the total basis set coefficient line integrals are the vector sum of the vectors for each individual material. This is illustrated in part (b) of Fig. 1.

4. COMPUTATION OF ENERGY SELECTIVE INFORMATION

Equation (8) shows that the data measured by a conventional system depend on two quantities, the basis set coefficient line integrals. There are two "unknowns" and a single conventional measurement cannot be used to compute them. However, under quite general circumstances (3) two measurements with different source spectra S1(E) and S2(E) can be used to uniquely compute the unknown line integrals. These two spectra can be generated by simply changing the x-ray tube voltage, but other more sophisticated techniques are available for generating the measurements with the difficulty of voltage switching.

Since the x-ray spectrum varies from tube to tube, techniques for calculating the basis set coefficient line integrals should not depend on knowing the spectrum. Such a technique has been developed. This technique is based on making calibration measurements on a phantom with known composition and thickness.

Normally, the calibration phantom attenuation coefficient must be known in order to calibrate. But by a clever choice of the basis functions of

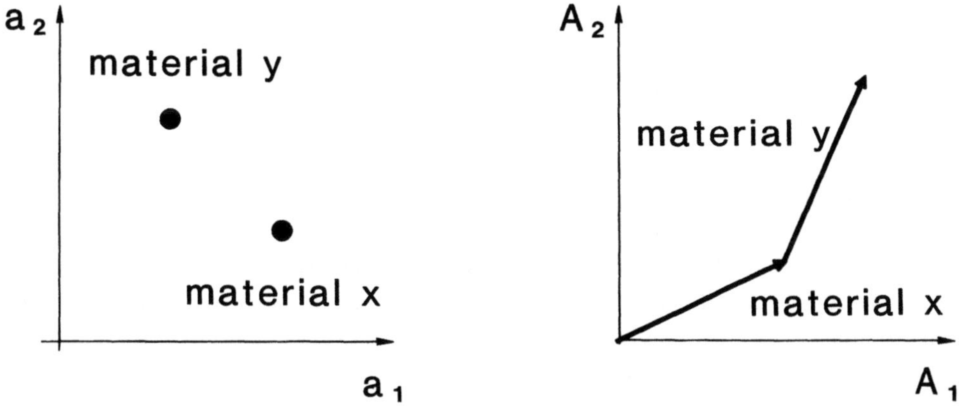

Fig. 1. Relationship of the basis set coefficients and their line integrals. The contribution of each material is a vector of fixed angle with length proportional to the amount of material. The total vector for a line through a material is the vector sum of the contribution of each material along the path.

the expansion of the attenuation coefficient, this can be avoided. Suppose the attenuation coefficients of the calibration materials themselves are used as the basis functions f1(E) and f2(E). Then the basis set coefficients for these materials will be either one or zero

$$\begin{aligned} &\text{material 1:} \quad a_1 = 1 \quad , \quad a_2 = 0 \\ &\text{material 2:} \quad a_1 = 0 \quad , \quad a_2 = 1 \end{aligned} \tag{10}$$

and the line integrals will be simply the lengths of the two materials along the x-ray beam path. Thus by using this basis set the calibration will rest only on length measurements which can, of course, be carried out accurately with relative ease.

The use of this basis set results in coefficients with interesting physical interpretations. They represent the amounts of the calibration materials which can be mixed to give a material with the same x-ray attenuation as the object at every energy. The basis set coefficient line integrals are the thicknesses (with units of length) of the two calibration materials which if placed along the path of the x-ray beam would yield an object with the same x-ray attenuation as the original object at every x-ray energy.

5. MATERIAL SELECTIVE IMAGES

Suppose that the object is composed of several materials so that along a line the vector sum of the basis set coefficients can be decomposed as shown in Figure 1. If at every point in the image we compute a quantity which consists of a "projection" perpendicular to the direction for one of the materials, the resultant image will not contain any contribution from that material. This is illustrated in Figure 2. The material will have been "cancelled" from the image by this generalized projection operation.

Since the body materials can be roughly decomposed into "bone" and "soft tissue", applying the material cancellation operation will yield images which correspond to anatomical features. Examples are shown in Figures 3 and 4. These are chest images with bone and soft tissue cancelled. The data were measured on a modified CT scanner which allowed the x-ray tube voltage to be rapidly switched while the images were measured on a line by line basis (4). The resolution of this electronic system is somewhat limited because of the relatively small number of channels (and therefore pixels in the horizontal direction) and the coarse collimation of the x-ray beam in the vertical dimension. However, the power of the technique is evident in the images.

6. TISSUE CHARACTERIZATION

Because of the effect of overlying tissue, it might seem that it is not possible to perform accurate measurements of the tissue characteristics in a single projection (i.e. the non-computed tomography system). However, this can be carried in some important cases by using the properties of generalized projections in the attenuation coefficient vector space.

The first step in understanding the characterization is to consider the representation of a single material in the vector space. The line integrals for a single material are simply

$$A_1 = a_1 t \qquad A_2 = a_2 t \tag{11}$$

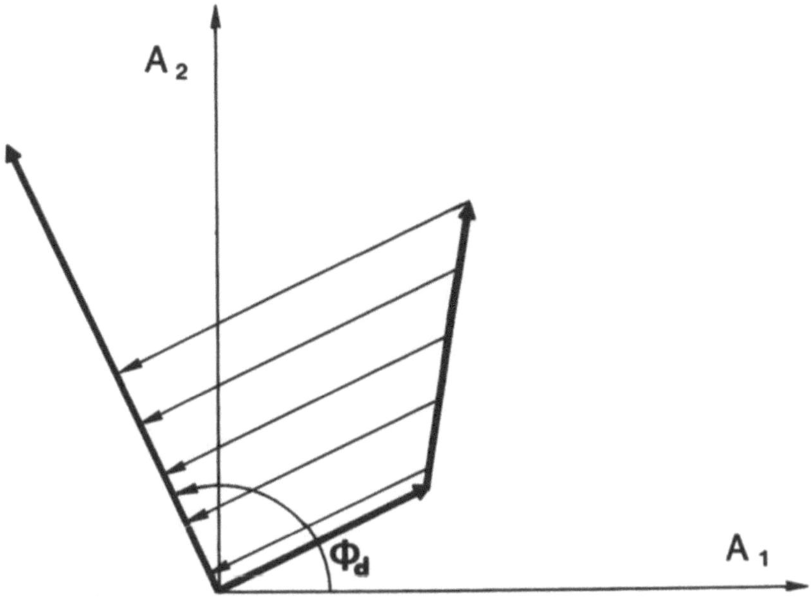

Fig. 2. Selective Tissue Cancellation. By forming a linear combination of the two quantities at every point in the image equivalent to the projection shown above, a material of a given composition can be cancelled from the image.

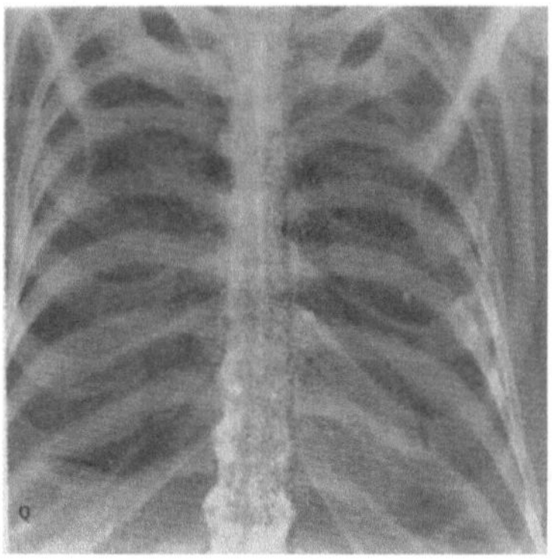

Fig. 3. Soft Tissue Cancellation Image. This is an image formed by performing a point by point linear combination of the basis set coefficient line integrals to cancel soft tissue. Photograph courtesy Stanford University Advanced Imaging Techniques Laboratory.

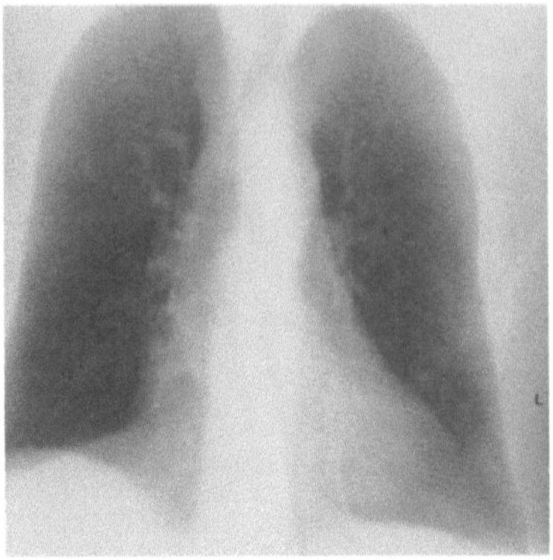

Fig. 4. Bone Cancellation Image. This image was formed from the same basis as Figure 3 but by using a different projection angle.

where t is the thickness of the material. Note that the ratio of line integrals is

$$A_1/A_2 = a_1/a_2 \tag{12}$$

which is a constant that depends only on the type of material. Thus the vector space representation of a single material is a straight line through the origin with length proportional to the thickness.

Now suppose the object consists, as shown in Figure 5, of a uniform background material with an unknown object of complex shape. The vector representation for this case is as shown in Figure 5. The background material contributes a constant vector while the unknown contributes a vector of variable length but fixed direction. The total vector will thus be along a line as shown in the figure.

If we make generalized projections with varying angle and observe the result in the image, at an angle perpendicular to the angle of the unknown material it will disappear. Furthermore the angle is independent of the composition or thickness of the background material. It depends only on the ratio of the basis set coefficients of the unknown material. This ratio, in turn, depends only on the average atomic number of the material (5).

Summarizing, by using a-priori knowledge of the geometrical features of the background material, the composition of an unknown can be determined by interactively measuring the generalized projection angle which causes

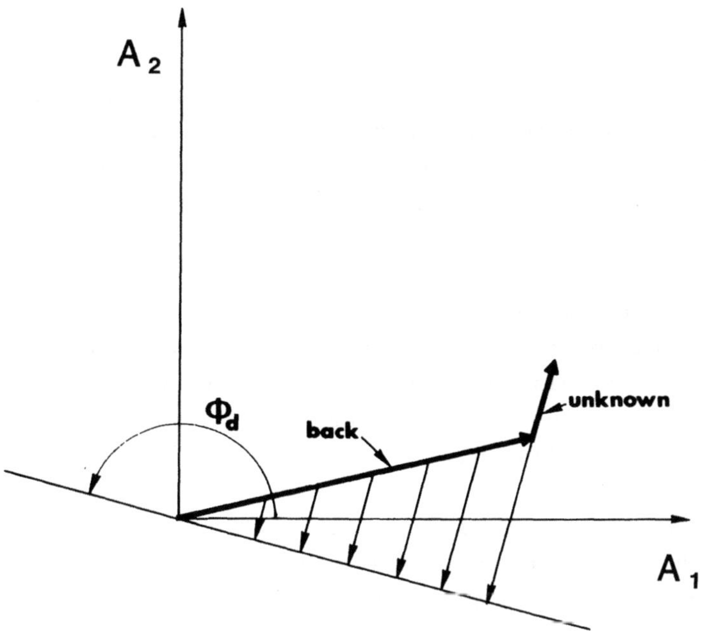

Fig. 5. Tissue characterization with energy selective data. The projection angle ϕd is varied until the unknown object disappears from the image.

its contribution to disappear from the image. This measurement can be quite accurate as illustrated in Table 3. This shows measured and computed angles for various materials. The data were measured on the modified CT scanner system described previously.

TABLE 3

MATERIAL CANCELLATION ANGLES
FOR AN ALUMINUM/PLASTIC BASIS SET
(data from Lehmann)(6)

MATERIAL	ANGLE (measured degrees)	ANGLE (calculated degrees)
Aluminum	90	90
Plastic	0	0
Water	15	15.6
Teflon	35	36.8
Bone	90	89.0

7. HYBRID SUBTRACTION

The most common "signal processing" technique used in digital radiography is simply subtraction of two images taken at different times. If something has changed (such as the perfusion of contrast agent into a region), then it will be enhanced while static (or nearly static) anatomical features will be suppressed.

While subtraction processing has been widely used, it has a major problem. Any anatomical motion will create subtraction artefacts which are large compared to the relatively small signals due to the contrast agent. The major problems are due to involuntary soft tissue motions such as bowel peristalsis or swallowing.

Hybrid subtraction is a combination of energy selective and subtraction processing which significantly lowers sensitivity to soft tissue motions (7). The technique involves subtracting soft tissue cancelled images taken at different times. The soft tissue subtracted images are formed from pairs of dual spectrum measurements taken nearly simultaneously. The sets of dual spectrum measurements are taken at times sufficiently separated so that the contrast agent has changed significantly.

The use of hybrid subtraction does give useful suppression of soft tissue motion artefacts. Figure 6 (a,b) shows temporal and hybrid subtraction images of phantoms taken at different times where a water-filled balloon has been inserted into the image. Conventional time subtraction shown in Figure 6a clearly shows the large artefacts. The hybrid subtracted image in Figure 6b has nearly eliminated the artifact.

8. SIGNAL AND NOISE IN ENERGY SELECTIVE RADIOGRAPHY

Examination of the energy selective images shows that they appear to have higher noise than the conventional images. From this one may (inaccurately) conclude that there is a noise "penalty" caused by extracting the energy dependent information. However, the energy selective and conventional images represent distinctly different physical quantities and care must be exercised in their comparison. In fact, there is a noise reward for using an energy selective system. From several points of view these systems extract more information for the same dose as conventional systems.

One form of comparison is the noise standard deviation. As discussed previously, a conventional system measures data which are related to the total x-ray flux. With digital radiography, the flux is used to compute the attenuation coefficient line integral at every point in the image. Equation (8) may be used to compute the line integral of the attenuation coefficient from the basic quantities calculated by an energy selective digital radiography system, the line integrals of the basis set coefficients. Note that this equation can be used to compute the line integral at any desired energy while a conventional system can only compute the line integral at a single average energy.

The noise in the line integral computed from energy selective data depends strongly on the energy used in the computation, the display energy (8). Figure 7 shows a plot of the noise standard deviation as a function of display energy. The standard deviation of the computed line integral has a minimum at an energy near the average energy of the two spectra used to measure the energy selective information. At this minimum, the line integral computed from the energy selective data has the same noise as the line integral computed by a conventional system using the same total dose as the energy selective system. But, since the energy selective data can be used to compute the line integrals at any desired energy in the region of applicability of Equation (8), the energy selective system extracts more information for the same dose as a conventional system (9).

Another way to compare x-ray imaging systems is from the point of view of signal-to-noise ratio. For this comparison an actual imaging problem may be used in order to precisely define a "signal". A common situation will be used as a basis for analysis. This involves the detection of the presence of a small object overlying a large uniform region of another object. An example of this situation would be the detection of a lung nodule with the overlying tissue in the rest of the thorax as a "background". The signal here is the difference in the measured quantity between a region containing only background material and a region containing the background plus the object of interest. The noise is the fundamental quantum noise found in any x-ray imaging system. In order to obtain a fair comparison, both systems must use the same dose. For analysis, this is achieved by assuming that the conventional system uses a spectrum equal to the sum of the two spectra used to compute the energy selective information.

Analyzing this situation leads to the following conclusion. The signal-to-noise ratio of the conventional system is always less than that of

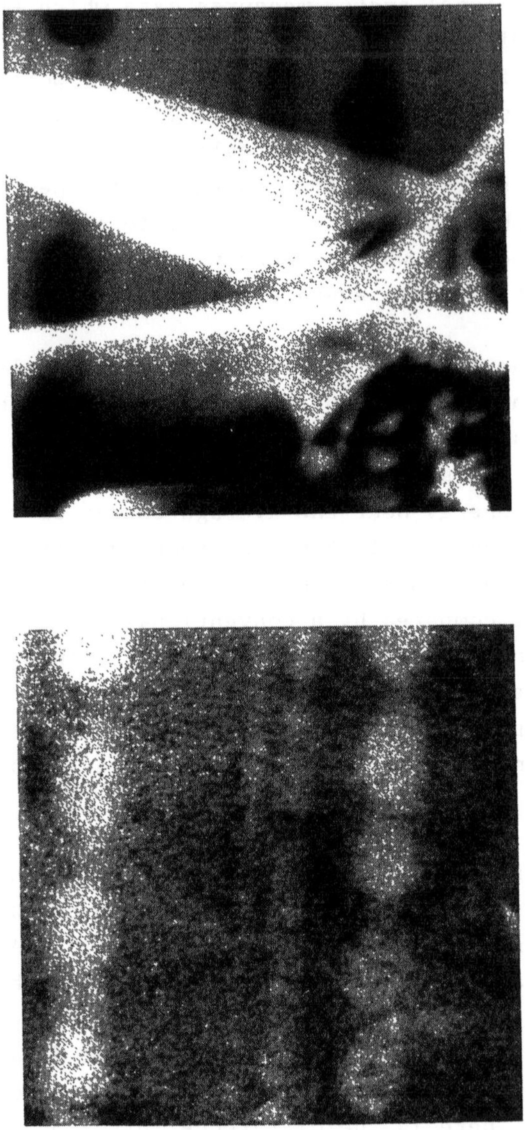

(a)

(b)

Fig. 6. Conventional Temporal and Hybrid Subtraction Images. Part (a) shows the temporal subtraction. A water-filled balloon has inserted into the field between the two times to simulate soft tissue motion. The temporal subtraction image shows large artefacts while the hybrid subtraction image which results from subtracting the soft tissue cancelled images shows little artefact. (Photograph courtesy Dr. William Brody, Stanford Advanced Imaging Techniques Laboratory).

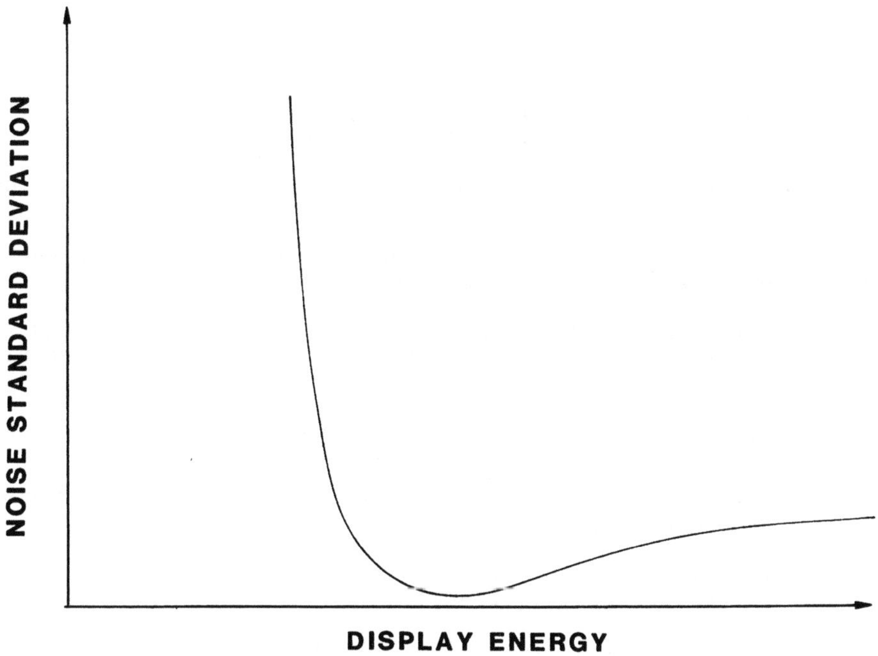

Fig. 7. Noise Standard Deviation versus display energy for conventional image calculated from energy selective data.

the energy selective system except for some physically unimportant situations. (The signal-to-noise ratios are equal only if the average energy of the two spectra used are equal so the energy selective system cannot compute energy dependent information or if the feature has zero attenuation so that the signal is always equal to zero for either system.) In all other cases the energy selective system is superior (9).

REFERENCES

1. Jackson DF and Hawkes DJ (1981). Attenuation coefficients of elements and mixtures. Physics Reports 70:169-233.

2. Alvarez RE (1983). Energy dependent information in x-ray imaging: Part I. The vector space representation. (To be published)

3. Alvarez RE and Macovski A (1976). Energy selective reconstructions in x-ray computerized tomography. Phys Med Biol 21:733-744.

4. Lehmann LA, Alvarez RE, Macovski A, et al (1981). Generalized image combinations in dual KvP digital radiography. Med Phys 8: 659-667.

5. Alvarez RE (1976). Ph.D. Thesis, Department of Electrical Engineering, Stanford University, Stanford, California.

6. Lehmann LA (1982). Ph.D. Thesis, Department of Electrical Engineering, Stanford University, Stanford, California.

7. Brody WR (1981). Hybrid subtraction for improved intravenous arteriography. Radiology 141:828-831.

8. Alvarez RE and Seppi E (1979). A comparison of noise in conventional and energy selective computed tomography. IEEE Trans Nucl Sci NS-26:2853-2856.

9. Alvarez RE and Lehmann LA (1983). Energy dependent information in x-ray imaging: Part 2. Information extraction and noise. (To be published)

10. Phelps ME, Hoffman EJ and Ter-Pogossian MM (1975). Attenuation coefficients of various body tissues, fluids and lesions at photon energies of 18 to 136 keV. Radiology 117:573-583.

11. Veigele WJ (1973). Cross sections from 0.1 keV to 1 MeV for elements Z=1 to Z=94. Atomic Data Tables 5:51-111.

MATCHED FILTERING FOR DIGITAL SUBTRACTION ANGIOGRAPHY

Robert A. Kruger

Department of Radiology, University of Utah
Medical Center Salt Lake City, Utah 84132

1. INTRODUCTION

Digital fluoroscopy methods are providing powerful diagnostic tools for imaging dilute amounts of contrast materials flowing through the cardiovascular anatomy. The predominate processing algorithms which have proven successful are temporal (1-6) and energy (7-9) subtraction schemes. The most common application of these subtraction methods has been in conjunction with intravenous angiography; although, intra-arterial studies also benefit from such processig (10). All these methods may be lumped together under the heading "digital subtraction angiography (DSA)".

To date, DSA imaging methods have been used almost exclusively for isolating vascular opacification due to a bolus injection of iodinated contrast medium. The preferred injection site is within the venous (rather than arterial) anatomy. Such an injection site results in safer and less time-consuming angiographic procedures. Such an injection site also results in severe dilution of the iodine concentration (approximately twenty-fold) prior to its arrival within arterial anatomy.

Because of this large dilution factor the iodine image contrast (signal) represents a small percentage (1-10%) of the image contrast that is produced by non-vascular anatomy (bone and fat tissue variation). The problem, then, which DSA techniques attempt to solve is the extraction of a weak time-dependent signal (iodine contrast bolus) from a much larger relatively static background in the presence of image noise (predominantly quantum). The ultimate performance of any DSA technique, assuming perfect extraction of the contrast bolus from background anatomy, will be limited by the magnitude of this image noise relative to the iodine signal.

One way of comparing DSA techniques, then, is to compare the resultant ratio of signal contrast to noise contrast (SNR) keeping other factors constant, i.e. x-ray exposure to the patient. Admittedly this type of comparison ignores image degradation factors which result from nonquantum noise sources - patient motion and vessel overlap - but such a comparison does not give an indication of relative ultimate performance. More importantly the establishment of an optimum DSA technique, in terms of SNR, provides a basis of comparison for all other techniques. When DSA techniques which may prove less sensitive to nonquantum image degradation factors than the SNR optimum technique are developed, the question "How big a price in terms of lost SNR does this technique cost?" will have an answer in hand.

2. SNR OPTIMUM TECHNIQUE - MATCHED FILTER

All temporal subtraction schemes can be thought of as subtractive temporal filters acting on a set of images acquired over some time period. Figure 1 shows how such images are acquired for a typical "mask mode" imaging scheme. In this example images are formed at a rate of one per

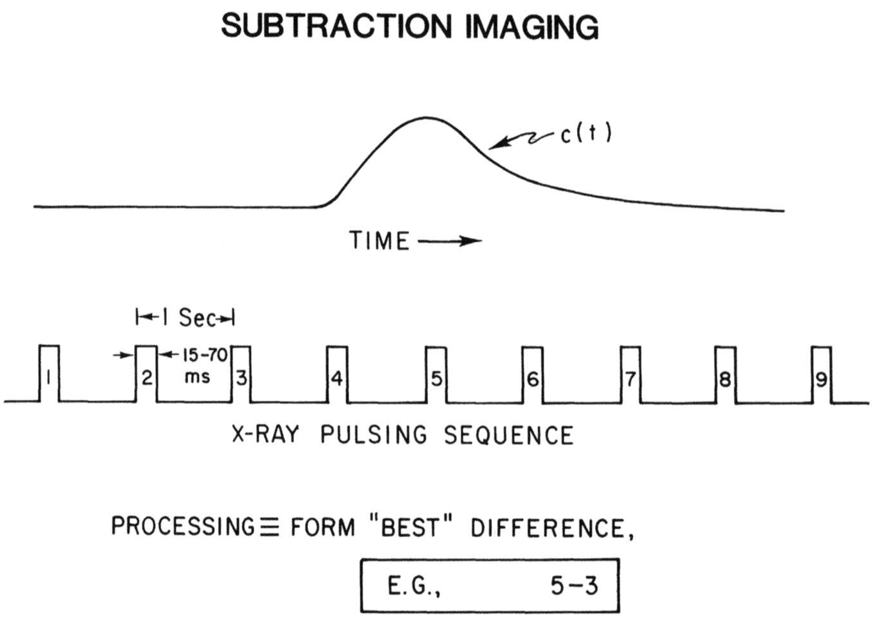

Fig. 1. Image acquisition scheme characteristic of an intravenous angiogram using mask mode subtraction. (Reprinted with permission from SPIE).

second and stored for later processing. The concentration of iodine with time, C(t), also is shown. The mask mode algorithm isolates iodinated vasculature by choosing a pair of images which exhibit a high differential in iodine contrast and a low differential in patient motion. Of all the images acquired, only two are retained. The others are discarded. In terms of filtering one can write the following equations which are descriptive of the above mask mode algorithm:

$$I = \sum_{i=1}^{9} k_i I_i , \tag{1}$$

$$k_i = \begin{cases} -1, & i = 3 \\ +1, & i = 5 \\ 0, & i \neq 3,5. \end{cases} \tag{2}$$

The set $\{k_i\}$ defines the finite impulse response (F.I.R.).

characteristic of this "high pass" temporal filter. Note that $\Sigma_i k_i = 0$. This is the condition imposed on all subtractive filters and insures that stationary image contrast is completely removed by the filtering process.

Assuming one has acquired the nine images as shown in Fig. 1, is there a better subtractive F.I.R. than that representative of a mask mode procedure? If one knows the shape of the contrast dilution curve C(t), the answer is "yes". In fact there exists an optimum filter which can be found by maximizing the iodine SNR associated with Eq. 1 subject to the condition that $\sum_i k_i = 0$.

The iodine signal is contained in I. The image noise, N, associated with I is given as:

$$N = \left[\sum_i k_i^2\right]^{1/2} \sigma , \tag{3}$$

where it is assumed that the noise associated with each image, I_i, is independent of i and has a standard deviation, σ. The SNR of the filtered image is given as the ratio of Eqs. 1 and 3:

$$SNR \equiv \frac{I}{N} = \frac{\sum_i k_i I_i}{\left[\sum_i k_i^2\right]^{1/2} \sigma} \tag{4}$$

If the contrast of each image is separated into two components, stationary background (B) and iodine contrast (C_i), $I_i = B + C_i$, then Eq. 4 can be rewritten:

$$SNR = \frac{\Sigma_i \, k_i \, . \, C_i}{\left[\Sigma_i \, k_i^2 \right]^{\frac{1}{2}} \sigma} \qquad (5)$$

Since $\Sigma_i k_i = 0$ the terms involving B cancel.

Maximizing Eq. 5 leads to a F.I.R. which reflects (matches) the shape of the contrast bolus (C_i). This F.I.R. has been derived by two investigators (11,12) and is given as:

$$k_i = \alpha \, (C_i - \overline{C}). \qquad (6)$$

Equation 6 gives a "matched filter" F.I.R., where α is an arbitrary constant and $\overline{C}$ is the mean of C_i . For this case the optimum SNR is given as:

$$SNR_{opt} = \frac{\sqrt{N} \, C_{rms}}{\sigma} \qquad (7)$$

Where N is the number of images acquired and C_{rms} is the standard deviation of the sampled contrast dilution curve about its mean value, $\overline{C}$.

This matched filter is determined retrospectively in a straightforward way. After a sequence of images is acquired and stored they are viewed one at a time or as subtraction pairs. A videodensitometric window within the field of view and overlying a vascular structure of interest is drawn with viewer interaction. The acquired image sequence (or a subset of the entire sequence) then is replayed, image-by-image. The integrated brightness within the window is measured and stored for each image, I_i . This time-dependent densitometric signal is used to form the matched filter F.I.R. The image sequence is again replayed, each image weighted by the appropriate term of the F.I.R., and summed to give the matched filter result, I. The entire procedure may take less than a minute to perform.

3. MATCHED FILTER PERFORMANCE

In order to quantitate matched filter performance, a reference must be established. Assume an imaging rate of 30 frames/sec (consistent with U.S. standard video), each frame is produced by a 1/30 sec x-ray exposure and results in a quantum noise level of σ per frame. The maximum iodine signal-to-noise ratio available from any single frame is C_{max}/σ. This single frame, corresponding to maximal vessel opacification, will serve as the reference image.

Defined in this way a signal-to-noise ratio improvement (SNRI) factor is defined as the ratio of the SNR after processing an image sequence to that of the reference image. For a matched filter the SNRI is given by

$$SNRI_{opt} = \frac{\sqrt{N}\ C_{rms}}{C_{max}} \tag{8}$$

which is the best SNRI available from any processing scheme given the same N images.

Consider for the moment the idealized case of a contrast dilution curve which is a box function; i.e. the iodine concentration is zero until some time at which the concentration jumps instantaneously to one unit, lasts say 5 seconds and then returns instantaneously to zero. Assume that 300 images are acquired over a 10-second period spanning the 5-second duration of the contrast bolus. The various quantities in Eq. 8 are N = 300, $C_{max} = 1$, $C_{rms} = 1/2$. The $SNRI_{opt}$ for the matched filter is easily computed to be

$$SNR_{opt} = \frac{\sqrt{300}}{2} = 8.7.$$

If a mask mode (MM) procedure were used instead, assuming an imaging rate of once per second, the SNRI would be:

$$SNRI_{MM} = \frac{\sqrt{60}}{2} = 3.9.$$

It is assumed that the x-ray exposure from each contiguous 30 frames is lumped together to give a once per second imaging rate and that the maximum opacification image is subtracted from a preopacification image. In actuality a box function contrast dilution curve is hard to come by. Rather, the contrast dilution curves encountered in clinical practice can be modeled by a family of gamma variate curves (13). The approximate shape of such a curve is shown in Fig. 2 and is represented by the parametric equation:

$$C(t) = A\,(t-T)^{\alpha}\,e^{-(t-T)/\beta} \tag{9}$$

The two parameters α and β determine the curve's shape, T determines its phase (appearance time) and A its magnitude. The quantity $\alpha\beta$ defines the time between contrast appearance and peak opacification. For typical intravenous exams $\alpha\beta \sim 5$ seconds and is about half as great for intraarterial injections.

The SNRI for a matched filter is both a function of the curve shape and the imaging time (number of acquired images). Furthermore, for any given imaging duration, matched filter performance will vary depending on when relative to the contrast appearance time, T, image acquisition begins. Figure 3 shows the matched filter SNRI for a particular contrast dilution curve ($\alpha\beta = 4.9$ seconds) and various imaging durations (TD). It is assumed that imaging begins at time T + TL. As can be seen, for every imaging duration there exists an optimum time relative to bolus appearance when imaging should begin. For a 10-second duration, imaging should ideally begin 4 seconds prior to bolus arrival and result in a SNRI $\sim$ 7.

Compared to the idealized "box" function the real contrast dilution curve produced a SNRI $\sim$ 7 compared to the previously calculated 8.7. The mask mode result, being the difference between two images taken at maximum opacification and preopacification still would give the same value of 3.9. The mask mode result is virtually independent of curve shape so long as $\alpha\beta$ is large enough (>2-3 seconds). The variation of the matched filtering result with curve shape is shown in Fig. 4. A fixed imaging duration of 10 seconds was assumed and the start of imaging was picked to coincide with contrast appearance time so as to give moderate values of SNRI rather than their maximal results. For all values of $\alpha\beta$ considered (0.5 to 11.25 seconds) the matched filter SNRI exceeded 3.9.

Since a matched filter results in the highest possible SNRI, its use also requires the least x-ray exposure. Rather than comparing SNRI factors, one alternatively can calculate the amount of x-ray exposure required during an imaging procedure in order to deliver a specified SNRI, given a particular contrast dilution curve. If one calculates the exposure requirement of an alternative processing scheme relative to the matched

$$C(t) = A\,(t-T)^{\alpha}\, e^{-(t-T)/\beta}$$

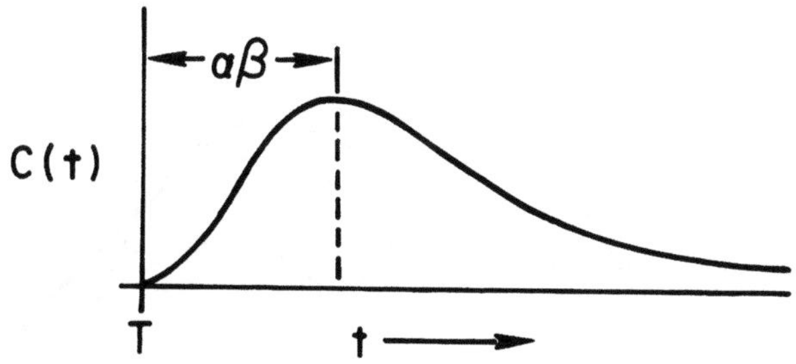

I.A. STUDY $\alpha\beta \sim 2$ sec

I.V. STUDY $\alpha\beta \sim 4-6$ sec

Fig. 2. Contrast dilution curve (C(t)) modeled as a gamma variate.

filter requirement, then a dose increase factor characteristic of that particular algorithm is defined. The dose requirement relative to a matched filter has been calculated for mask mode imaging (one image per second) (11). Imaging durations of from 5 to 40 seconds were considered and a contrast dilution curve for which $\alpha = 2.68$ and $\beta = 1.84$ was assumed. The results are plotted in Fig. 5. The values of α and β used to calculate the relative doses in Fig. 5 are characteristic of an intravenous carotid angiogram. The results suggest that if one employs a once per second mask mode imaging procedure, then one must use 3-4 times the x-ray dose of the SNRI optimum technique, i.e. matched filtering.

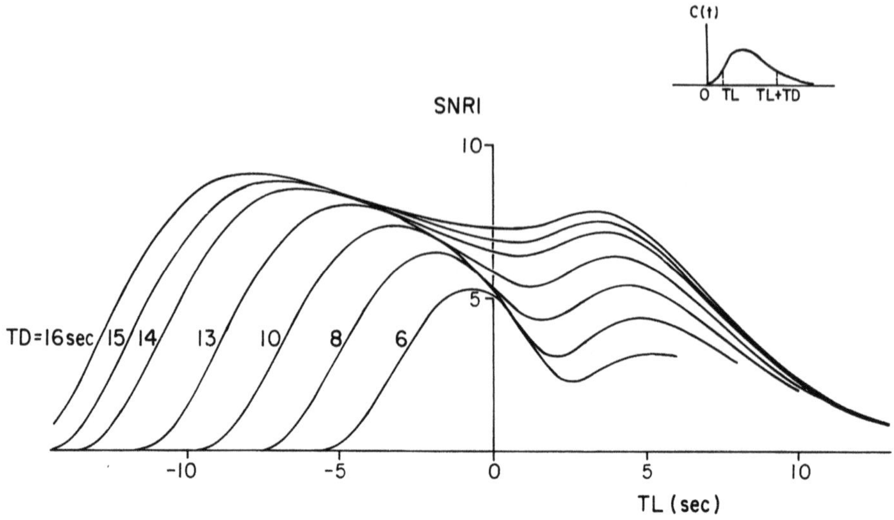

Fig. 3. Relationship between SNRI, TD and TL for a matched filter $\alpha = 2.68$, $\beta = 1.84$.

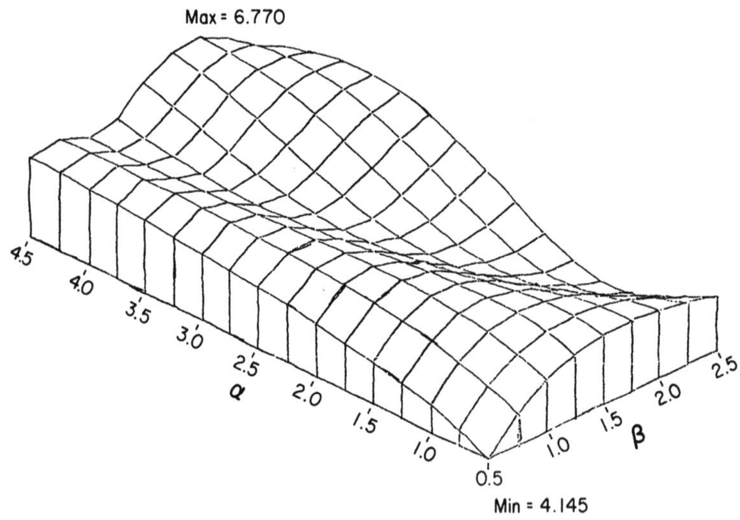

Fig. 4. SNRI of a matched filter as a function of α and β (TD = 10 sec, TL = 0).

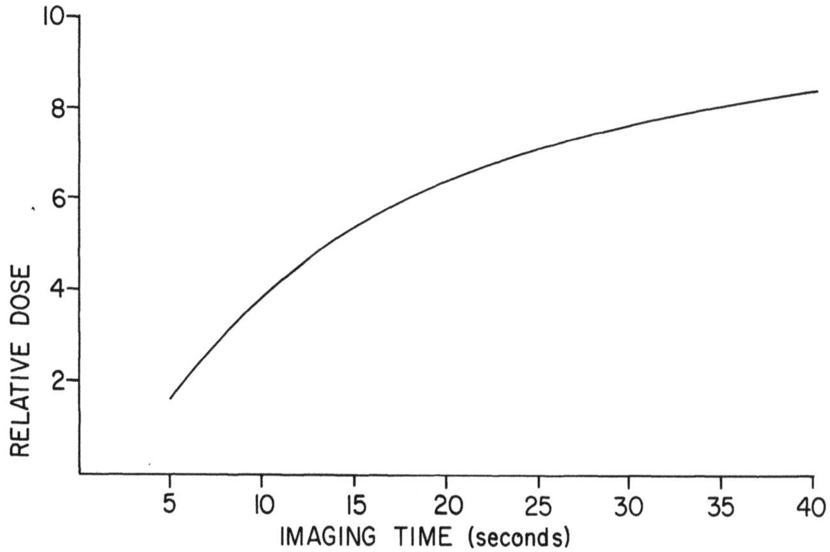

Fig. 5. Relative dose required for mask mode imaging vs. matched filtering as a function of imaging time. Intravenous angiography of the carotid bifurcation was considered. (Reprinted with permission from IEEE Trans. Med. Imaging).

SUMMARY

Digital subtraction angiography (DSA) algorithms share the common goal of attempting to isolate a weak, time-dependent signal from large image variations which results from bone and soft tissue variations in the body. Mask mode subtraction has been used as the technique of choice in most exams to date. All temporal subtraction schemes, including mask mode, are part of a larger class of subtractive temporal filtering algorithms. Within this class of algorithms an optimum technique, in terms of maximizing iodine SNR given an initial set of images spanning a measurable contrast bolus, exists. This algorithm is a filter "matched" to the shape of the contrast dilution curve. Thought of in another way, a matched filter algorithm can deliver a specified SNR for a minimum of x-ray dose.

Were the SNR or x-ray dose the only criteria for choosing a DSA algorithm, matched filtering would be preferred. However, patient and arterial motion introduce artifacts into any imaging technique which may compromise SNR advantages. The decision to use mask mode imaging or any other technique should be based on the possible "advantages" of the technique which may offset its dose inefficiency. These "advantages" depend on the relative ability to avoid vessel edge blurring and the ability to remove patient motion artifacts compared to matched filtering. For mask mode such "advantages" are as yet undocumented, but may offset the necessity of increased dose if they are significant.

For any other technique such as energy subtraction or time-energy (hybrid) subtraction SNRI factors (or relative dose requirements) can and should be calculated so that one knows the severity of the SNR penalty being paid. After documentation of the "advantages" of such techniques a decision can then be made as to whether the "advantages" outweigh the cost (in terms of lost SNR, or increased patient dose) for implementing a particular technique.

REFERENCES

1. Kruger RA, Mistretta CA, Houk TL, et al (1979). Computerized tomography in real time for non-invasive visualization of the cardiovascular system. Radiology 130:49-57.

2. Brennecke R, Brown TK, Bursch J et al (1977). Computerized videoimage processing with application to cardioangiographic roentgen image series. (In) Digital Image Processing, (Nagel HH, ed), Springer, New York, p 244.

3. Strother CM, Sackett JF, Crummy AB, et al (1980). Clinical applications of computerized fluoroscopy. The extracranial carotid artery. Radiology 136:781-783.

4. Ovitt TW, Christensen PC, Fisher HD, et al (1980). Intravenous angiography using digital video subtraction: X-ray imaging system. AJNR 1:387-390.

5. Meaney TF, Weinstein MA, Buonocore E, et al (1980). Digital subtraction angiography of the human cardiovascular system. SPIE 233: 272-278.

6. Kruger RA, Mistretta CA, Houk TL, et al (1979). Computerized fluoroscopy techniques for intravenous study of cardiac chamber dynamics. Invest Radiol 14:279-287.

7. Kruger RA, Mistretta CA, Crummy AB, et al (1977). Digital K-edge subtraction radiography. Radiology 125:243-245.

8. Riederer SJ, Kruger RA and Mistretta CA (1980). Three-beam K-edge imaging of iodine using differences between fluoroscopic video images: Experimental results. Med Phys 8:480-487.

9. Brody WR (1981). Hybrid subtraction for improved arteriography. Radiology 141:828-831.

10. Crummy AB, Strother CM, Lieberman RP, et al (1981). Digital video subtraction angiography for evaluation of peripheral vascular disease. Radiology 141:33-37.

11. Kruger RA and Liu P-Y (1982). Digital angiography using a matched filter. IEEE Trans Med Im MI(I):16.

12. Riederer SJ, Hall AL, Keyes GS and Pelc NJ (1982). Contrast sensitivity of digital fluorographic systems. Proc IEEE International Workshop on Physics and Engineering in Medical Imaging.

13. Thomson HK, Stormer RE, Whalen RE and McIntosh HD (1964). Indicator transit time considered as a gamma variate. Circ Res 14:502-515.

FUNCTIONAL ANALYSIS OF ANGIOGRAMS BY DIGITAL IMAGE PROCESSING TECHNIQUES

Rüdiger Brennecke and Joachim H. Bürsch

Departments of Pediatric Cardiology
and Biomedical Engineering
University of Kiel, FRG

Digital image processing techniques are increasingly applied in clinical angiography. Their main application is to increase the conspicuity of morphological details provided by subtraction imaging. However, the availability of angiographic image sequences in digitized form offers numerous possibilities to add the computerized evaluation of functional parameters to the spectrum of angiographic techniques. Earlier developments of functional analysis such as angiodensitometry form the basis for new approaches in digital imaging. We will discuss the progress obtained in the functional analysis of angiograms in the context of these earlier developments.

1. IMAGING OF STRUCTURE AND FUNCTION

Radiological diagnosis is primarily based on the visual assessment of images by the experienced observer. Anatomical details are usually providing clues for the diagnosis. In angiographic imaging, the application of x-ray opaque contrast material is primarily motivated by the need to improve the visualization of details of vascular and cardiac structures otherwise superimposed by shadows of tissue and bone. Important information such as the extent and position of stenotic obstructions of vessels or septal lesions of the heart can then be extracted by viewing a selected image from the picture sequence obtained during angiocardiography. Convincing documentation is possible by saving this image. Often this assessment of morphology - possibly combined with ECG, blood pressure and oxygenation data - is sufficient for the diagnosis and as a basis for the decision, which kind of correction or therapy is to be applied.

There remain, however, a large number of diagnostic questions not adequately answered by the observation of morphological peculiarities:

A. The observation of motion anomalies, e.g. in cardiac contraction, provides important clues for the diagnosis of cardiac diseases.

When the assessment of motion forms the basis of a diagnostic decision, the evaluation and documentation of deviations from normal function is much more difficult than in the case of morphological lesions, since the functional information is spread over a time sequence of images.

B. There are numerous situations where the extent of a morphological obstruction is difficult to assess quantitatively. Here, it is generally more efficient to ask for a quantitation of the effect of the obstruction on the hemodynamic function than to try to improve the estimation of the primary deficiency. An example is to perform flow measurements in the affected vessel when it is not clear from viewing the narrowing if the (three-dimensional) occlusion is severe. Peripheral vascular disease in an organ such as the lungs and kidney is another illustrative example. Here, the blood flow in the supplying artery may be decreased due to the increased peripheral resistance.

C. Even if a prominent morphological deficiency is visualized, a compensatory mechanism may exist so that the overall function of the organ under study is not severely impaired, e.g. collateral blood flow may compensate for the primary reduction in myocardial perfusion occurring after development of a coronary artery occlusion. Therefore, the assessment of regional myocardial blood flow adds essential information to the morphological findings. Even more generally, one should know to what extent the pump function of the region of musculature supplied by an obviously obstructed coronary is impaired.

This is a partial categorization of diagnostic situations requiring functional analysis. The translation of these problems into procedures of image analysis can sometimes be difficult. We have to define relevant parameters of the function of an organ or a vessel and to extract them from the angiographic image sequence. The definition of parameters of function is difficult when detailed pathophysiological models are not available. An example of this situation is the controversial question of the quantitation of regional ventricular wall motion.

The preceding discussion on the functional analysis of angiograms reminds of the fact that in this field the details of the questions to be asked are sometimes in an early state of development. In this situation we will try to define the goals more precisely and give an overview on tools for the assessment of functional parameters which are available from disparate fields such as angiodensitometry, filter theory and digital image processing.

2. MOTION ANALYSIS AND FUNCTION

In the preceding paragraph we have discussed indications for the assessment of functional parameters by x-ray imaging. Motion is a phenomenon connected in a straight-forward way with the function of the heart and with the primary function of the vessels, the transport of blood flow. Injected radiopaque contrast material is used as an indicator for the assessment of these types of motion by x-ray imaging. Therefore, angiographic motion analysis forms the basis for practically all types of functional evaluation from x-ray image sequences.

The motion and distribution of the mixture of dye and blood in the body presents a complex pattern. In x-ray imaging, usually only one projection of the time varying three-dimensional density distribution is available. This fact introduces considerable ambiguity into the interpretation of image data. Three-dimensional cross-sectional reconstruction from numerous projections (CT imaging) does not provide the temporal resolution and the volume coverage required for detailed motion analysis. Dynamic spatial reconstruction (1) should in principle offer an adequate temporal resolution, but numerous technical improvements are still needed. On the other hand, projection imaging at a high frame rate can generally be adequate for a majority of cases if the mode of contrast injection, the projection of the area of interest, and the timing of image acquisition are chosen carefully. Under these conditions the following idealized types of motion can be approximated (Fig. 1):

A. Propagation of a bolus of contrast material in a vessel that is not moving with respect to the x-ray system. The distribution of dye concentration along the bolus produces a contrast inhomogeneity moving with the blood flow. This is the typical situation approximated in many studies of arterial blood flow.

B. Motion of the heart walls while one or several cavities of the heart are filled by blood with a more or less homogeneous contrast solution. In this way the total outline of the cavity is visualized simultaneously and density measurements in different regions of a cavity become comparable.

C. Distribution of arterially or peripherally injected contrast material in tissue regions that are not moving with respect to the x-ray system. An example is a perfusion study of renal parenchyma. In the case of blood perfusion of the myocardium, ECG gating of image analysis can be used to avoid motion blurring by the movement of the ventricular walls. Even then, however, perfusion studies are frequently complicated by the superposition of functionally different regions of the organ in the projection image.

In the following report on image analysis techniques we will assume that one of these conditions prevails in the region of interest. Venous blood flow has not yet been studied in detail by digital techniques.

In conclusion, functional analysis primarily requires the definition of relevant parameters of object motion as discussed in the previous paragraph. By careful choice of the conditions fo image acquisitions, some of the relevant parameters of motion can be observable in a single angiographic projection plane (Fig. 1).

The following report will review techniques for the extraction of these parameters from the projection image sequence. This report will include a description of relevant analog video techniques for functional analysis. Computer algorithms, although frequently used for static two-dimensional image processing tasks, have to be refined for this type of studies. The contributions to a book edited by Huang (2) and especially the careful overview by Nagel (3) on numerous areas of image sequence analysis showed that this field of image processing does not yet provide general tools for application oriented approach. Thus it is still necessary to study solutions to specific problems, as will be done in the following paragraphs.

3. DEVELOPMENT OF VIDEOANGIOGRAPHIC IMAGE ANALYSIS

Quantitative analysis of cine and videoangiograms had been developed by a small number of groups for more than twenty years when digital video techniques began to find application in angiography in the years since 1976. Fig. 2 gives a schematic overview on these developments. "Classical" angiographic and angiocardiographic procedures are still the main sources of information. Diagnosis is still predominantly based on the visual assessment of cine film recordings by the experienced observer. However, as an application of analog video techniques, two quantitative

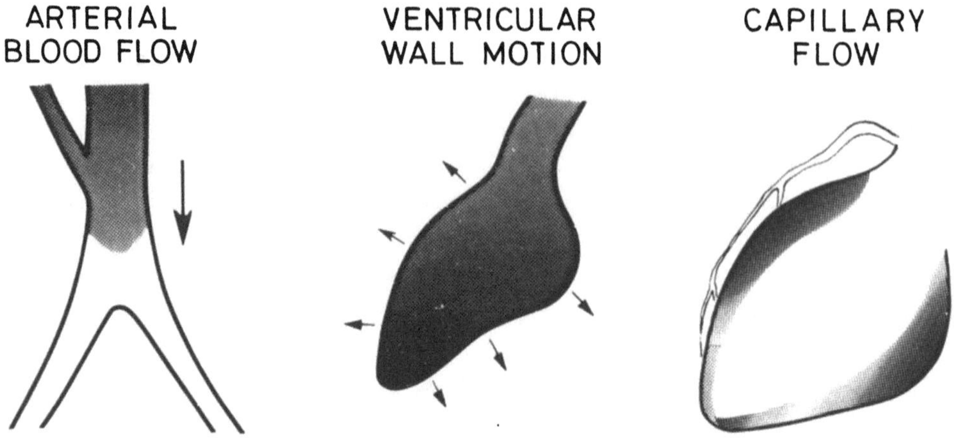

Fig. 1. Three types of physiologic motion that can be analyzed by digital angiographic techniques. These are blood flow in the arterial system (left), contraction of the cardiac ventricles (middle) and organ perfusion (right).

analysis methods found increasing applications: videometry and videodensitometry. Overviews on early developments in the field were given by Rockoff (4) and in a book edited by Heintzen (5).

In the most important application of videometry, the dynamic volume, shape or diameter changes of a ventricular chamber are determined by interactively outlining the opacified ventricular silhouette from stop action videoangiographic images. Volume determination is based on models of the three-dimensional shape of the ventricle. A number of semiautomatic and automatic methods for the detection of heart borders in angiocardiograms have been described. A book edited by Heintzen and Bursch (6), and reviews by Nagel (3), and Spiesberger and Tasto (7) have presented overviews. In typical semiautomatic systems the ventricular boundary has to be traced in the first frame from the sequence. This is then used as a contour estimate for the next frame. Even these semiautomatic systems generally require operator interaction to correct errors introduced by the system in the evaluation of subsequent frames. This special field of computerized contour detection and segmentation will not be discussed in detail in the following article, since most recent developments in functional analysis are primarily based on the quantitative evaluation of regional x-ray transmission instead of edge features and dimensional measurements. Video- and cine-densitometry (6,8,

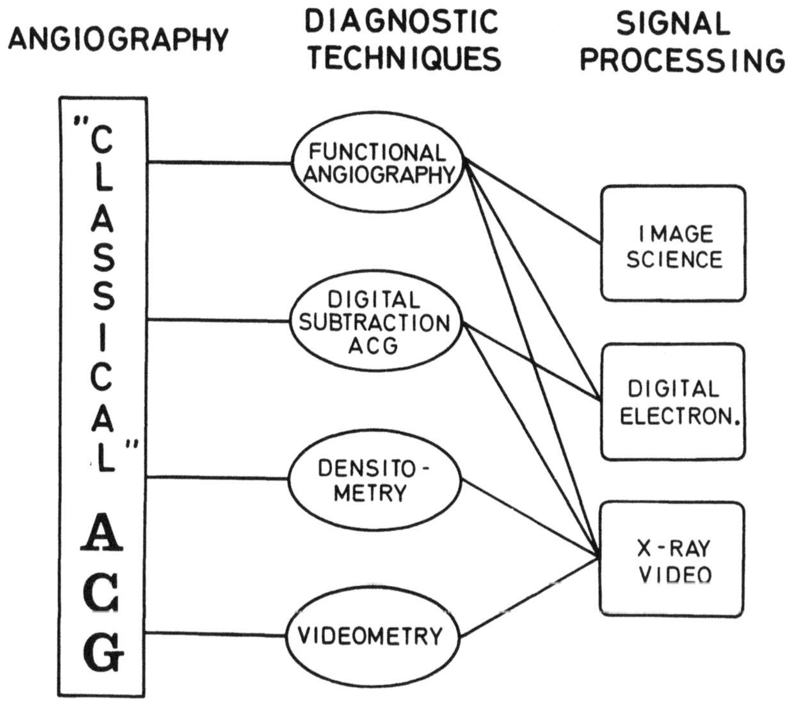

Fig. 2. Schematic overview on development of diagnostic techniques in angiography. "Classical" angiography (left) is still based on the visual assessment of films, but technical and scientific achievements (right) can be utilized to develop quantitative and automatic modes of image sequence analysis (middle) for functional diagnostic.

9,10), the second group of diagnostic techniques derived from angiography (Fig. 2), are primarily based on the evaluation of temporal changes of x-ray transmission occurring regionally during passage of a contrast material solution. The principal approach can be discussed using the simplified problem shown in Fig. 3. In a projection region of infinitesimal area da the x-ray transmission changes from i_1 (before opacification) to i_2 (after opacification), due to the passage of contrast material. From Lambert Beer's law we obtain approximately:

$$\ln i_2 - \ln i_1 \sim cd = z \tag{1}$$

where z is the product of concentration c and thickness d of radiopaque material transradiated in this area. Multiplying Eq. (1) by the area element da, one determines the infinitesimal mass element (concentration times volume) dm in the infinitesimal volume element 1×da. A density or mass signal m proportional to the total mass of contrast material in the

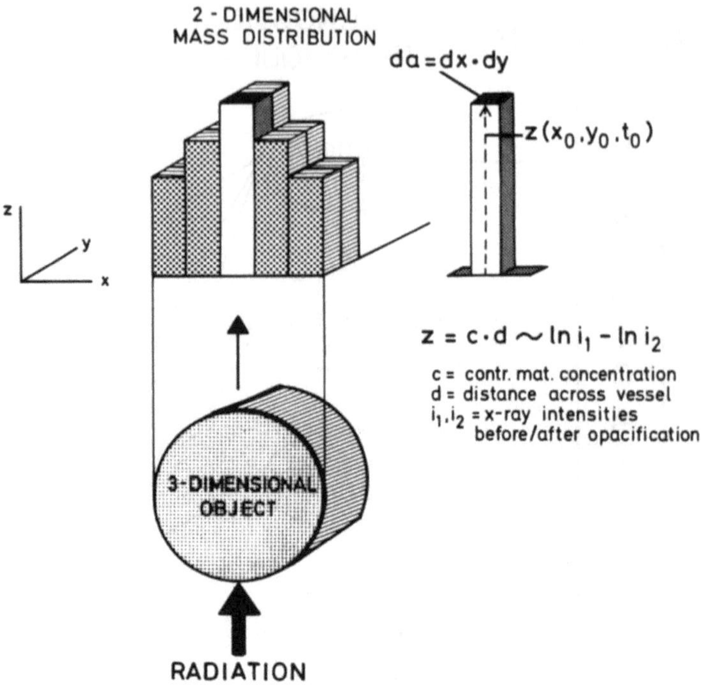

Fig. 3a. Principle of densitometric techniques. The spatial mass distribution in the projection image (upper part) is derived from the absorption distribution of the three-dimensional object as descibred by Lambert Beer's law. The background intensity (f_1) is subtracted.

region of interest with area A can be derived by integration of the contributions of all infinitesimal elements:

$$m = k_A \int dm = k_A \int \ln (i_2/i_1) da \qquad (2)$$

The constant k accounts for the specific mass absorption coefficient of the contrast material and the magnification. The influence of image degradations on this relation is discussed in another chapter of this book. Numerous approaches were developed using cine- and videodensitometry to extract from Eq. (2) data such as the dynamic volume change of ventricular chambers and the volume flow in vessels passed by the contrast material. It has always been considered an advantage of these densitometric volume measurements that they do not require geometric models of the three-dimensional shape of the object (11). These models have to be

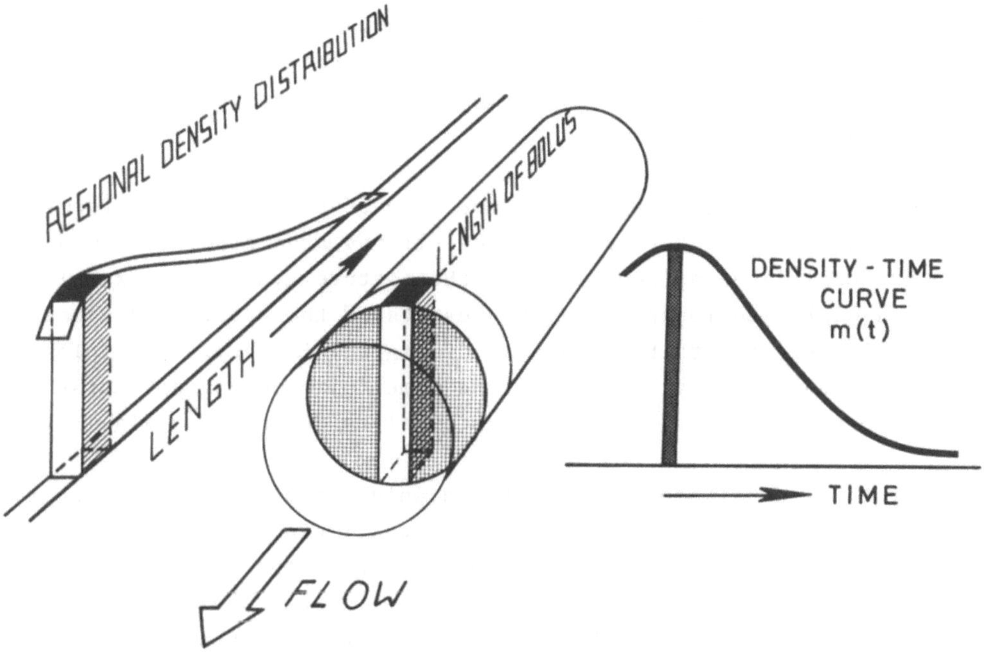

Fig. 3b. The motion of the spatial contrast material distribution (left) produces temporal changes in the brightness of regions traversed by the bolus projection (right). These changes depend on amplitude and shape of the spatial mass distribution and on the velocity of movement or flow. If the area of interest is confined to a single pixel, the corresponding mass-time curve is called "pixeldensogram".

applied in videometric volume analysis based on the borders of the silhouette of the object.

Measurements of time parameters related to the motion of a dye bolus are also accessible by densitometric techniques since the mass m is usually a function of time t (Fig. 3b). For example, the mean time of passage of the contrast material bolus (mean transit time) can be defined by:

$$\bar{t} = \int_T tm(t)dt / \int_T m(t)dt \qquad (3)$$

where T is an integration interval comprising the relevant part of the bolus signal. In this way time delays of bolus passage through two regions of interest can be estimated. A book edited by Heintzen and Bursch (6) and a recent article by Bursch (12) presented numerous examples of the development and application of these principles. Careful experimental work evaluated these videodensitometric results by comparison with more conventional methods such as electromagnetic flow probes and indicator tecniques.

Angiodensitometry provides the basis for many of the recent developments in digital functional analysis. In fact, if the area elements shown in Fig. 3a are considered as representing the picture elements (pixels) of sampled and digitized image data, the results of the previous discussion can be easily extended to the analysis of digital imagery. This close relation is underlined by the use of the name "pixeldensogram" for the mass-time curve $m(x_o,y_o,t)$ derived at a pixel as depicted in Fig. 3b.

Details of those densitometric operations which find application in digital imaging will be discussed later. Digital implementations of densitometric techniques are expected to reduce some of the problems which have over many years interfered with a wider clinical application of the analog implementations of these quantitative methods. Easier use, the digital restoration of image degradations and the presentation of quantitative results in pictorial form are among the expected advantages of digital densitometry.

As shown in Fig. 2, it was digital subtraction angiography that followed the development of videometry and videodensitometry, when digitization of the whole angiographic image sequence became technologically feasible (13,14,15).

The primary application of the new technique was to improve visibility of regions of the vasculature opacified by contrast material. However, soon several groups (16,17,18,19) developed methods for the extraction of functional information from digitized angiographic image sequences. From the original angiogram this approach extracted both: new types of synthetic images derived from motion parameters and condensed quantitative data such as regional volume flow. This new field of angiographic image analysis is depicted in the overview (Fig. 2) as the last development: digital functional angiography or functional imaging. In the refinement of the algorithms applied, high-level image processing techniques will play an increasing role (Fig. 2, right part). The discussion of this development will be covered in the following paragraphs.

Another field of medical picture analysis with some influlence on the development of functional analysis of angiograms by digital techniques is nuclear imaging. The computer was quite early used for the analysis of the manageable amount of data acquired in examinations by nuclear emission imaging and the basic concept of functional imaging was developed in this field (20). Overviews have been given by Alpert (21), Pretschner (22), and Adam and coworkers (23). In principle, digital angiography offers a higher spatial and temporal resolution and avoids most of the problems associated with long-term gated (ECG-synchronized) studies. This partially compensates for disadvantages of digital angiography such as the more invasive mode of administration of the indicator and the larger risk of artefacts due to motion of superimposed background structures.

Another competing technique will be nuclear magnetic resonance imaging with its potential to assess regional blood flow (24). Here, however, both the temporal resolution and the volume coverage may remain insufficient to make this method applicable to a larger part of the diagnostic fields depicted in Fig. 1. In the computerized derivation of functional data from 2D-echocardiograms, one of the main problems is the low signal-to-noise ratio of the primary data (25), and the mode of image data acquisition which is difficult to standardize.

4. IMAGE PROCESSING TECHNIQUES FOR MOTION EXTRACTION

As discussed in the previous paragraphs, practically all approaches to functional analysis by x-ray imaging depend on the assessment of motion and are based on angiographic images from a single projection plane. From the description of the development of diagnostic techniques in angiography given in the previous paragraph, we can deduce the three basic approaches to motion extraction that will be discussed below.

4.1 Digital Subtraction Angiography

This technique relies on the logarithmic conversion and subtraction of two projection images characterizing different phases of the motion (Fig. 4a). Regions in the subtraction image corresponding to increasing or decreasing densities in the original image pair will appear as dark or bright relative to non-moving parts of the image. A certain type of functional image can be created by displaying the changes occurring between an end-diastolic and an end-systolic image of the heart, thus simplifying in some cases the observation of akinetic regions of the wall (Fig. 5).

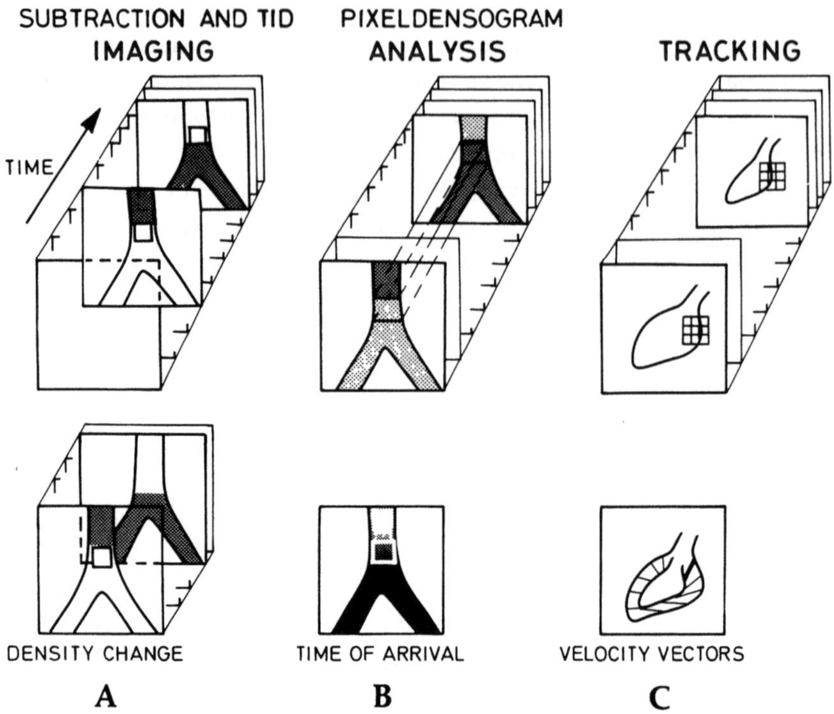

Fig. 4. Three main techniques for change or motion extraction from an angiographic image sequence. Digital subtraction imaging produces a new image series (A), while parametric imaging based on pixeldensogram analysis can be used to compress relevant information into a single picture (B). Tracking (C) is mainly applicable to motion analysis at a few selected sites in the projection plane.

For visual assessment of these density changes, usually a subtraction image sequence is produced from the original angiographic series as shown in Fig. 4a (lower part). For a pixel in the subtraction image with coordinates (x_o, y_o), the temporal variation of brightness m(t) is then given by:

$$m(x_o, y_o, t) = \ln i(x_o, y_o, t) - \ln i\ (x_o, y_o, t_r) \tag{4}$$

The values of m are obtained by logarithmic conversion and subtraction of the intensities i obtained in the unprocessed x-ray images which are taken at time t_r and t, respectively (Eq. 1). Typically, a fixed reference image is chosen (t_r = const) corresponding to the state before opacification by the contrast material. The process of logarithmic subtraction described by Eq. (4) derives a time-varying subtraction image sequence from the original angiogram which is then usually visualized by an experienced observer for diagnostic evaluation.

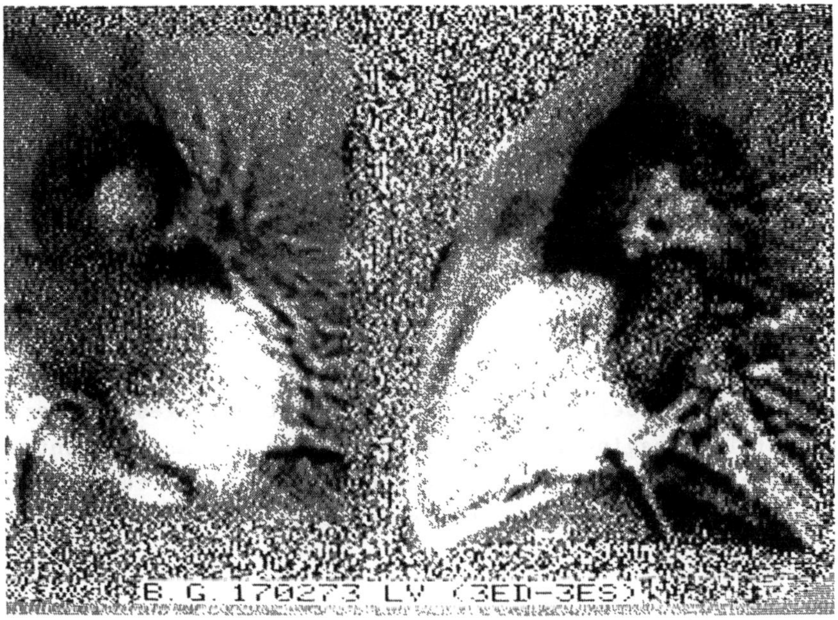

Fig. 5. Digital subtraction angiocardiography. Subtraction of an end-diastolic from an end-systolic image gives an indication of regional ventricular contraction.

A quantitative application of subtraction imaging was described in 1972 by Trenholm and coworkers (26). Their goal was the measurement of dynamic left ventricular cardiac volumes by digital densitometry. A venous injection was used to provide mixing of the contrast material before opacification of the left ventricle. A value proportional to the total mass of contrast material in the region of interest (slightly larger than the end-diastolic left ventricle) was calculated for each TV-frame and was based on the approach described by Fig. 3a and Eq. (2). In order to calibrate these measurements (const k in Eq. 2), the measured mass distribution along a diameter across the ventricle was fitted to an idealized projection of the corresponding cross-section (assumed to be circular). Similar attempts on analog (28) and digital (29) densitometric methods have been reported later. At present it seems difficult to reduce sufficiently all errors of these techniques that are due to inhomogeneous contrast material distribution, superposition of contrast material in the opacified lungs, and the effect of radiation scatter. Therefore, still the videometric approach based on dimensional measurements is nearly exclusively used for left ventricular volume determination although it requires exact boundary detection and it depends on models of the three-dimensional shape of the ventricle. Digital subtraction of the background can help in these dimensional measurements also (30).

A special subtraction mode (14) mainly used for qualitative imaging is time-interval differencing (TID). It is also described by Eq. (4). However, here t_r is variable with:

$$t_r = t + \Delta t \tag{5}$$

where Δt is a fixed delay between the subtracted images. Several applications of TID imaging for the visualization of cardiac contraction have been described (14,31). This technique has also been used to make visible the displacement of the bolus front in arterial blood flow analysis (32).

The TID-signal obtained at a given pixel location can be analyzed in a straight-forward way when assuming a translational motion of the bolus. If the bolus has the local gradient (dm/dx, dm/dy), the subtraction signal can be approximately derived from an expansion of the mass-time function m (x,y,t) into a Taylor series:

$$-\frac{\partial m}{\partial t} \equiv v_x \frac{\partial \ln i}{\partial x} + v_y \frac{\partial \ln i}{\partial y} + \ldots \tag{6}$$

Thus, the TID-signal is approximated by:

$$\Delta m = - v_x \frac{\partial \ln i}{\partial x} \Delta t - v_y \frac{\partial \ln i}{\partial y} \Delta t \tag{7}$$

This dependence of the TID-amplitude Δm at a given location (x_o, y_o) on both the spatial gradient of the object and the velocity of edge motion makes its interpretation difficult. However, in principle the components of the spatial gradient can be calculated by image processing techniques. In this way, from Eq. (7) regional values of velocity could be calculated (2,33).

4.2 Parametric Imaging

By analysis of the mass versus time curve or pixeldensogram (Figs. 3b, 4b), the value of a single feature is calculated for each pixel. Typical features are the area under the curve or the relative time at which the maximum of the curve occurs. After extraction of feature values for each pixel of the image matrix, these values can be displayed in a single image (Fig. 4b, lower part). This parametric image then represents information otherwise spread over an image sequence.

Using the same symbols as in Eq. (4), we describe the determination of a feature value $f(x_o,y_o)$ from a pixeldensogram $m(x_o,y_o,t)$ (Eq. 4) by:

$$f(x_o,y_o) = \mathcal{F}\ (m(x_o,y_o,t) \qquad (8)$$

Thus, the function $\mathcal{F}$ replaces the time series $m(t)$ by a single parameter value. Figure 6 shows schematically the evaluation of densograms from two pixels located over a vessel.

Applying Eq. (8) to each pixeldensogram constituting the image sequence produces a parametric image matrix $f(x,y)$. The selection of   is sometimes based on phenomenological features (e.g. "maximum of $m(t)$") or on physiological or angiodensitometric models, e.g. the area under the curve $m(t)$ correlates with transradiated thickness of the vessel passed by a bolus. Often, a set of pixeldensogram parameters is derived from each angiogram. An example from nuclear medicine is the extraction of a set of Fourier coefficients by ECG-gated analysis of each time-function $m(t)$ and their display in a series of images each representing the amplitude of one coefficient (34). The choice of functions will be discussed in detail in the next paragraph.

4.3 Tracking or Matching

A moving object in a two-dimensional subregion centered at (x_o,y_o) is searched for in a second frame (Fig. 4c). In a typical procedure, this subimage or two-dimensional mass distribution $m^+\ (x_o,y_o,t_r)$ is detected or interactively marked (target acquisition) in a first reference image. m^+ is then compared with the mass distribution in the neighborhood of (x_o,y_o) in a second image $m\ (x,y,t_1)$ by shifting m^+. A function measuring similarity is applied to find the shift resulting in an optimum match. In this way, we can define a two-dimensional function of the shift $(\Delta x,\ \Delta y)$:

$$R(\Delta x,\Delta y) = \mathcal{G}(m(x-\Delta x,y-\Delta y,t_1),\ m^+(x,y,t_r)) \qquad (9)$$

If $\mathcal{G}$ is the cross-correlation function, a maximum in $R\ (\Delta x,\ \Delta y)$ indicates

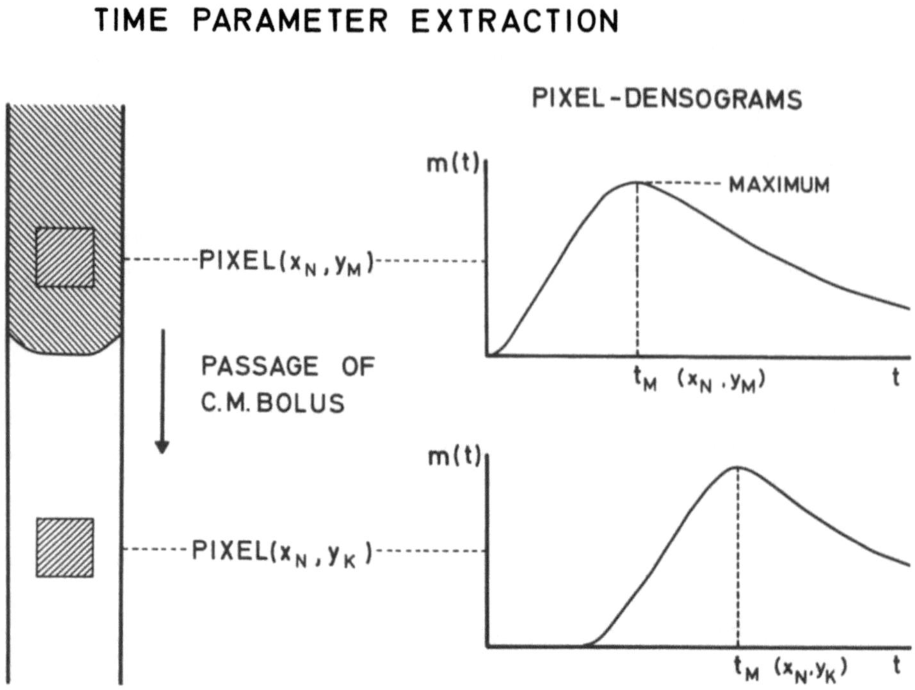

Fig. 6. Feature extraction from pixeldensograms. In this example the parameter "time of maximum opacification" is computed for each of typically 64000 pixels. This process is shown for only two pixels located over a vessel.

the optimum match. Under simplified conditions, the vector (Δx, Δy) then describes the motion of the features imaged in the subregion m^+.

The principle described by Eq. (9) has not yet been directly applied in angiographic subtraction imaging. A simplified approach, however, was used for tracking the horizontal component of the movement of the border between heart and lungs. Heart motion video tracking, radarkymography (35,36) or videokymography result primarily in time curves describing an aspect of the motion of heart borders, but a pictorial display similar to that shown in Fig. 4c could also be obtained. Some of the methods for automatic extraction of ventricular borders from angiocardiograms use matching as an essential part of the algorithm (37). There has also been a report on the transient appearance of landmarks along the silhouette of the opacified left ventricle (38). These could be used in tracking schemes that are not confined to horizontal motion.

4.4 Comparison of Motion Extraction Techniques

Digital subtraction angiography (4.1) and parametric analysis (4.2) are primarily imaging methods while video tracking or matching (4.3) would usually be applied to relatively few selected objects and thus typically provides results in the form of numerical data and time curves. Subtraction (and TID-) imaging is mostly used for improved visual assessment of image sequences. Their potential for quantitative imaging has been mentioned above. The basis for most of the recent progress in functional analysis by angiographic image processing techniques is parametric imaging. Correspondingly it will be discussed in detail in the following paragraphs. Tracking has been included into the comparison shown in Fig. 4 in order to remind of the fact that it is better adapted to certain problems in motion extraction than the principle of pixeldensogram analysis. The latter technique does not take into account relations between the pixeldensograms from a spatial neighborhood. This makes it more noise sensitive. Moreover, features such as "velocity" or "direction of motion" are not directly accessible to pixeldensogram analysis if not very simplifying assumptions are made. These complex features can, however, be extractable on the basis of a refined analysis of one or of a group of feature images, each representing a different parameter of the same angiographic scene. Examples of this approach will be discussed in the paragraph on the analysis of blood flow.

We end this comparison of basic motion extraction techniques by pointing out that subtraction imaging (Eq. 4) should not be considered as a special case of parametric imaging (Eq. 8) as is sometimes done in the literature. An essential distinction between those methods is directly visible from Fig. 4 and from these equations. Subtraction imaging produces a new time-varying sequence of images while parametric imaging can be considered as a projection of the original image sequence on a single feature image. In this process information on temporal context is concentrated. In most cases, this "projection", however, also results in an irreversible loss of information that does not occur in subtraction imaging.

5. PARAMETER EXTRACTION FROM ANGIOGRAMS

As defined above, parametric imaging relies on the separate analysis of the temporal intensity variation of each picture element or pixeldensogram. The resulting local values for a given feature (Eq. 8 and Fig. 4b) are then displayed as a new type of synthetic image. A critical comparison with other motion extraction techniques has been given above. Here, we will present an overview on the feature extractors applicable to mass-time curve analysis. These can be categorized on account of their principle:

A. Descriptive indices. An example is the calculation of the time of arrival of the contrast material bolus. The values of this feature do not have an immediate pathophysiological significance but with some experience, "normal values" might be acquired.

B. Physical and physiological quantities. Examples are cardiac output or the value of flow through a kidney artery relative to cardiac output.

Another more formal distinction can be made between parameters that characterize temporal aspects and those that characterize amplitude parameters of the pixeldensograms. It is a rather unexpected result of the extraction of these parameters that some of the techniques originally designed for functional analysis provide also enhanced morphological information. Finally, an important criterium for the comparison of parameter extracting functions (Eq. 8) is their sensitivity to random noise such as quantum fluctuations.

5.1 Image Acquisition

Reports on the digital acquisition of video image sequences for subtraction imaging have been published recently (39,40). In the field of quantitative functional analysis, several additional requirements have to be considered:

A. The analysis of blood flow in small arteries (e.g. coronaries) generally requires a higher spatial resolution than that provided by present digital video systems. Therefore, usually cine film is used as primary storage medium. A region of interest is then optically magnified and scanned by a TV system (41).

B. Numerous degradations of image quality result from effects such as beam hardening, radiation scatter and veiling glare of the image intensifier. These are discussed in another chapter of this book. Their influence in quantitative videodensitometric measurements is sometimes not as critical as might be expected, e.g. when relative measurements are performed (42).

C. The mode of contrast material injection has to be chosen carefully in order to reduce the influence of injection flow on the measured parameters. Lantz and coworkers (43) have given some rules which are quite generally applicable.

The following are minimum requirements on the x-ray TV system for parametric imaging that are sufficient for blood flow studies in the greater arteries and for estimation of capillary flow in organs:

A. Use of highly stabilized x-ray generator; the automatic exposure control must not change settings during image digitization.

B. Application of pulsed radiation and of a TV camera with low temporal lag to avoid motion blurring (42), high linearity of the camera tube.

C. Prehardening of radiation using a copper filter (d = 0.2 mm) between x-ray tube and patient.

D. Use of an image intensifier with low veiling glare. The x-ray beam should be collimated as much as possible.

E. Use of a 256x256 image matrix, 25 to 60 fields per second.

5.2 Preprocessing Techniques

Feature extraction by pixeldensogram analysis can be enhanced by preprocessing the image data. A first essential step is the reduction of random noise. Since motion feature extraction depends basically on the estimation of temporal intensity changes, fluctuations in image intensity due to random noise (quantum noise, electronic noise) represent an essential limitation of the performance of these techniques.

Random noise can be reduced by applying temporal median filtering or temporal recursive low-pass filters (44,45) to each pixeldensogram constituting the image sequence. The filter parameters have to be chosen carefully to avoid biasing of the process of feature extraction. Two-dimensional filtering is also applicable. The resulting change in spatial gradients will generally have an effect on the shape of the pixeldensograms as is clear from Eq. (7). Temporal filtering with motion compensation (46) can reduce noise while preserving image edge sharpness.

Another interfering effect of basic importance to parametric imaging is the temporal change of background absorption. Usually, a fixed background image (or the average over a short background image sequence) is subtracted logarithmically (Eq. 4). Note, however, that for some of the following operations (e.g. determination of the time of maximum opacification) this logarithmic subtraction may not be essential. Movement of the background due to normal physiologic action (respiratory movement of lung structures, ribs, the diaphragm and bowel movement) and involuntary movement of the patient will usually not be discriminated from bolus or heart movement. Respiration gated subtraction (44) can be applied to reduce some of the resulting motion artefacts. Temporal high-pass or band-pass filtering have a similar effect on slow background motion. In local measurements on small objects such as opacified coronaries superimposed to a slowly varying background, the reference level for the estimation of density changes is sometimes derived from measurements in small windows positioned close to the contrast structure (47). This compensates also for the gradual increase in background absorption due to tissue opacification. This can also be reduced by subtraction from each pixeldensogram of a smooth baseline curve interpolating between "early" and "late" values of the pixeldensogram (16). Temporal high-pass filtering can have a similar effect, however, presumable at a higher computational cost.

5.3 Time Parameter Extraction

The primary goal of these feature extraction techniques is to characterize the time of arrival of the injected contrast material for each pixel that is passed by the bolus. The essential information, however, is not the local time of arrival but the comparison of time intervals for bolus passage between regions of interest (Fig. 6). Therefore, an additive error in the determination of time of arrival will generally not be critical.

The definition of a characteristic feature of each pixeldensogram is somewhat ambiguous due to the extended shape of the pixeldensograms. Moreover, the shape of the spatial distribution of contrast material changes while a bolus moves through the circulation. Therefore, the comparison of arrival times obtained at different sites is difficult. Finally, even if the shape of the spatial distribution of contrast material were constant, the translational movement with varying velocities will modify the shape of the pixeldensograms recorded (Fig. 3 and Eq. 7). This is critical in pulsatile flow measurements. Especially difficult conditions prevail in organ perfusion when functionally different structures are superimposed.

The following parameters have been described in the literature (17,48-51):

A. Time of maximum amplitude

This parameter (t_{max}) is noise sensitive. A small random noise amplitude superimposed to the relatively flat region around the maximum (m_{max}) of a pixeldensogram will be transformed into a large scatter in the estimation of the "time of maximum amplitude.

B. Appearance time

The time when the signal exceed a small fraction of the maximum amplitude (e.g. $0.1 \cdot m_{max}$). This parameter is also noise sensitive.

C. Time of half-maximum amplitude

As shown in Fig. 7a, in this case the determination of the time parameters is performed in a region of the pixeldensogram with a steep slope so that the noise problem is reduced.

D. Mean transit time

This is an integral measure of time of arrival provided by calculation of a line of gravity in the pixeldensogram (Fig. 7b and Eq. 3). The influence of the "tail" of the pixeldensogram due to layering of the contrast material and tissue opacification should be excluded.

E. Mean time of bolus front

The mean time of passage of the increasing part of each pixeldensogram (17) is computed (Fig. 7c):

$$\bar{t}_F = t_{max} - \frac{1}{m_{max}} \int_o^{t_{max}} m\,dt$$

F. Time of maximum correlation

Each pixeldensogram is compared to a "template". This is usually ano-

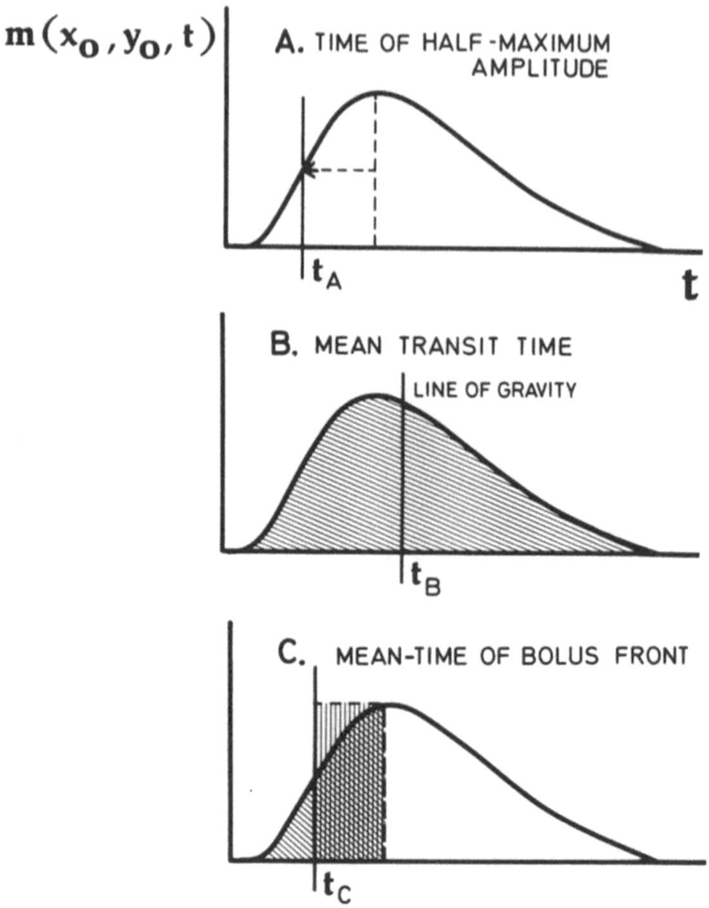

Fig. 7. Three examples for time parameter extraction from pixel-densograms. From each densogram representing typically hundred and more samples of a local mass-time curve a single parameter is extracted as indicated.

ther pixeldensogram typically derived by averaging over a neighborhood of several pixels which are passed by the bolus. The template curve is shifted in the time direction over the measured curve while calculating for each shift value the covariance or the correlation coefficient. The temporal shift resulting in a maximum value of correlation is stored as "time of maximum correlation" (50). This process is shown schematically in Fig. 8. A typical parametric image of type "maximum correlation" obtained from an abdominal arteriogram is shown in Fig. 9.

The performance of four of these parameters (C,D,E,F) in the measurement of time delays (arterial circulation) has been systematically evaluated by videodensitometry in animal experiments (12,17). The last two techniques (E,F) reduced errors due to the physical effects inherent in contrast dilution methods such as layering of the contrast material as well as effects of pulsatile flow.

We have also experimented with the application of time parameter imaging to studies of left ventricular motion (18). Figure 10 is an example of a display of the parameter "time of half-maximum amplitude" extracted from a selective angiocardiogram. The parameter image was masked by a binary image showing the end-diastolic projection area of the ventricle to suppress noise that is otherwise visible outside the ventricular silhouette.

5.4 Amplitude Parameter Extraction

In addition to the temporal features discussed above, amplitude parameters are extracted by similar techniques. The main goal has been the determination of parameters related to the volume or thickness of vessels. The quantitative basis for these measurements is discussed in the following paragraphs. Two typical amplitude parameters are (17,49):

A. Maximum value of opacification

Usually the logarithmically converted signal is analyzed. Since the maximum amplitude is stored for each pixel, random noise amplitudes are "rectified".

B. Temporal integral of pixeldensogram ("volume")

This is a measure of the relative thickness of the vessel under the corresponding pixel. When the temporal integration is extended over a region of the pixeldensogram not containing the bolus signal, the signal-to-noise ratio of the resulting parameter image is relatively low. Fig. 11 shows an example of a parameter image of type "volume".

A quite different task is the detection of the bolus signal in the presence of noise. In this context the concept of parameter extraction is also applicable. Here, however, the primary goal is not the estimation of a characteristic value of the signal, but a high probability of correct decision if a pixeldensogram contains contributions due to the passage of contrast material. This is essentially a binary (yes/no) decis-

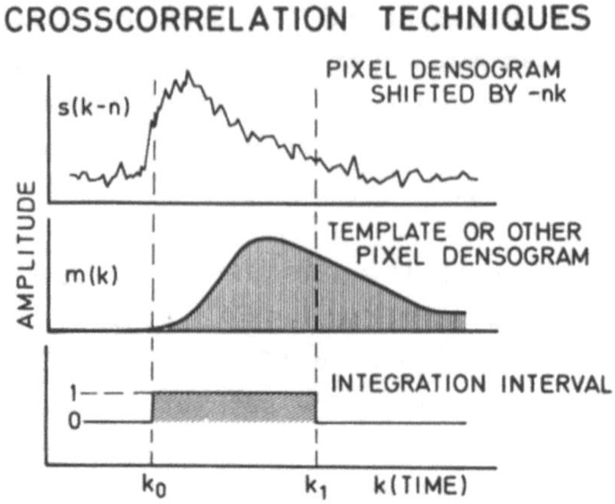

COVARIANCE FCT: $COV(n) = \sum_{k_0}^{k_1} (s-\bar{s})(m-\bar{m})$ — n : temporal shift

CORRELATION FCT: $COR(n) = \frac{COV(n)}{\left[\sum (s-\bar{s})^2 (m-m^2)\right]^{1/2}}$ — $-1 \leq COR \leq +1$

MAX. CORRELATION FCT: $MCOR = \max_n [COR(n)]$ — for all n

TIME OF MAX. OF CORRELATION FCT: $TCOR = n_0$ — n_0: temp.position of MCOR

Fig. 8. Cross-correlation techniques for parameter extraction. The signal (pixeldensogram) is compared to a template (ideal pixeldensogram) in an integration interval. This comparison is repeated with different values of the delay between densogram and template. Different implementations of this principle are defined in the lower part of the diagram. Results are shown in Figs. 9 and 42.

ion. This operation is required when regions containing contrast material have to be separated from the non-specific surroundings by automatic techniques (segmentation). A different application is noise reduction for visualization of the arborization of vessels using low amounts of contrast material. A parameter applicable is:

C. Maximum of correlation function

As described above, the shape of each pixeldensogram is compared to a template by cross-correlation (Fig. 8). The maximum value of the correlation function (function of the relative time shift of the curves) is stored as a parameter (Fig. 12). An extension of this technique uses the pixeldensograms from a spatial neighborhood as "templates" (49).

5.5 Applications of Parametric Imaging

Fig. 9. Time parameter image resulting from correlation analysis of abdominal arteriogram. The gray level of each pixel is proportional to that delay between the pixeldensogram and the template that produced the maximum of the covariance function (Fig. 8). Thus, early/late arrival times to bright/dark gray levels.

The described parameter extraction techniques derive for each pixel a feature such as the time of maximum opacification thus creating new types of images. The primary application that is discussed below is the visual assessment of time parameter images by the radiologist or cardiologist. Another point of view is to consider the feature data calculated by feature extraction as the basis for a more refined quantitative analysis. The typical goal of these procedures is the determination of regional volume flow as will be discussed in the following paragraph.

The evaluation of the performance of the described parametric imaging techniques in the diagnosis of diseases seems to lag behind their technical development. Hohne and coworkers (48) reported on experience with some 500 cases (selective angiograms from the kidneys and the liver). They found that it was difficult to extract from the parameter images conclusive information on a normal or abnormal state of the organ under

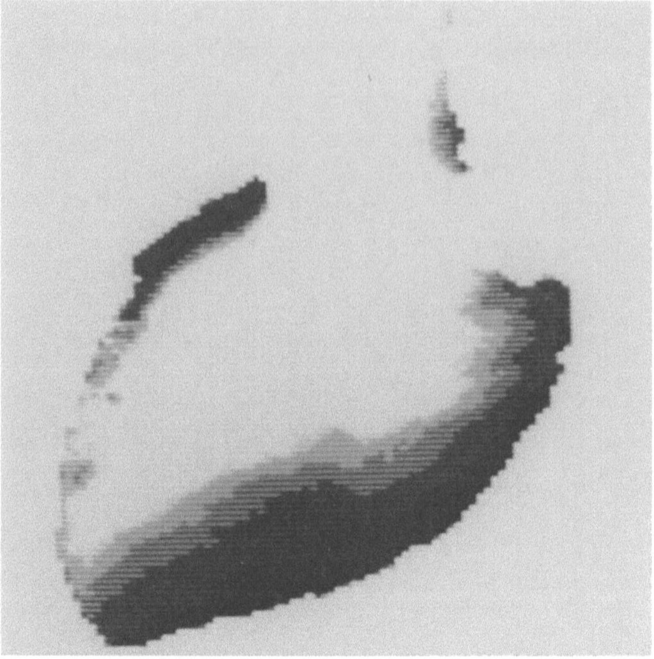

Fig. 10. Time parameter image of type "time of half-maximum opacification" derived from a selective angiocardiogram (18).

study. They attributed this partial failure to the lack of physiologic models which would predict certain types of contrast material kinetics. Another reason could be seen in the rather simple nature of feature analysis of time density curves applied to date. Potentially important features such as multiple peaking are only very indirectly imaged by the described techniques (51). Hohne and coworkers, however, stressed that time parameter imaging can be a good basis for the determination of relative effects. Examples are the assessment of times of arrival in a given region before and after medication or surgery.

Gallagher and coworkers (52) studied the selectivity of some of the time parameters in renal transplant evaluation. Intravenous injections of contrast material were used. In this investigation, parametric images of type "time of maximum amplitude" were found to distinguish normal from abnormal renal function.

A more basic result of parametric imaging is the concentration of some of the information on the regional temporal behavior into one or a few images thus providing an overview on the morphology of vascular struc-

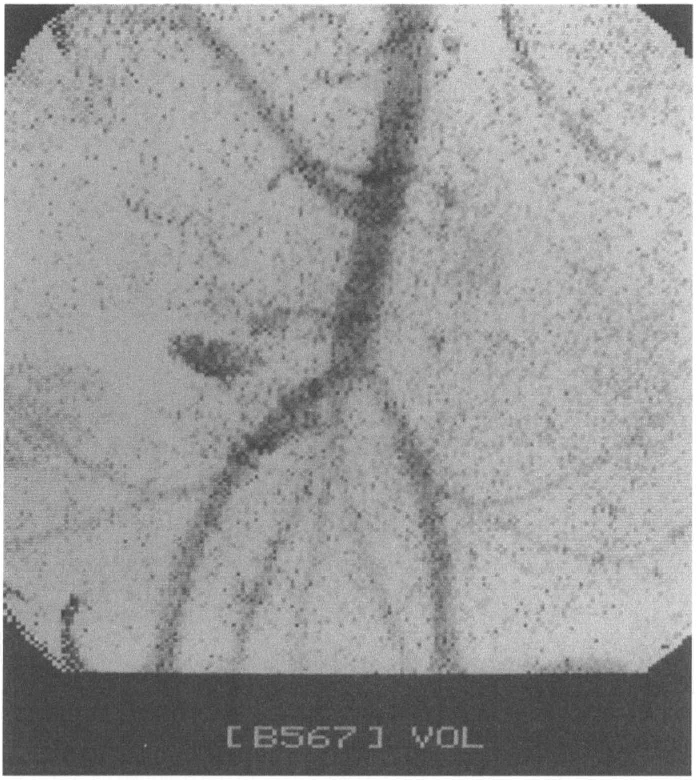

Fig. 11. Amplitude parameter image of type "volume" derived from abdominal arteriogram.

tures passed by contrast material. Vessels that are superimposed to each other can often be discriminated on account of their coding into different colors or gray levels which correspond to different times of arrival. Thus, functional imaging can be considered in some applications as another computerized image enhancement technique complementing the potential of digital subtraction angiography to selectively display structures opacified by contrast material.

6. QUANTITATIVE VOLUME FLOW MEASUREMENTS

A main goal in applying digital image processing techniques to the analysis of angiograms is the extraction of diagnostically relevant quantitative parameters. Parametric imaging, as described above, derives for each pixel location quantitative values related to motion of the heart and the blood. However, parameters such as local time of arrival of contrast material are often difficult to apply as diagnostic criteria in an interindividual comparison. Therefore, much interest exists in the der-

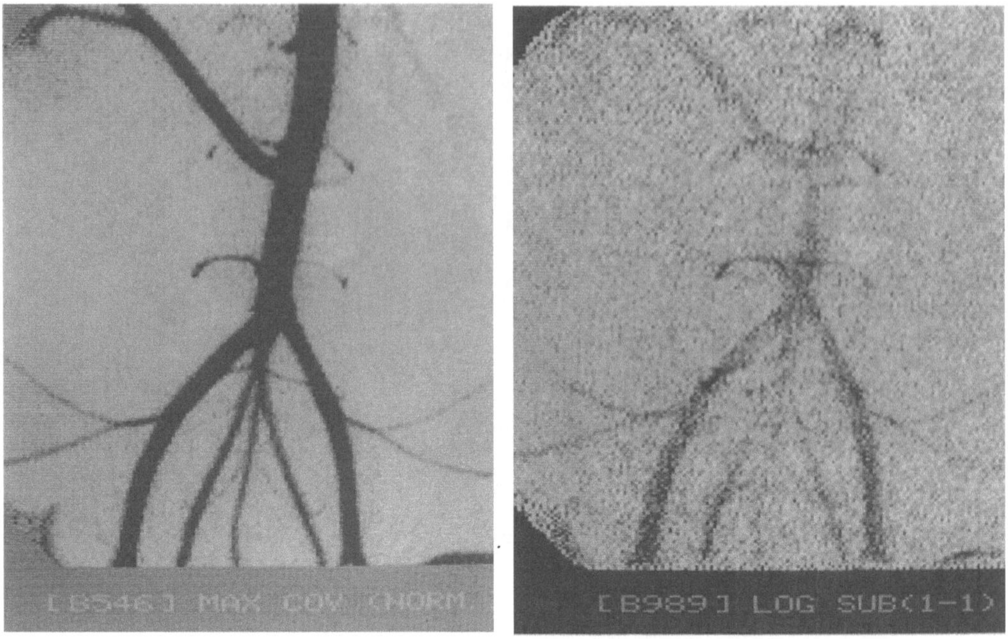

Fig. 12. Left: Amplitude parameter image resulting from correlation analysis of abdominal arteriogram. The maximum of the covariance function was obtained as indicated in Fig. 8. Right: For comparison, a digital subtraction image from the same arteriogram is shown.

ivation of parameters having a more general diagnostic significance. Regional volume flow, especially when measured relative to the flow in a main artery, is a functional parameter with a direct physiologic interpretation. Techniques for the interactive determination of local flow that are known from (analog) videodensitometry (see section 3) can be improved by their implementation using digital hardware. A further extension is the derivation by digital image processing techniques of relative flow values for all regions traversed by the bolus. This analysis is based on a group of parameter images derived, as described above.

The most direct approach to the measurement of flow is derived from the definition of flow as a function of front velocity v_F of the contrast material bolus and the radius r of the vessel:

$$Q = \pi T^2 v_F \tag{10}$$

Both the velocity and the radius have to be estimated by angiographic techniques. This calculation assumes a circular cross-section of the vessel and its orientation parallel to the projection plane.

For front velocity measurements, several methods have been applied. In conventional densitometry, two windows are positioned over the vessel at a distance 1. The regional arrival times t_1, t_2 are determined (section 5.3) and the bolus velocity is computed from:

$$v_F = 1/(t_1 - t_2) \tag{11}$$

The arrival time is often simply determined as the time of half-maximum opacification. This estimation contains an error due to the finite time interval Δt between imaging the sampling of TV fields (Fig. 13a). A typical value is $\Delta t = 20$ ms. A large relative error in v_F is caused by this timing error especially when measuring fast flow in short segments of vessels.

The technique shown in Fig. 13b avoids this timing error by measuring the distance traversed by the bolus during a fixed multiple of the field sampling period Δt (Fig. 13b) by image processing techniques. This estimation can in principle be performed by subtraction of subsequent TV fields (TID-imaging, Eqs. 4,5,) and interactive delineation of the positions of the bolus front. The velocity v_F is then again calculated as shown in Eq.(11) (32,53).

The absolute error of this <u>dimensional</u> measurement is determined by the spacing of pixels Δs. This is usually of the order of 0.7 mm (image matrix 256x256 pixels). The timing error of the conventional method (<u>time</u> measurement) can be compared by calculating the equivalent dimensional error Δs^+ produced by the timing error:

$$\Delta s^+ = v_F \, \Delta t \tag{12}$$

Equating Δs^+ and Δt and inserting typical values ($\Delta t = 20$ ms, $\Delta s =$ 0.7 mm) we find that the time interval measurement is less accurate than the displacement measurement for velocities of the order of 35 mm/s and more. However, even at lower velocities the displacement measurement (Fig. 13b) offers an important advantage since one can use a reduced rate of field sampling ($1/\Delta t$) without decreasing the accuracy of the velocity measurement. Spiller and coworkers (32) used this technique for the measurement of displacement versus time diagrams describing phasic flow in coronary arteries. For consistent flow measurements in these arteries, the additional problems of the superposition of bolus and heart motion is still not solved.

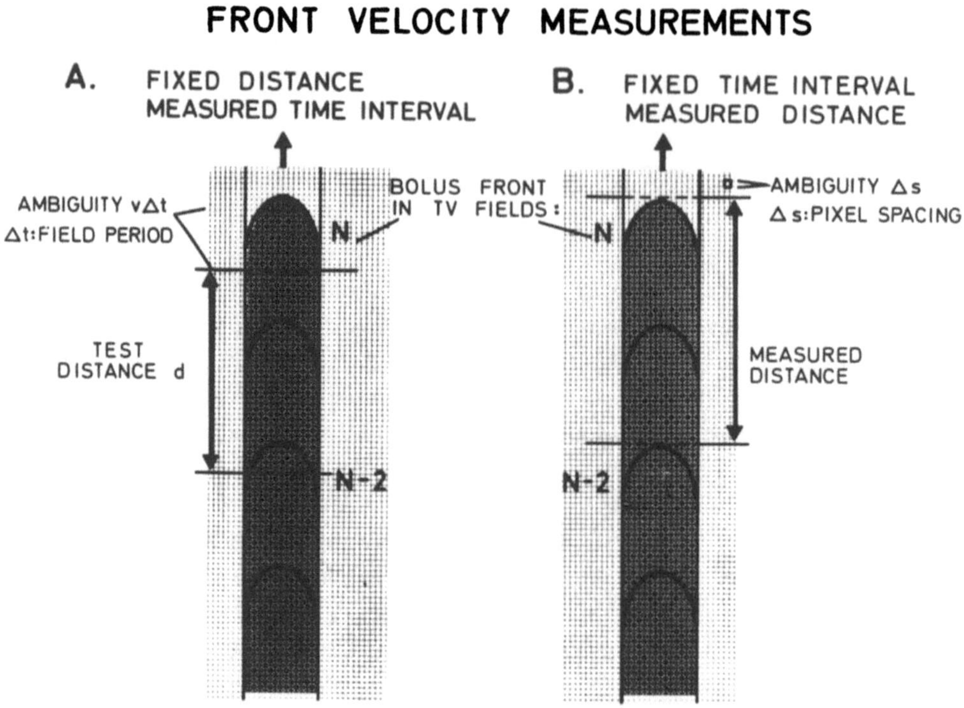

Fig. 13. Comparison of front velocity determinations based on time measurements (left) and distance measurements (right). The position of the bolus is shown in several TV fields. At high bolus speed, velocity measurements obtained by estimation of the distance traversed in a given number of TV field periods is more accurate, if accurate edge detection of bolus front is assumed.

The determination of bolus front displacement is complicated by the problem of an unambiguous detection of the edge of the contrast material front in the TID images. Motion tracking (Fig. 4c) may be more appropriate for this analysis of the spatial mass distribution than the interactive evaluation of TID images. Another approach to this problem which is based on the analysis of parametric images is described below.

Besides velocity measuremens, the estimation of flow values (Eq. 10) requires the determination of vessel radius r. Image processing techniques such as the correlation of the regional mass distribution corresponding to the vessel with templates characterizing the assumed cross-sectional projection (13,54,55) can be applied for accurate measurements even with noisy data. Another statistical approach especially suited for images sampled at a relatively low spatial resolution has been proposed by Shmueli and coworkers (56).

While the approach to flow measurements described above uses distance measurements only (Fig. 14a), a method for imaging of relative flow distributions developed by Bursch and coworkers (57) is based exclusively on densitometric principles (Eqs. 1,2). In this method, (relative) flow is defined as the (relative) volume V_N of blood contained in a region of the vessel that is traversed by the bolus in a given time interval. Both the determination of this region ("displacement segment") and of the corresponding vessel volume is performed by densitometric techniques.

This calculation begins with the determination of the projection area A_N of the vessel segment traversed in a given time interval (Fig. 14b). This displacement segment A_N comprises the areas of all contiguous pixels (coordinate x,y) with pixeldensograms characterized by a time of arrival t_N falling into the interval defined by:

$$N\Delta t \leq t_N (x,y) < (N+1)\,\Delta t \qquad N = 1,2,... \tag{13}$$

where Δt is again the period of video field sampling (or multiple). This arrival time t_N is derived from the pixeldensograms m (x,y,t) by one of the time parameter extraction techniques described in section 5.3. Usually, the further analysis is in addition confined to those pixel locations that correspond to high values of one of the "amplitude" parameters (e.g. "maximum of correlation", section 5.4).

After having defined the displacement segment A_N as described above, this definition of (relative) regional volume flow requires the calculation of the (relative) densitometric volume V_N of the part of the vessel corresponding to the displacement segment A_N (projected area, see Fig. 14b). This volume V_N is computed by temporal and spatial integration of pixeldensograms m (x,y,t):

$$V_N \sim \int_{A_N} \left(\int_T m\,dt \right) da \tag{14}$$

This estimation of V_N follows from:

$$\int_{A_N} \left(\int_T m\,dt \right) da = \int_{A_N} \left(\int_T d(x,y)\; c(x,y,t)\,dt \right) da \tag{15}$$

(see Eqs. 1,2) and by applying the Stewart-Hamilton principle in the form:

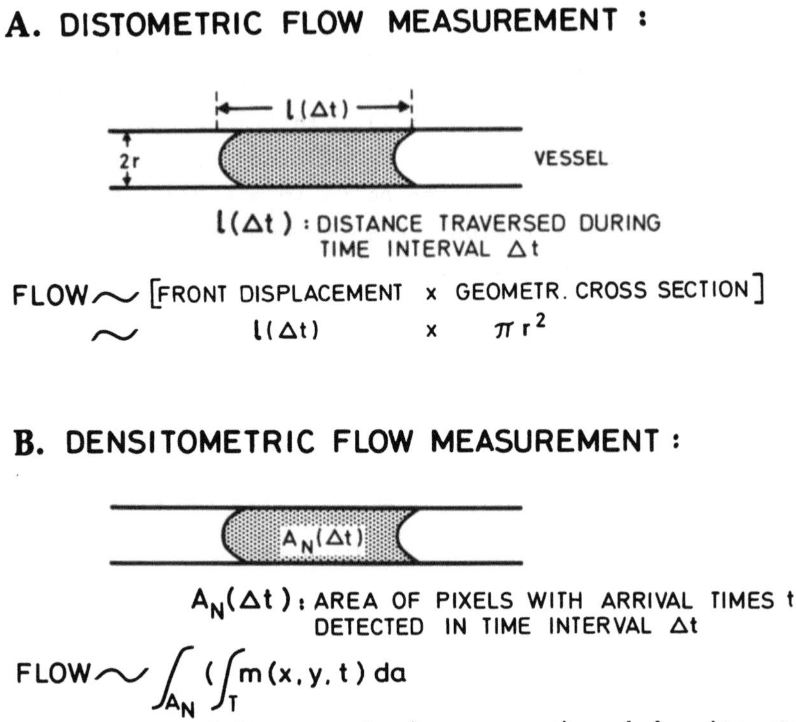

Fig. 14. Comparison of distometric (upper part) and densitometric (lower part) flow measurements. Distometric measurements rely exclusively on edge detection and distance measurements, while the densitometric technique is based on the analysis of pixeldensograms obtained in displacement segments.

$$\int_{0}^{\infty} c(x,y,t)dt = \text{const} \qquad (16)$$

to all parts of the circulation traversed by the contrast material injected centrally.

Image processing techniques were used to implement this principle. As shown in Fig. 15, the determination of flow distribution was based on three parametric images generated from the original angiogram. The image showing the parameter "similarity to signal" was computed by correlation techniques (Figs. 8,12). This image is used for the segmentation of the image plane into those areas containing vessels and those with non-specific signals (spatial segmentation). Usually, this process must be supported by manual corrections. The second image shows for each pixel one of the time parameters (section 5.3) and was typically also determined by correlation techniques (Fig. 8). Based on this information, the vessel segments detected in the first step (spatial segmentation) were partitioned into the (temporal) displacement segments defined by Eq. (13). Finally, the relative volume of the transradiated vessel segment is derived by local summation of amplitudes in areas of the "volume image" (parameter "volume", section 5.4, Fig. 11) corresponding to the displacement segments (Eq. 14). The results can be shown as a flow distribution image (Figs. 15,16) or as numerical data in the form of a table (Fig.15).

When compared to flow estimations based on distance measurements (Eq.10), this method has the following advantages:

A. The values of relative flow determined by the densitometric technique do not depend critically on the orientation of the vessels relative to the projection plane. Even though the displacement segment varies with the angle between vessel and projection plane, the volume as determined by densitometric technique keeps essentially constant. In addition, one has not to assume a circular cross-section of the vessel.

B. The determination of the displacement segments defined above is a process supported by conventional two-dimensional image segmentation techniques (58,59) and by experience from the analysis of image sequences (3,60).

Both methods of flow measurement described have not yet been applied extensively. Their main problem is the assessment of pulsatile flow. Measurements in different displacement segments along the vessel will generally correspond to different phases of this periodic flow variation (61). Only the volumes measured in segments corresponding to the same phase can be directly compared. Several attempts can be made to solve this problem:

A. The time period chosen for the calculation of segments is made equal to one heart cycle. This results in a coarse spatial resolution of flow measurements.

B. The flow measurements are averaged over several injections with different delays with respect to the R-wave of the ECG (62).

Work is in progress to obtain a correction for the locally varying flow velocity by image processing techniques using information contained in the original angiographic image data.

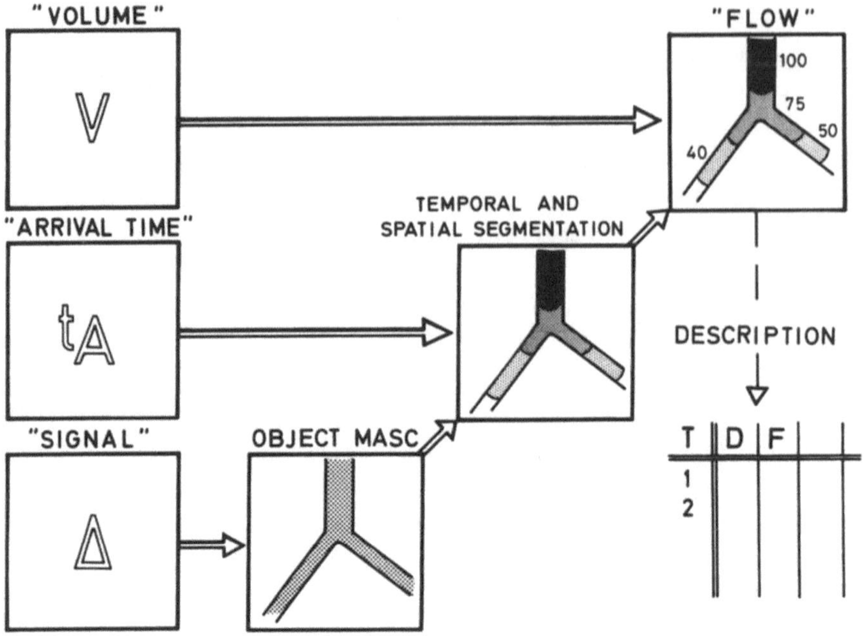

Fig. 15. Schematic diagram showing the derivation of a functional "flow" image from three parametric images. The data shown in this image can alternatively be listed (T = mean time of arrival, D = distance from injection site, F = relative flow). More complex analysis can be based on these reduced data which provide a simple image sequence description.

7. CONCLUSION

Functional studies of the heart and circulation can be based on digital image processing techniques for the extraction of motion parameters. The development of these tecehniques can profit from earlier experience with videometry, videodensitometry, target tracking and from other field of digital image sequence processing. Different approaches to the central problem of motion estimation are required, depending on the area of application. Ventricular contraction studies are still the domain of interactive evaluation of edge information, but digital subtraction imaging is increasingly applied for the preprocessing of intravenous angiograms. Details of organ perfusion can be visualized by computerized extraction of local arrival time parameters and their display as parametric images. The quantitative assessment of regional blood flow in the arterial system requires more complex techniques since the spatial variation of temporal features has to be evaluated. Dimensional or angio-

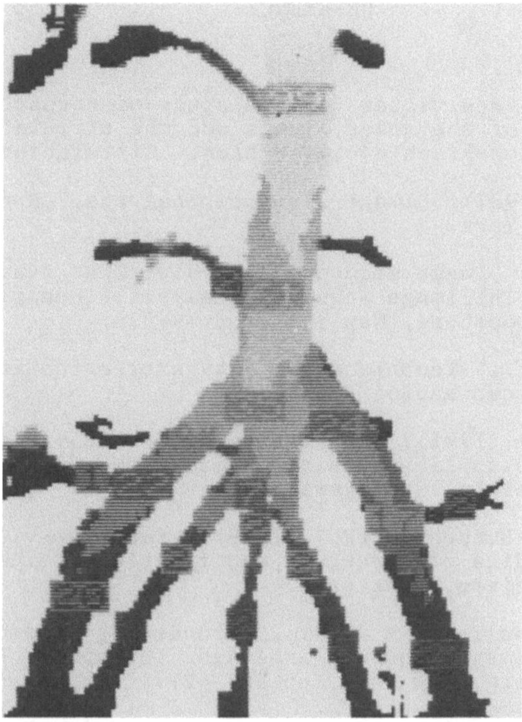

Fig. 16. Regional blood flow distribution computed by techniques based on densitometric principles. The numbers indicate the percentage of the flow in the abdominal aorta measured in regional displacement segments.

densitometric measurements can be used to construct functional images that display the spatial distribution of this hemodynamic parameter. Improved methods for edge and region segmentation have to be applied to automate this analysis.

The main unsolved problems in this field arise from the fact that three-dimensional motion is imaged in a single angiographic projection plane. Angiodensitometric concepts can often reduce the necessity to rely on greatly simplified models of the object under study as is necessary in projection image analysis that is based on edge features only.

ACKNOWLEDGEMENTS

Numerous discussions with Professor Heintzen and with Dipl. Math Hans-Jürgen Hahne have greatly helped to develop the concept of this overview. The assistance of Mrs. U. Bürsch in the preparation of diagrams and photo prints is gratefully acknowledged. We would also like to thank Mrs. F. Liepner for her work in preparation of the manuscript.

REFERENCES

1. Wood EH (1977). New vistas for the study of structural and functional dynamics of the heart, lungs and the circulation by noninvasive numerical tomographic vivisection. Circulation 56:506-520.

2. Huang TS (Ed) (1981). Image Sequence Analysis. Springer, Berlin, Heidelberg, New York.

3. Nagel HH (1981). Image sequence analysis: What can we learn from applications? (In) Image Sequence Analysis, (Huang TS, Ed), Springer, Berlin, Heidelberg, New York, pp 19-228.

4. Rockoff SD (1972). Techniques of data extraction from radiological images. Invest Radiol 7:206-217.

5. Heintzen PH (Ed) (1971). Roentgen-, cine- and videodensitometry. Fundamentals and applications for blood flow and heart volume determination. Thieme, Stuttgart.

6. Heintzen PH and Bursch JH (Ed) (1978). Roentgen video techniques for dynamic studies of structure and function of the heart and circulation. Thieme, Stuttgart.

7. Spiesberger W and Tasto M (1981). Processing of medical image sequences. (In) Image Sequence Analysis, (Huang TS, Ed), Springer, Berlin, Heidelberg, New York, pp 381-421.

8. Heuck F (1973). Densitometrie in der Radiologie. Thieme, Stuttgart.

9. Rutishauser W (1969). Kreislaufanalyse mittels Rontgendensitometrie. Huber, Bern.

10. Wood EH, Sturm RE and Sanders JJ (1964). Data processing in cardiovascular physiology with particular reference to roentgen videodensitometry. Mayo Clin Proc 39:849-865.

11. Trenholm BG, et al (1974). Automated ventricular volume calculation from single plane images. Diagnostic Radiology, pp 299-304.

12. Bursch JH (1983). Principles and methods of roentgen densitometry for circulatory studies. (In) Angiocardiography, (Just H, Ed), Springer, Berlin, Heidelberg, New York (in press).

13. Brennecke R, Brown TK, Bursch JH and Heintzen PH (1977). Computerized video-image preprocessing with applications to cardioangiographic roentgen-image series. (In) Digitale Bildverarbeitung, (Nagel HH, Ed), Springer, Berlin, Heidelberg, New York, pp 244-262.

14. Kruger RA, Mistretta CA, Houk TL, Riederer SJ and Shaw CG (1979). Computerized fluoroscopy in real-time for non-invasive visualization of the cardiovascular system. Radiology 130:49-57.

15. Ovitt TW, Christenson PC, Fisher HD, Frost MM, Nudelman S, Roehrig H and Seeley G (1980). Intravenous angiography using digital video subtraction: X-ray imaging system. AJR 135:1141-1144.

16. Hohne KH, Bohm M, Erbe W, Nicolae GC, Pfeiffer G and Sonne B (1978). Computer angiography: A new tool for x-ray functional diagnostics. Med Progr Technol 6:23.

17. Bursch JH, Hahne HJ, Brennecke R, Hetzer R and Heintzen PH (1979). Funktions-Angiogramme als Ergebnis der densitometrischen Analyse digitalisierter Rontgenbildserien. Biomedizinische Technik Suppl 24:189-190.

18. Brennecke R, Hahne HJ, Moldenhauer K, Bursch JH and Heintzen PH (1979). A special purpose processor for digital angiocardiography. Design and applications. Proc Comp Cardiol, IEEE Computer Society, Long Beach, pp 343-346.

19. Crummy AB, Strother CM, Sackett JF, Ergun DL, Shaw CG, Kruger RA, Mistretta CA, Turnispeed WD, Lieberman RP, Myerowitz PD and Ruzicka FF (1980). Computerized fluoroscopy: Digital subtraction for intravenous angiocardiography and arteriography. AJR 135:1131-1140.

20. Kaihara S, Natarajan C and Maynard D (1969). Construction of a functional image from spatially localized rate constants obtained from serial camera and rectilinear scanner data. Radiology 93:1345.

21. Alpert NM (1977). Functional imaging. (In) Computer Methods, the Fundamentals of Digital Nuclear Medicine, C.V. Mosby, St. Louis.

22. Pretschner DP (1981). Planar imaging and picture analysis in nuclear medicine. (In) Digital Image Processing in Medicine, (Hohne KH, Ed), Springer, Berlin, Heidelberg, New York, pp 149-194.

23. Adam WE, Tarkowska A, Bitter F, Stauch M and Geffers H (1979). Equilibrium (gated) radionuclide ventriculography. Cardiovasc Radiol 2:161-173.

24. Battocletti JH, Halbach RE, Salles-Cunha SX and Sances A (1981). The NMR blood flowmeter. Med Phys 8:435-443.

25. Brennecke R, Hahne HJ, Wessel A and Heintzen PH (1981). Computerized enhancement techniques for echocardiographic sector scans. Proc Comp Cardiol, IEEE Computer Society, Long Beach, pp 7-11.

26. Trenholm BG, Winter DA, Mymin D and Landsdown EL (1972). Computerized determination of left ventricular volume using videodensitometry. Med Biol Eng 10:163.

27. Bursch JH, Heintzen PH and Simon R (1974). Videodensitometric studies by a new method of quantitating the amount of contrast medium. Europ J Cardiol 1:437-446.

28. Kedem D, Kedem DR, Smith CW and Brill AB (1978). Blood flow and velocity distribution measurements by means of digital densitometry. (In) Roentgen Video Techniques, (Heintzen PH, Bursch JH, Eds), Thieme, Stuttgart, pp 58-64.

29. Shaw CG, Ergun DL, Van Lysel MS, Peppler WW, Dobbins JT, Zarnstorff WC, Myerowitz PD, Swanson DK, Lasser TA, Mistretta CA, Dhanani SP, Strother CM and Crummy AB (1981). Quantitation techniques in digital subtraction videoangiography. Digital Radiography, Proc SPIE 314:121-129.

30. Carey PH, Slutsky lRA, Ashburn WL and Higgins CB (1982). Validation of cardiac output estimates by digital video subtraction angiography in dogs. Radiology 143:623-626.

31. Engels PHC and Ludwig JW (1982). Digital subtraction arteriography of the left ventricle using time interval difference mode. (In) Digital Subtraction Arteriography, (Mistretta CA, Crummy AB, Strother CM, Sackett JF, Eds), Year Book Med Publ, Chicago, pp 123-124.

32. Spiller P, Jehle J, Politz B and Schmiel FK (1982). A digital image processing system for measurement of phasic blood flow in coronary arteries. Proc Comp Cardiol, IEEE Comp Soc, Long Beach (in press).

33. Limb JO and Murphy JA (1975). Measuring the speed of moving objects in television signals. IEEE Trans Communications 23:474-478.

34. Budinger RF and Harpootlian J (1975). Developments in digital computer implementation in nuclear medicine imaging. Comp Biomed Res 8:26-52.

35. Schuette WH and Simon AL (1968). A new device for recording cardiac motion. Med Res Eng 7:25-27.

36. Kazamias TM, Gander MP, Ross J and Braunwald E. Detection of left ventricular wall motion disorders in coronary artery disease by radarkymography. New Engl J Med 285:63-71.

37. Barrett WA, Clayton PD and Warner HR (1975). A system for the evaluation and refinement of left ventricular border recognition algorithms. Proc Comp Cardiol, IEEE Comp Soc, Long Beach, pp 251-253.

38. Slager CJ, Hooghoudt TEH, Reiber JHC, Schuurbiers JHC, Booman F, and Meester GT (1979). Left ventricular contour segmentation from anatomical landmark trajectories. Proc Comp Cardiol, IEEE Comp Soc, pp 347-350.

39. Kruger RA (1982). Basic physics of computerized fluoroscopic difference imaging. (In) Digital Subtraction Angiography, (Mistretta CA, Crummy AB, Strother CM, Sackett JF, Eds), Year Book Med Publ, Chicago, pp 16-22.

40. Brennecke R and Heintzen PH (1983). Digital acquisition and processing of video-angiocardiograms. (In) Angiocardiography, (Just H, Ed), Springer, Berlin, Heidelberg, New York (in press).

41. Reiber JHC, Slager CJ, Schuurbiers JHC, den Boer A, Gerbrands JJ, Troost GJ, Kooijman CJ and Serruys PW (1983). Transfer function of the x-ray-cine-video chain applied to digital processing of coronary cineangiograms. (In) Digital Imaging in Cardiovascular Radiology, (Heintzen PH, Brennecke R, Eds), Thieme, Stuttgart (in press).

42. Brennecke R, Hahne HJ, Bursch JH and Heintzen PH (1983). Digital videodensitometry: Some approaches to radiographic image restoration and analysis. (In) X-ray Contrast Media in Vascular Functional Diagnosis (Heuck F, Ed), Springer, Berlin, Heidelberg, New York (in press).

43. Lantz BMT, Foerster JA, Link DP and Holcroft JW (1980). Determination of relative blood flow in single arteries. AJR 134:1161-1168.

44. Brennecke R, Hahne JH, Moldenhauer K, Bursch JH and Heintzen PH (1978). Improved digital real-time processing and storage techniques with applications to intravenous contrast angiography. Proc Comp Cardiol, IEEE Comp Soc, Long Beach, pp 191-194.

45. Kruger RA (1981). A method for time domain filtering using computerized fluoroscopy. Med Phys 8:466-470.

46. Huang TS, Hsu YP (1981). Image sequence enhancement. (In) Image Sequence Analysis, (Huang TS, Ed), Springer, Berlin, Heidelberg, New York.

47. Schmiel FK, Huber H, Neuhaus KL and Spiller P (1978). Densitometric measurements of coronary blood flow. Methodological improvements. (In) Roentgen Video Techniques, (Heintzen PH, Bursch JH, Eds), Thieme, Stuttgart, pp 49-55.

48. Hohne KH, Bohm M and Obermoller U (1981). X-ray functional imaging - Evaluation of the properties of different parameters. Digital Radiography, Proc SPIE 314:224-228.

49. Brennecke R, Hahne HJ, Bursch JH and Heintzen PH (1983). Optimization of generalized subtraction operations for digital fluorography. (In) Digital Imaging in Cardiovascular Radiology, (Heintzen PH, Brennecke R, Eds), Thieme, Stuttgart (in press).

50. Rosen L and Silverman NR (1973). Videodensitometric measurements of blood flow using cross-correlation techniques. Radiology 109: 305-310.

51. Hohne KH, Bohm M and Nicolae GC (1980). The processing of x-ray image sequences. (In) Advances in Digital Image Processing, (Stucki P, Ed), Plenum, New York, pp 147-160.

52. Gallagher JH, Meaney TH, Flechner SM, Novick AC and Buonocore E (1981). Parametric imaging of digital subtraction angiography studies for renal transplant evaluation. Digital Radiography, Proc SPIE 314:229-234.

53. Sonne B and Hohne KH (1978). Vergleichende Untersuchungen verschiedener MeBmethoden zur BlutfluBbestimmung aus digitalen Angiogrammen. Biomed Techn(Berlin) 23:208-215.

54. Alderman EL, Berte LE, Harrison DC and Sanders W (1981). Quantitation of coronary artery dimensions using digital image processing. Digital Radiography, Proc SPIE 314:273-278.

55. Barth K, Epple E, Irion KM, Faust U and Decker D (1981). Quantifizierung von Stenosen der Herzkranzgefasse durch digitale Bildauswertung. Biomed Technik(Berlin) 26(Suppl):67.

56. Shmueli K, Brody WR and Macovski A (1981). Estimation of blood vessel boundaries in x-ray imagse. Digital Radiography, Proc SPIE 314:279-286.

57. Bursch JH, Hahne HJ, Brennecke R, Gronmeyer D and Heintzen PH (1981). Assessment of arterial blood flow measurements by digital angiography. Radiology 14:39-47.

58. Rosenfeld A and Kak AC (1976). Digital Picture Processing. Academic Press, New York.

59. Pratt WK (1978). Digital Image Processing. Wiley & Sons, New York.

60. Thomson WB (1980). Combining motion and contrast for segmentation. IEEE Trans Pattern Anal Machine Intell 2:543-549.

61. Bursch JH, Hahne HJ, Brennecke R, Eicker C and Heintzen PH (1983). Arterial blood flow analysis by digital angiography. (In) Digital Imaging in Cardiovascular Radiology, (Heintzen PH, Brennecke R, Eds), Thieme, Stuttgart (in press).

62. Decker D, Faust U and Heuck F (1980). Hydraulischer Hochdruck-Injektor mit digitalem Injektionsprozessor.

ACOUSTICAL IMAGING: HISTORY, APPLICATIONS, PRINCIPLES AND LIMITATIONS

Glen Wade, Hua Lee and Carl Schueler

Department of Electrical and Computer Engineering, University of California, Santa Barbara, CA ; Santa Barbara Research Center, Goleta, CA

1. INTRODUCTION

"Seeing" with sound instead of light seemed far-fetched a generation ago. However, after more than twenty years of successful ultrasound use in clinical medicine (1-10), many persons have seen acoustically-produced images of the organs in their bodies and are accustomed to the concept of acoustical imaging. An example of such imaging is shown in Fig. 1. The ultrasonic instruments in our hospitals are the most generally known of all acoustical imaging systems, but, in addition to medical diagnosis, there are many other applications for sonic images. For example, acoustical imaging is being employed for oceanic search, seismic exploration, non-destructive evaluation, and microscopy (11). The systems involved use mechanical and electronic scanning and pulse-echo ranging, as well as focusing with acoustic lenses and digital processing (12). The more detailed descriptions in this chapter will be restricted to systems employed for microscopy, diagnostic medicine and non-destructive evaluation.

These systems are all relatively recent because modern technology is absolutely essential to a successful man-made acoustical-imaging system. Such systems do exist in nature, but we have only recently become aware of them. Humans are not naturally endowed to process acoustic information and obtain a mental image. For example, no human could go into a dark room and shout or hum and listen to the echoes to identify and accurately locate the objects in the room.

Although nature has not equipped man to "see" with sound, bats and dolphins do it with ease. Bats can fly about in a dark room and avoid hitting each other, the walls, the floor, the ceiling or any object in the room in spite of lack of light. The evolutionary development of the bat has provided that animal with a high level of sophistication in its acoustical-imaging capability. The bat's talent in using sound in this way is comparable to that of a human in using light. A large portion of

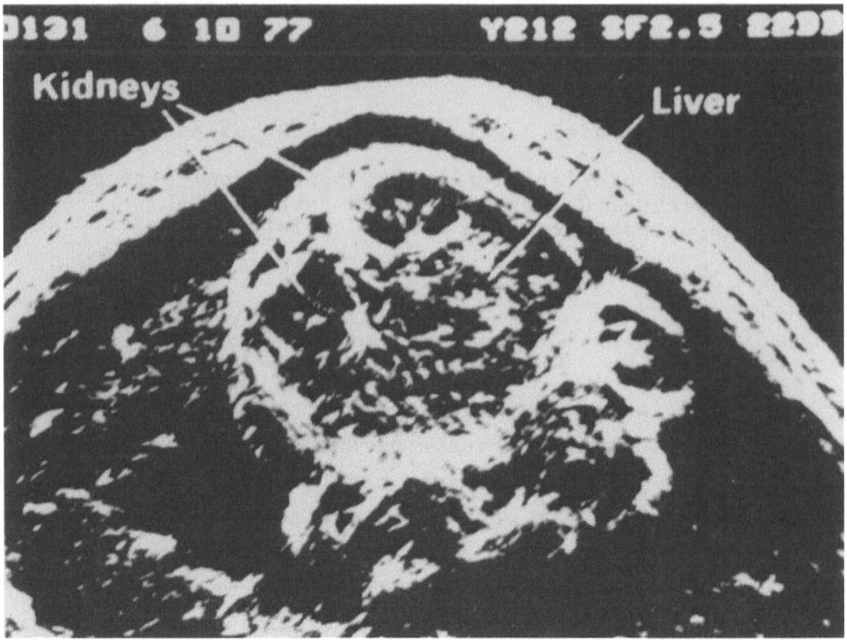

Fig. 1. Cross-section longitudinal-scan image from a Picker instrument showing the kidneys and liver of a human patient at the Seoul National University Hospital.

the bat's brain is devoted to acoustical vision, just as a considerable part of the human brain is used for optical vision.

It is well known that dolphins also have excellent capability to navigate and find food by means of sonic energy. The dolphin has very good eyes, but nevertheless depends primarily on acoustic echoes to locate food. A dolphin may have almost no objection to a blindfold for its eyes, but would strongly resist having its sense of hearing impaired.

The single audible indication to a human that an acoustic phenomenon is taking place when a bat or a dolphin is near is a clicking sound. The clicking from a dolphin can be detected underwater and is similar to the noise from a squeaking gate. The clicking from a bat in flight is easily heard as the bat passes by.

Only recently have these facts been fully appreciated. The Italian scientist Spallanzani wrote in 1793 that he was inclined to believe that bats have "a new organ or sense" that humans do not have and of which no one was at that time even aware (13). Nevertheless Langevin's artificial acoustical imaging system called SONAR evolved independently of any knowledge about bats during World War I (14,15). Only after the war was the bat's "new sense" understood by Hartridge to be somewhat similar to SONAR in its operation (16). When Spallanzani described the bat's acoustical imaging ability as a "new sense", he was entirely right from his 1793 point of view. Until Galton invented the ultrasonic whistle ninety years later, the existence of sound at frequencies above 20kHz or so was generally known. Even now, without electronic equipment, one is hard put to say if an ultrasonic whistle is working unless the family dog responds. For two hundred years we have been able to appreciate the fact that bats have special imaging ability. However, not many details are known to this day about the way these animals process acoustic information to guide their movements. Although humans now have remarkable artificial ability to image acoustically, this ability has been wrought through independent technological achievement rather than by attempts to imitate animals that possess natural acoustical imaging ability.

This chapter initially provides a review of the history, principles of operation, and applications of artificial acoustical imaging systems that have evolved since Paul Langevin's pioneering work seventy years ago. The interesting history of the development of acoustical imaging cannot be told without simultaneously introducing key concepts upon which these systems operate. Moreover, applications of acoustical imaging will necessarily be brought to the forefront as we tour the melange of enticing, and creative ideas proposed by early and mid-twentieth century investigators.

Upon reaching the modern era, we will find that successful new applications of acoustical imaging have depended in very large measure upon advances in signal processing, and in particular, upon digital processing. Digital techniques have special relevance for acoustical imaging (17) where, in general, the wavelength of the radiation being used to produce the image data is the same size as the fine structure to be resolved. It is far more important to employ sophisticated signal processing for acoustical imaging than for optical imaging where, except in microscopy, the wavelengths are invariably much shorter than the structural variation being observed. Therefore, section 3 of this chapter is dedicated to a review of applications of digital processing to acoustical imaging.

Finally, no technical review is complete without an attempt to discern the most likely future course of the pertinent technology. Section 4, entitled "Potential, Limitations and Tradeoffs", explores key physical restrictions in the art of acoustical imaging, and provides an optimist's view of the best that we can expect of this far from fully-exploited technology.

2. HISTORY, PRINCIPLES AND APPLICATIONS

2.A The Early Pioneers

With the advent of the submarine as a weapon of war in World War I, the French navy contracted with Paul Langevin to develop an effective means to spot German U-boats underwater. Langevin's solution achieved a place in Hollywood history via such submarine films as the 1957 classic "Run Silent, Run Deep", with Clark Gable and Burt Lancaster. Indeed, Langevin's solution involved the ingenious use of sound to measure the distance and speed of approach or recession of an underwater target. The technique, based on projecting a directed pulse of sound and timing the returned echo, is reminiscent of a backpacker's method to estimate a canyon's depth by listening to the echo from a shout projected over a precipice. Both Lord Rayleigh and O.P. Richardson had thought of employing ultrasonic waves many years before, and M.C. Chilowski developed an ultrasonic device for the French navy prior to World War I, but its acoustic intensity was much too weak to be practical. Langevin succeeded in providing high ultrasonic intensity by means of a piezoelectric transducer operating at resonance (14,15). He continued important work after World War I, and as late as 1940 was invited by the French navy to direct a research program on ultrasonic depth-finders.

Known as SONAR, Langevin's technique, when applied repetitively over a range of azimuthal angle, and provided with a capability to respond to targets at various distances, produces a very useful image of objects in the area of interest. SONAR, an acronym for sound navigation and ranging, predates its electromagnetic counterpart known as RADAR, an acronym for radio detection and ranging.

Langevin's acoustical imaging technique, though simple in concept, to this day remains the most generally useful means for producing acoustical images. The fundamental pulse-echo principle upon which SONAR was founded forms the basis of a wide range of modern acoustical imaging technologies.

The well-known ultrasonic instruments used by hospitals to produce television images of an unborn fetus in a mother's womb are some the progeny from the simple SONAR concept (6). Equipped with complicated transceiver arrays and electronic timing circuits, these instruments listen to echoes scattered by the tissues in the mother's body, and produce a picture of the intensity profile of these echoes. The major feature that distinguishes these modern imaging instruments from Langevin's original ranging device, is the ability to focus and direct an ultrasonic beam using a transceiver array, and to time the echoes by means of sophisticated circuits that match the spatial location of a scattering point in the acoustic field of view of the transceiver array to the corresponding point on the television raster.

A development going considerably beyond Langevin's basic ideas involved taking advantage of acoustic phase information. Work in this area, well after Langevin's pulse-echo intensity methods had been developed, was started a decade or so later with S.J. Sokolov's pioneering efforts at the University of Leningrad. Sokolov's major practical goal was to use sound to detect flaws in metal objects as a quality inspection tool for industrial manufacturing. Indeed, commercial ultrasonic imaging systems now exist that provide excellent non-destructive evaluation of welds in metal pipe flaws in concrete structures, as well as micro-flaws in tiny integrated circuits (18,19). Many of these devices are based on principles enunciated by Sokolov in the early part of the twentieth century.

He was one of the first persons to recognize and systematically explore the usefulness of ultrasound for "imaging" internal structures in optically opaque objects. His productive work extended over three decades beginning in the 1920's. He devised several techniques to obtain optical patterns corresponding to the structure within metal objects using information acquired by insonifying the objects with acoustic energy in the megahertz region. Many of his schemes were proposed to detect inhomogeneities such as flaws and voids within the objects.

In one of his systems, the inhomogeneities were made "visible" by reflecting collimated light from a liquid surface that carried acoustic information in a fashion similar to modern liquid-surface holography, to be described later. In another system light was diffracted after being passed through a glass container filled with turpentine in a manner reminiscent of modern Bragg-diffraction imaging, also to be discussed later. The recent work on liquid-surface holography and Bragg-diffraction imaging began in earnest less than twenty years ago. But the ideas of Sokolov were quite similar in nature to these modern ideas and were enunciated more than thirty years earlier.

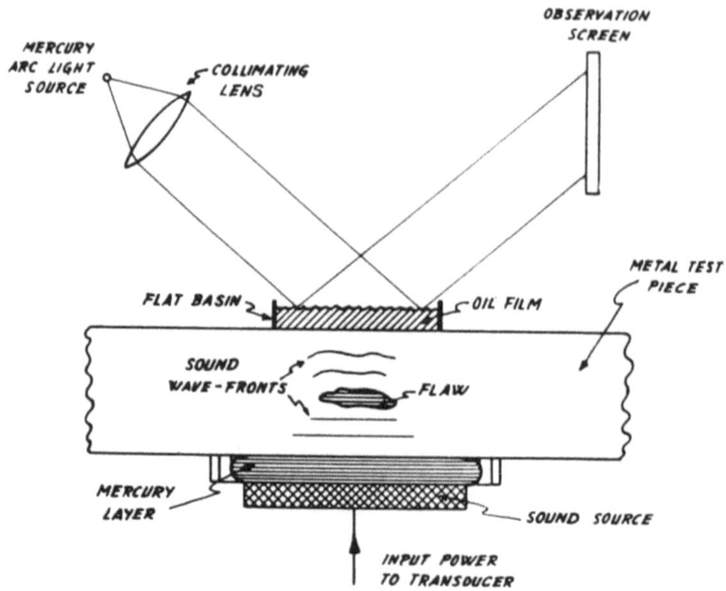

Fig. 2. The liquid-surface system of S.J. Sokolov for detecting voids and other flaws in metal manufactured pieces. (Courtesy of Acoustic Imaging, Plenum Press).

The liquid-surface system he constructed for detecting flaws in metal test pieces is illustrated in Fig. 2 and was first proposed in 1929. Later Sokolov patented this system and published a paper concerning its operation (20). The system has all the basic elements necessary to produce the acoustic equivalent of the "Gabor-type" hologram of optical holography. Conceivably Sokolov can be regarded as making the first hologram, and doing so almost twenty years before Dennis Gabor's invention of holography in 1948! It is therefore possible to argue that, from a chronological standpoint, acoustical holography preceded optical holography. Of course Sokolov had no way of understanding holographic principles as we understand them today. His images were not of high quality. Not only did he operate without the use of a laser, but he made no attempt to eliminate either the undiffracted light which carries no image information or the diffracted light corresponding to the conjugate image.

In another of Sokolov's early attempts to detect flaws with ultrasound, he used a bulk system (21) strikingly similar in overall appearance to that of the present Bragg-diffraction system (22). A diagram of Sokolov's instrument is shown in Figure 3. Sokolov was interested in detecting inhomogeneities, such as cracks and casting errors in manufactured metallic parts, rather than in forming images per se. His system utilized the arrangements of Debye and Sears (23) and of Lucas and Biquard (24) to display the diffraction spectra of the light as produced by interaction with the sound. Thus, strictly speaking, Debye-Sears diffraction rather than Bragg-diffraction was involved. Sokolov provided no spatial filtering to eliminate unwanted beam components in the reconstruction process and no actual "image" was produced.

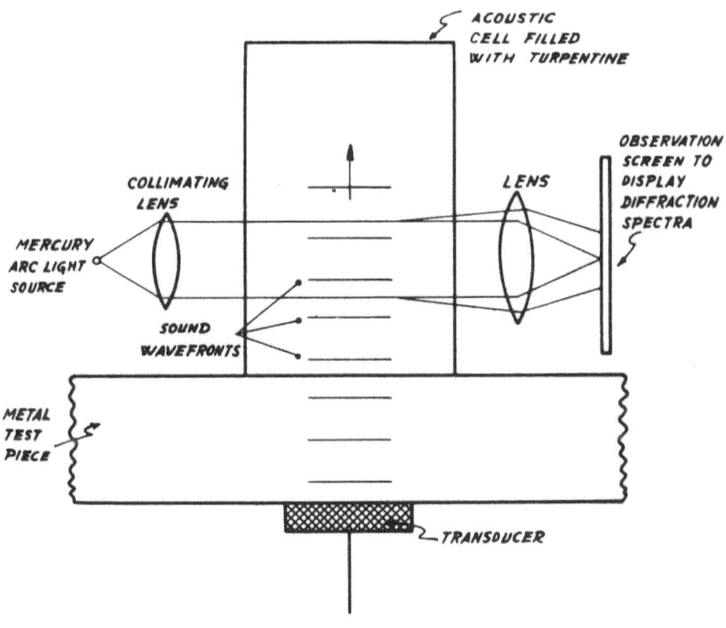

Fig. 3. Sokolov's bulk-diffraction instrument for flaw detection.

Although we have emphasized Sokolov's ideas on the use of phase information for imaging, he also contributed to the development of systems which measured intensity. He originally proposed using sound at 3 GHz where the wavelength of the sound in water is very short (half a micrometer) and capable of resolving truly minute objects. The name he gave to the corresponding device was the "ultrasonic microscope" (25). Technological impracticalities prevented Sokolov from generating sound at sufficiently high frequencies for the purposes of microscopy. However, the principle he put forth (that of reading out localized electronic charges developed on a piezoelectric crystal in response to an acoustic input) has since become embodied in a well-known instrument called the Sokolov tube which is even now used for low-frequency acoustic imaging (26).

Practically every modern acoustic instrument has its roots in, or is at least related to, the ideas of Langevin and Sokolov. Furthermore, the fundamental principle upon which almost any present acoustical imaging instrument is based is traceable to intensity-imaging techniques, to pulse-echo concepts, or phase-amplitude measurements. As a result, we will review the development of modern acoustical imaging from these three perspectives. Naturally, some systems apply a combination of the three to enhance the resulting image.

In the next three subsections, we shall consider each fundamental approach in turn. We will start with intensity mapping because it is conceptually the simplest of the three approaches. We shall discover that even the most elementary intensity-mapping techniques can provide a very powerful acoustical imaging tool. Following the discussion of instruments developed for intensity mapping, we present two subsections which consider pulse-echo and phase-amplitude concepts, respectively. As we might expect, the imaging capacity inherent in these latter approaches has already permitted the development of powerful imaging instruments. The research on these approaches is still going on and we can anticipate even greater achievement in the future.

2.B Intensity-Mapping Systems

Perhaps the simplest conceptual approach to forming an image with acoustic energy is to take advantage of the transparency of many objects to sound, by mapping the intensity of the acoustic waves transmitted through the objects. Though elementary in concept, and similar to the principle upon which a standard X-ray picture is made, intensity-mapping was not the first method used to make acoustical images. Langevin's pulse-echo approach, and Sokolov's phase-amplitude schemes were proposed and used several years before the first practical intensity-mapping technique was developed. However, imaging by mapping transmitted acoustic intensity, forms the basis of the very successful modern acoustic microscope developed by C.F. Quate and colleagues at Stanford University within the last dozen or so years (27,28). We shall discuss this instrument in some detail shortly.

But first let us look at another approach of this type which was developed more than thirty years earlier by Reimar Pohlman (29). It was based on a special technique that Pohlman invented for making acoustic intensity visible with light. This device, called the Pohlman cell, was the first practical, high-resolution, acoustical imaging tool that employed intensity-mapping. The Pohlman cell acts as an acousto-optical energy converter. We avoid the use of the word transducer because of the peculiar way in which the device operates.

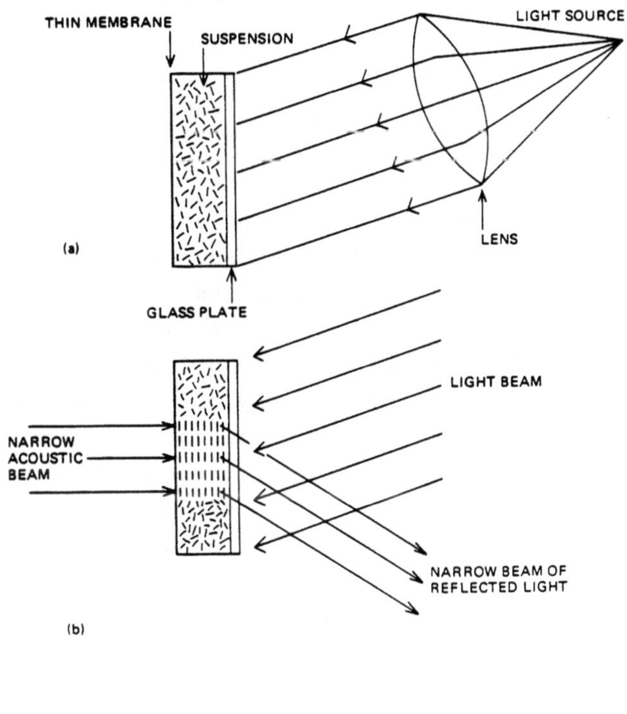

Fig. 4. Acoustic intensity mapping with the Pohlman Cell. (a) Cell with no insonification, (b) Cell with localized insonification. (Courtesy of Acoustic Imaging, Plenum Press).

The cell consists of a sandwich-like structure containing a suspension of fine metallic flakes in a suitable liquid as shown in Fig. 4. One side of the sandwich is a glass plate and the other side is a thin membrane which is acoustically transparent. Because of thermal motion, the metallic flakes will normally be oriented randomly in the medium. If the suspension is illuminated by light projected through the glass plate, the light reflected from the metallic particles, as seen by an observer, will have a diffusely scattered, matte appearance. However, when sound, passing through the stretched thin membrane, is incident on the suspension, the metallic flakes will tend to orient themselves locally with their broad surfaces parallel to each other. Each flake will present the maximum area to the sound field. Thus in the regions of high acoustic intensity, the planar surface of each particle tends to become perpendicular to the direction of the acoustic propagation. If the suspension is then irradiated by light, the reflected light delineates the pattern of acoustic intensity impinging upon the cell through the membrane on the other side. The regions of high acoustic intensity, with their aligned metal flakes, will present a more reflective surface to the incident light than other regions with their randomly-oriented flakes. Under these conditions, an observer will see areas of varying brightness corresponding to spatially varying acoustic intensity, superimposed on the gray matter-appearing background.

The Pohlman cell has relatively high intrinsic resolution because the size of the metallic particles used in the suspension and the thickness of the membrane through which the sound passes can be very small compared with a wavelength. The chief disadvantage of this imaging technique is that the response time for the particles to orient themselves when insonification is present, or to become randomly oriented in the absence of insonification, is too long for real-time operation. Nevertheless the idea is intriguing and has attracted much attention over the years since it was first proposed. The Pohlman cell continues to be studied in various research laboratories for a variety of specialized applications.

Perhaps the most spectacular of the acoustical images are those from ultrasonic microscopes. Owing to the long wavelength of ordinary sound, it may be difficult to imagine that sonic waves could possibly be useful to obtain highly resolved images of minute objects. It is easy to do this with ordinary light whose wavelengths are well known to be very short or with electrons whose size is negligibly small. Sound might at first glance appear to be entirely too cumbersome. This would be true if we were referring to sound in air, where the frequency spectrum is severely limited. However, in water the acoustic spectrum is much larger. Because sound velocity is extremely slow compared to light, sound wavelengths for a given frequency are extremely short compared to those of electromagnetic waves for the same frequency. For example, at 1.5 MHz the electromagnetic vacuum wavelength is 200 meters. The sound wavelength is one millimeter. At 3 GHz, a realizable acoustic frequency, the wavelength in water is down to that of visible light! Figure 5 illustrates the quality of images obtainable when the frequency is 600 MHz, with sound wavelengths equivalent to that of shortwave infrared. The pictures were furnished by Professor Quate at Stanford and show various images of unstained human red blood cells. The photograph on the left is an optical image and the other three are acoustical images with different magnifications. The acoustical images have a resolution of about 2 micrometers. Quate's system is of the lens type and reads image information with a piezoelectric transducer. It is the acoustical equivalent of the scanning electron-beam microscope (SEM), and is generally referred to as the SAM (for scanning acoustic microscope).

A sound beam from a transmitting transducer is focused by an acoustic lens to a small spot in the object plane. The spot is made to scan out a raster pattern over the object by rapidly moving the object through such a pattern within its plane. The scattered sound is then gathered by another lens and projected onto a receiving transducer. The time variation of the output from this transducer, after suitable processing, is converted into a corresponding spatial variation of intensity on the screen of a cathode-ray tube by means of synchronous scanning, as described in more detail elsewhere (27,28). The acoustical images of red cells in Fig. 5 compare favorably with the optical one. Each blood cell averages about 7 micrometers in diameter. Obviously, for objects of this size the resolution is adequate in both systems. However, the contrast is much better acoustically than optically, and this fact illustrates one advantage of acoustic microscopy over optical microscopy.

In the next subsection we discuss the application of pulse-echo concepts to modern acoustical imaging instrumentation.

2.C Pulse-Echo Systems

Modern medical ultrasonic diagnostic equipment is built upon Langevin's pulse-echo technique. Many of these instruments operate in real-time, producing so-called B-scan gray-scale images in the plane of the insonification by scanning a focused acoustic beam rapidly over azimuth and slowly over range to obtain a television display. Although the concept dates back to World War I, the technology necessary to steer and focus the acoustic beam and to reproduce the televised image, was not readily available until about twenty years ago. In fact, the more sophisticated of these instruments have been in commercial use for only about ten years

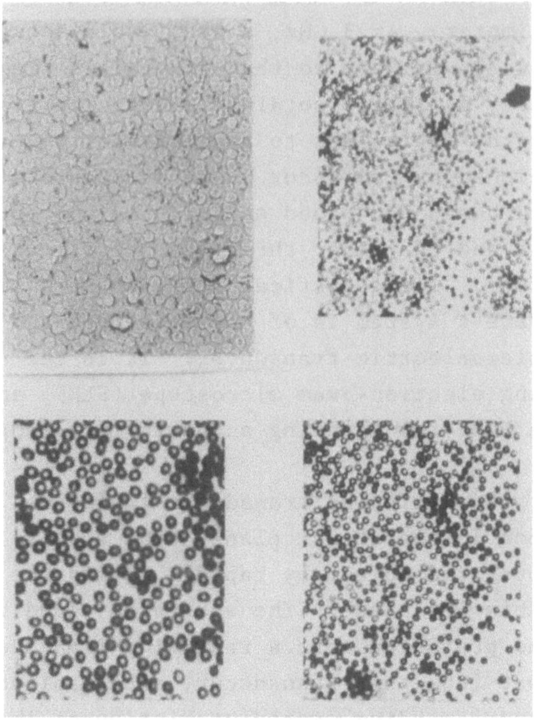

Fig. 5. An optical image (upper left) and three acoustical images from Stanford's scanning acoustic microscope. The images are of unstained human red blood cells. (Courtesy of IEEE Trans. Sonics and Ultrasonics).

with the first laboratory version of a digitally controlled B-scan system reported in 1968 (3).

In 1970 a prototype system was reported that used a commercial ultrasound B-scanner and a transducer mounted on a rectilinear drive and immersed in a water tank (29). A programmable timing and control unit supervised the motion of the rectilinear scanner and the transmission of the acquired data to an external four K-byte data buffer. The data could then be transferred to a computer or to digital magnetic tape. The echo amplitude was digitized with 10-bit resolution. Only one sample was taken from each resolution cell and the transducer was restricted to moving in a straight line at constant speed.

The first digital scan converter for B-scan ultrasound was reported in 1974 (30). This system could store the echo amplitude acquired by the random motion of a hand-held transducer in a contact B-scan examination and then display the image in TV format. The system was limited to 128 x 128 pixels. Due to computer speed limitations, 10 to 15 seconds were required to complete a scan.

Commercial digital scan converters were reported as early as 1976 (31), with many others introduced shortly thereafter. The commercial scan converters of today possess image quality that rivals that of analog machines. Typical pixel grid size is 512 x 512, with either 16 or 32 gray shades. These devices offer flicker-free operation, high reliability, and excellent capability for image processing. In addition, the modern digital scan converter can provide good spatial resolution which is independent of brightness contrast to improve image sharpness.

Another very important application of the pulse-echo technique is for industrial non-destructive evaluation (NDE). The simplicity of the pulse-echo concept lends itself well to field instrumentation for detecting flaws in manufactured parts. Although for pedagogical reasons we classify modern pulse-echo NDE equipment as descending from Langevin's original work, it was F.A. Firestone's Reflectoscope built at the University of Michigan that spurred the development of today's industrial pulse-echo test equipment. The Reflectoscope operated by generating a succession of ultrasonic pulses of short duration and detecting their echoes from sub-surface discontinuities within a solid. The Reflectoscope made it possible to examine an object from one surface only to determine, within limits, the size of any internal inhomogeneities and to measure the depth of the inhomogeneities below the surface.

Ever since Firestone's day, the pulse-echo approach has had substantial success in non-destructive evaluation. As with SONAR and RADAR, the systems based on this approach provide a map of the regions under investigation, producing cross-sectional images of the internal structure of the object, similar to those of the B-scan medical instrument.

The latest stage in the application of pulse-echo techniques in acoustical imaging has followed the advent of X-ray computer-aided tomography. The measurement of acoustic echo intensity from one direction only produces a B-scan image that is somewhat similar to a tomogram in that it is a slice image in the plane of insonification. Thus it was natural that computer-aided X-ray tomography, developed in the early 1970's, would stimulate research on computer-aided acoustic tomography. Nevertheless, the results of this research so far have been disappointing. The key reason that acoustic tomographic systems have not replaced X-ray scanners in the hospital, in spite of advantages in terms of low patient irradiation, is the lack of sufficient resolution of detail for the purposes to which these tomographic systems are applied. However, the research is continuing and we shall describe the accomplishments briefly. To appreciate the principles involved, we shall give a short review of the more familiar X-ray tomographic systems and then treat the acoustic systems.

In conventional radiology, X-rays diverge from a single source through a patient's body to project onto film a shadowgraph of the structure along the path of the rays. The structural elements, cleanly separated in the three-dimensional object, often overlap in the final two-dimensional image in such a way as to make the structure difficult to discern.

In computer-aided tomography, there is no overlap. The images are computed from a large number of projections. The image has the form of a two-dimensional mapping of discrete non-overlapping structural elements in a single plane of the body. Ordinary X-ray technology is combined with sophisticated computer processing to make this possible.

The X-ray source and detector move around the body and in effect hundreds of X-ray pictures are made. Instead of being recorded on film, the information is sent to a computer to be processed into a tomogram. With this approach it is possible in principle to obtain the image of any cross-section within the body.

But X-rays are not the only kind of energy for which computer-aided tomography is feasible. Ultrasound, microwaves, electron beams, fast subatomic particles from accelerators, gamma rays from such sources as positron annihilation, and even magnetic fields can be used (32).

Exposure to X-rays, no matter how low the level, is regarded as being harmful to a patient. Hence ultrasound is particularly attractive to computer-aided tomography. It can be completely non-invasive. In addition, acoustic energy can often give a view of a cross-section not available with X-rays or other types of radiation. A mapping of acoustic and elastic variation can be expected to provide a basically different pattern than a mapping of variations of X-ray absorption and scattering coefficients. Moreover, a mapping of one kind of acoustic parameter yields quite a different picture than that of another. So far two acoustic parameters have received the most attention in research; attenuation and acoustic refractive index.

Such workers as Greenleaf et al (33), Carson et al (34), and Jakowatz and Kak (26) were the first to produce tomograms involving these parameters. Mappings of variations of either parameter require transmitting ultrasound through the object. But ultrasound can also be reflected from objects. Inhomogeneities within an object provide echoes, and research is going on to produce computer-reconstructed tomographic mappings of variaions in ultrasonic reflectivity. In addition, tomographic extensions of Doppler processing (9) are being pursued.

Thus, starting with the simple backpacker's concept of pulse-echo ranging, man has wrought ever more sophisticated acoustical imaging tools. Langevin's early work seventy years ago and Firestone's work three decades later have been combined with creative application of television and digital electronics. These innovations have spawned a still-developing technology for "seeing" into optically opaque objects, including a mother's womb or a concrete structure, to produce slice images of complex structure.

But, as we shall discover in the next subsection, Sokolov's ideas on the use of acoustic phase information had their own interesting evolutionary consequences. Because of concepts of this sort, acoustical imaging appears to offer future performance which may be unrivaled by any other imaging modality for certain applications.

2.D Phase-Amplitude Systems

Sokolov's original work to develop a liquid-surface system that transformed acoustic energy into an optical replica of the acoustic wavefield resulted in what we can think of as a type of "holographic" imaging that predated Gabor's invention of optical holography by nearly twenty years. However, it was the success of optical holography in the early sixties, after the advent of the laser, that impelled researchers to develop the field of acoustical holography and to produce practical instrumentation based on the measurement of the acoustic wavefield, both in phase and amplitude. In fact, many of the systems that evolved from acoustical holographic research employ lasers for the acoustic wavefield readout.

Although Sokolov can be credited with being the first to produce holograms, he was probably unaware that he had done so, and the first worker to purposely set out to produce an acoustical hologram and publish that effort in the literature was Pal Greguss in 1965 (35). Greguss' method was fundamentally different from any of the approaches that have developed into useful concepts. However, his unique idea illustrates the kind of open-minded creative energy that contributes to progress in research.

To record the hologram, Greguss used a "sonographic plate" whose exposure depended upon a "sonochemical" reaction rather than a photochemical one. If such a plate is developed in the presence of a sound field, the pattern of standing-wave components in the sound will be converted into a black-and-white image suitable for use in making a hologram. The drawback that limited this method to a laboratory curiosity is that the sound used for the recording process must be very intense.

The type of holographic system which, from publication dates in the research literature, has been investigated over the longest sustained period of time, is the kind that uses scanning receiving transducers moving through a raster pattern in the holographic plane. The procedure involved was first described in an elementary form by Thurstone in 1966 (36), and is illustrated in Fig. 6. Normally water is used as the medium for acoustic propagation in the embodiment shown. A sound source is placed on one side of the object and a scanning hydrophone on the other. The effective diameter of the hydrophone should be less than or, at most, equal to the fringe spacing to be recorded.

A second holographic approach was also the subject of much sustained research over a number of years. The principles of operation were enunciated by Mueller and Sheridon in their classic paper of 1966 (37). The system uses a liquid-air surface as the recording medium and this is similar to the Sokolov system of Fig. 2. Figure 7 shows the arrangement. The scattered wave fronts of the object beam propagate upward toward the surface of the water in the tank, as do the relatively well-ordered wavefronts of a reference beam. In this system, two different transducers provide the object beam and the reference beam, whereas only one source was used in Sokolov's system. This gives rise to a sideband hologram as opposed to the Gabor hologram of the Sokolov system. The wavefronts from the reference and object beams undergo reflection at the liquid-air interface above the membrane shown in the figure. This reflection produces radiation pressure on the liquid surface. The pressure causes the surface to deform until gravity and surface tension achieve a new balance. The deformation appears as a stationary ripple pattern corresponding to the interference between the two beams. The ripples can be used directly to produce spatially-modulated, first-order diffraction sidebands on a beam of laser light. Thus the ripple pattern on the liquid surface becomes an embossed hologram of the object. The zero-order light and one

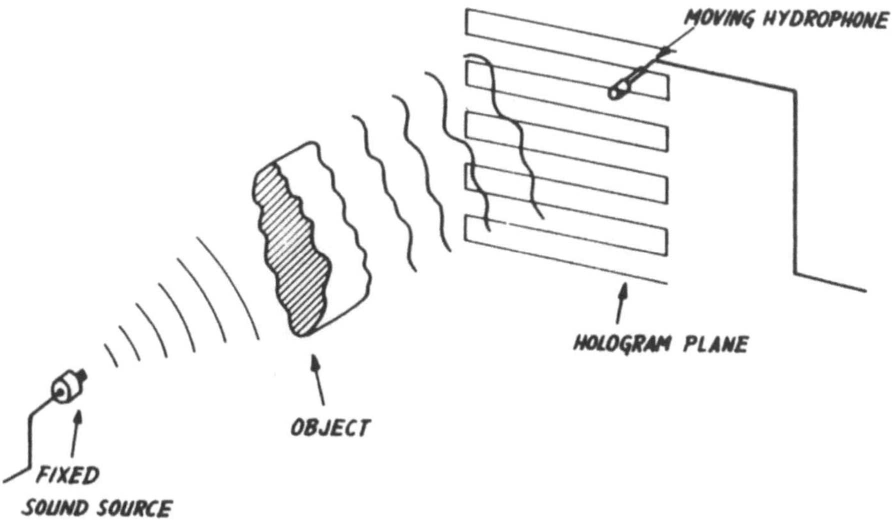

Fig. 6. Acoustic holography with a fixed transmitter and a scanning receiver. No acoustic reference beam is needed since the reference can be simulated electronically. (Courtesy of Acoustic Imaging, Plenum Press).

of the diffraction sidebands are eliminated by an optical spatial filter before the image is further processed for viewing.

In this type of holography, frequently referred to as "liquid-surface" holography, image reconstruction takes place instantaneously. The best images are achieved when an acoustic lens is placed between the object and the surface, so that an acoustical image is formed directly on the hologram plane, as illustrated in Fig. 7. Under these circumstances, the reconstructed visual image can be of such high quality that images produced this way have been among the best obtained to date by any of the orthographic acoustical imaging systems. A hologram made in this fashion is equivalent to the "focused-image" hologram or the "image-plane" hologram of optical holography (38).

A third acoustic imaging system that employs laser-beam readout and operates in real-time, utilizes a solid surface from which to detect the

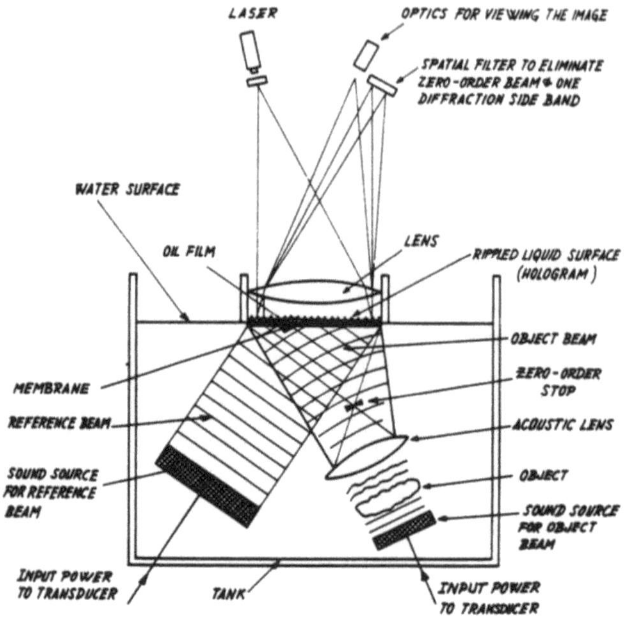

Fig. 7. Liquid-surface acoustic holography. Static ripples on the liquid-air interface constitute a record of the interference between the object and the reference beams and permit the rippled liquid surface to operate as a hologram in real time. (Courtesy of Acoustic Imaging, Plenum Press).

sound field. The technique involved is sometimes referred to as "solid-surface" holography. The system is illustrated in Fig. 8. The sound waves are first scattered from the object to be viewed into the water bath in which the object is placed. The scattered waves then impinge upon a solid membrane with a shiny surface which serves as one of the sides of the tank containing the bath. As the waves strike the solid surface, they generate a moving ripple pattern on the interface between the membrane and the water. Even though no reference beam is used in the system, the surface records what might be called a dynamic embossed "hologram". The image is read out by using a scanning, focused laser beam which produces a spot on the surface. The spot scans through a raster pattern on the "holographic" plane, and is reflected from the shiny surface. In the process, the beam becomes angularly modulated by the dynamic ripple on the surface. The image information is converted into weak electrical signals by passing the reflected laser beam through

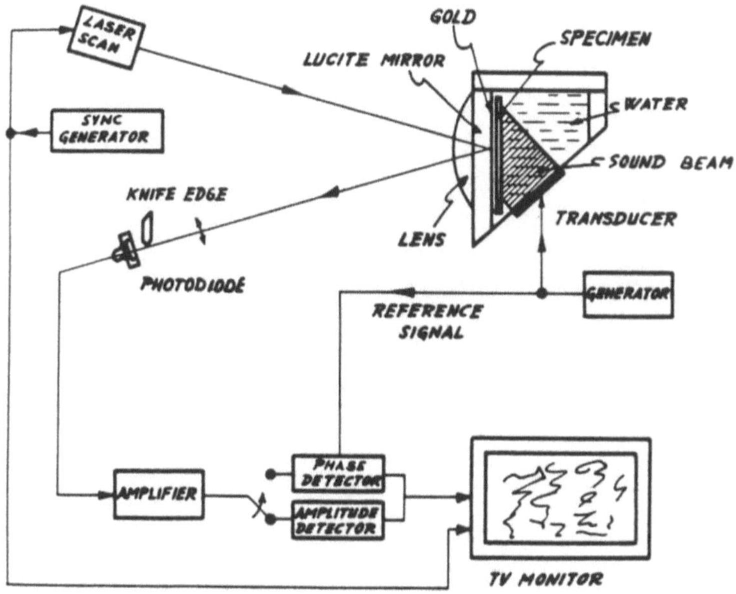

Fig. 8. Solid-surface system. Image information contained in the dynamic ripples on the surface of the lucite mirror is read out by a scanning laser beam. (Courtesy of Acoustic Imaging, Plenum Press).

a knife-edge-and-photodiode combination. The signals are amplified and decoded, resulting in an acoustic shadowgraph displayed on a synchronously-scanned television screen. Such a system has been profitably packaged by Sonoscan. It is presently commercially available for acoustic microscopy (39), and is known as the scanning laser acoustic microscope (SLAM). This microscope was developed by Dr. Lawrence W. Kessler and colleagues while Dr. Kessler was still with Zenith Radio Corporation.

The SLAM system has the unique feature that it can be adjusted to give simultaneously both an acoustic and an optical image of the same object. The photographs in Fig. 9 show real-time, simultaneous optical and acoustical images from this system. The object in this case was a fruit-fly larva, and the images were displayed simultaneously on the screens of three color television monitors. Fig. 9(a) presents the optical image of the larva, 9(b) the acoustical image, and 9(c) a superposition of the preceding two images. Thus 9(c) displays the simultaneous sum of the

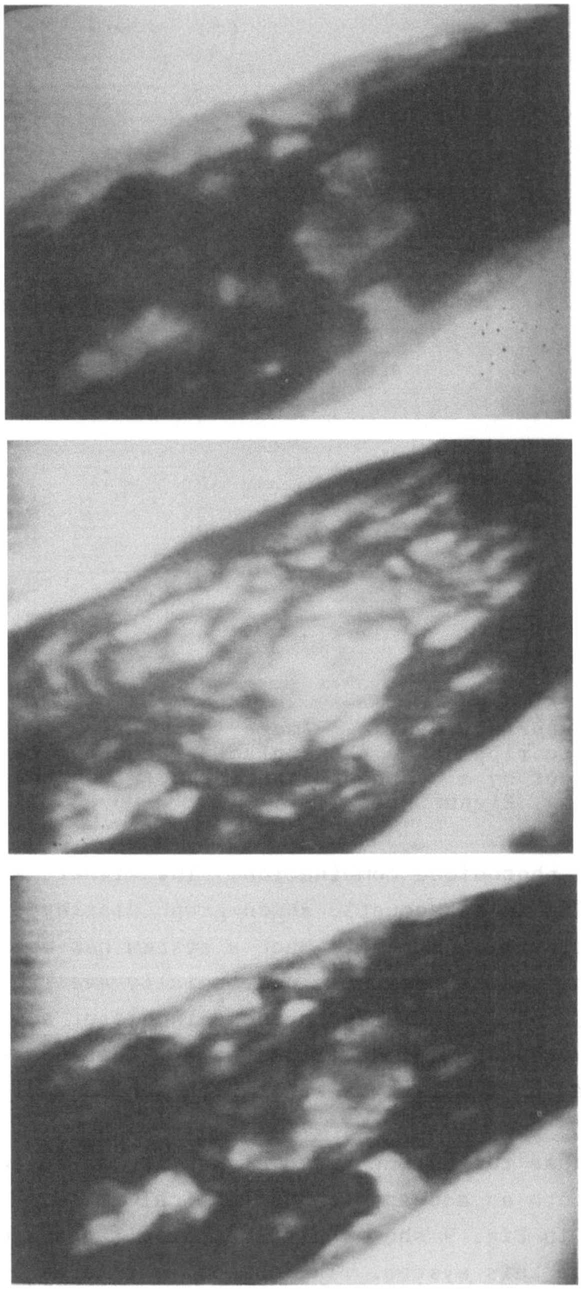

a

b

c

Fig. 9. Optical and acoustical images of a section of a fruit-fly larva obtained by a SLAM microscope.
(a) Optical image, (b) Acoustical image, (c) Superposition of (a) and (b). (Courtesy of IEEE Trans Sonics and Ultrasonics).

optical and acoustical information. The optical image is shown in yellow and the acoustical in blue, so that the bicolored representation maintains the identity of the source. This is more useful, as well as more spectacular, than a black-and-white presentation would be.

The larva shown in the displays of Fig. 9 was about 1 millimeter in diameter and 4 millimeters long. The length is considerably greater than the uniform region of the sound field. Therefore only a portion of the specimen could be imaged at any one time. In each of the images, the larva head is located beyond the display to the right. More fine detail is visible in the acoustical image (Fig. 9(b)) than in the optical one (Fig. 9(a)). Extending back from the head is a system of tubes through which oxygen is delivered to the cells, and many small bubbles are present. There are two main branches of the system, which are more clearly delineated acoustically than optically. One would expect this to be the case because the impedance mismatch for sound passing through a region containing bubbles is much greater than for light. Also, the optical opacity of the specimen is generally greater than the acoustical opacity, so that much internal detail is masked from view in the optical image but not in the acoustical image. Even with X-rays, where opacity is not a problem and where differentiation between hard, calcified tissue and soft tissue is easy to produce, it is still difficult to distinguish between various soft tissues. Soft-tissue differentiation is readily apparent, however, in acoustic micrographs without the need for chemical staining. For some experimental situations, considerable optical detail as well as acoustic detail is present, and one image tends to complement the other. Under these circumstances, the simultaneous superposition of both the optical and acoustic signals may be greatly advantageous and can be accomplished with striking effectiveness as demonstrated in Fig. 9(c).

As remarkable as the images from acoustic microscopes have been, there are still problems that make their use in non-destructive evaluation less than ideal. When an acoustic microscope operates in the transmission mode, the micrograph is simply a shadowgraph of all the structure encountered by the paths of the acoustic rays passing through the object. The resultant image is a two-dimensional mapping of three-dimensional internal structure and is particularly difficult to comprehend in the case of specimens of substantial thickness and structural complexity. For example, Fig. 9 shows a two-dimensional image of a three-dimensional object. It is very difficult to ascertain the depth of the scattering center within the larva corresponding to any of the image elements. The third (out-of-the-paper) dimension is completely unspecified.

The principles of computer-aided tomography can be applied to acoustical microscopy to solve this problem. As was previously discussed, tomograms are unambiguous slice images. A tomographic microscope is one that produces slice images of the various layers of a tiny three-dimensional object. If the object is an integrated circuit, each interior layer can be imaged separately by a tomographic microscope. The interior structure of an opaque microscopic specimen, such as a fruit-fly larva, is unambiguously imaged in a two-dimensional format without thin sectioning. Research is presently underway on this type of microscope and it has been given the acronym STAM for scanning tomographic acoustic microscope.

In principle, STAM is capable of producing microscopic cross-sectional images to overcome the difficulties referred to in connection with microscopic shadowgraphs. It is particularly attractive for complex objects with multiple planar structure. STAM uses elements from the scanning laser acoustic microscope (SLAM) in which the acoustic wave field is detected by a scanning laser beam. STAM incorporates digital signal processing in the reconstruction step to obtain microscopic tomograms of the interior structure of the specimens to be evaluated.

The tomograms are produced in a fashion somewhat reminiscent of commercial X-ray computer tomography. In STAM, however, laser-beam scanning of the energy of acoustic waves transmitted through the object constitutes the data acquisition and produces the projections. These projections are then digitally processed by means of a method which takes sound-wave diffraction into account. Such a method, called "back-and-forth propagation", was published in 1982 by the authors of this article, in collaboration with Dr. Gail Flesher at UCSB (40).

Preliminary evidence suggests that successful use of the back-and-forth propagation technique, an amalgam of tomographic and holographic concepts, may successfully produce tomograms using the SLAM system coupled to a digital computer.

The fourth acoustical imaging system using laser-beam readout makes use of the principle of Bragg-diffraction of coherent light from ultrasonic wavefronts in water (22). Although somewhat similar to the Sokolov system illustrated in Fig. 3, Bragg-diffraction imaging is different in important ways. The concepts were initially enunciated and experimentally studied by Korpel in 1966 (41). The ideas were independently thought of by Hance, Parks and Tsai (42) at Lockheed Research Laboratories, and by one of the present authors (43) at UCSB. Research efforts similar to Korpel's at Zenith were mounted at about the same time by these workers. First, a wedge-shaped beam of laser light is caused to traverse the acoustic cell shown in Fig. 10 rather than a collimated beam from a mercury source as in the Sokolov system. Second, Bragg-diffraction is used rather than Debye-Sears diffraction. Third, a masking stop is inserted in the system to prevent unwanted light from interfering with the image.

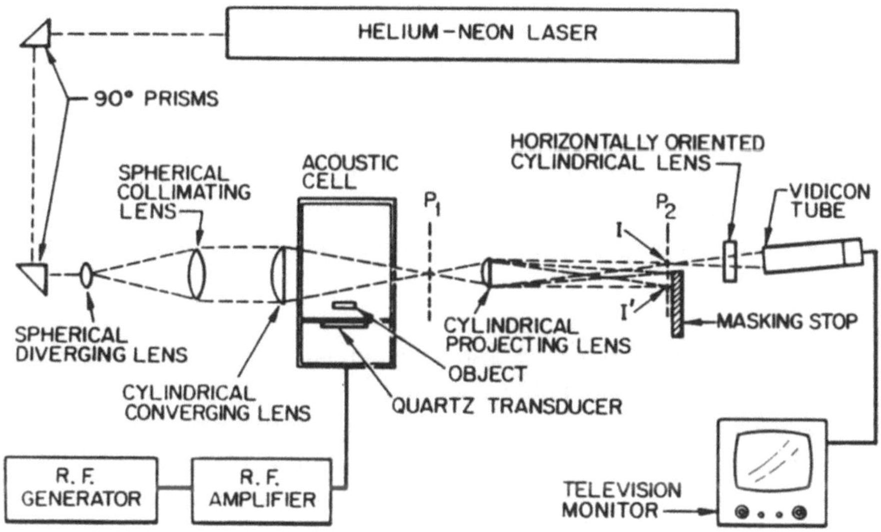

Fig. 10. Bragg-diffraction imaging system. (Courtesy of Acoustic Imaging, Plenum Press).

Phase information is retained by the Bragg-diffracted light that leaves the sound cell and this fact suggests a way of looking at Bragg-diffraction imaging that is reminiscent of holography where phase information is also retained. As described in reference (12), the analogue between a Bragg-diffraction imaging system and a thick optical hologram is very close.

Bragg-diffraction imaging systems produce images in real-time. They are easy to operate and inexpensive to construct. However, they have some severe drawbacks, including low sensitivity and limited field of view. In an era when patient radiation is a major concern, this lack of sensitivity produces a requirement for high insonifying power and makes Bragg-diffraction imaging unacceptable for clinical diagnostic purposes. However, this approach does have potential in the area of non-destructive evaluation, and perhaps in microscopy for the in vitro study of tissues where radiation hazard is not an important factor (22).

2.E Discussion

The most useful implementation of acoustical imaging that employs phase-amplitude information is Sonoscan's SLAM system, and this system operates without benefit of digital signal processing. Of the pulse-echo applications of acoustic imaging (from subsection 2.C) that use only intensity information, the ultrasonic B-scan instruments are the most widely employed, and they also became popular without benefit of digital signal processing. However, digital methods have by now been used to enhance the images from the B-scan instruments, and future B-scanners that do not take advantage of digital techniques are probably doomed in the competitive marketplace.

Calvin Quate's intensity mapping SAM and Sonoscan's phase-amplitude SLAM are now in a competitive position, although the SLAM has enjoyed the most commercial success. This is partly due to the fact that SLAM has been around longer and uses phase as well as amplitude information. As we have discussed, the next step in the development of SLAM may be a tomographic version, using digital signal processing to implement the conversion.

Although they have not gained as much application, as the above system based on solid-surface "holography", the liquid-surface methods have been shown to yield excellent images, and it may be possible that digital signal processing could correct the orthographic character of these images that mitigates against them. Finally, the holographic techniques involving scanning transducers comprise a data-acquisition approach with a resolution too limited for typical medical or NDE applications. Therefore, these methods are restricted to larger-scale operations such as oceanic or seismic exploration. Discrete transducer acquisition methods are commonly used in seismic work, and the data-processing techniques employed to interpret received data are very well-developed and quite sophisticated, as discussed, for example, in Robinson and Treitel's comprehensive text (44).

Clearly, digital signal processing has a crucial role to play in the further development of acoustical imaging for medical, microscopic, and NDE applications, and in the next section we review several of the most useful techniques and some promising new ones.

3. DIGITAL PROCESSING OF ACOUSTICAL IMAGES

Linear models can be easily applied to acoustical-imaging systems to simplify the analysis and processing. If the size of the receiving aperture is significantly large, space-invariance can also be assumed. Data acquisition can then be regarded as a linear filtering process and characterized by a convolution integral. The object distribution and the received wavefield can therefore be treated as the input and output, respectively, of a linear system.

The amplitude and phase of the resultant wavefield is detected at the receiving aperture by the data-acquisition system. A reference wave is employed to remove the carrier frequency and can be realized by electronic synchronization. The detected complex wavefield can then be stored and filtered to reconstruct the images.

Using a discrete receiving array is equivalent to sampling the wavefield. The spacing of the detector contained in the sensor array is directly related to the sampling rate. Due to the attenuation of the wave components exceeding the spatial frequencies that can be propagated, called evanescent-waves, the resultant wavefield has a low spatial-frequency pass band (45). It is circularly bounded with a cut-off frequency of $1/\lambda$ where λ is the wavelength. For this reason a half-wavelength can be considered as the optimal array spacing in order to avoid aliasing as required by the Nyquist sampling theorem. In two dimensions, hexagonal geometry is optimal for the array structure because it results in a 15% saving in terms of the array density due to the circularly bounded spatial-frequency spectrum (46).

When the receiving aperture is in the far field, Fraunhofer or Fresnel approximations can be used to simplify the formulation of the wave propagation (45). The relationship between the object distribution and the detected wavefield then becomes a Fourier transform which can readily be inverted by using the FFT algorithm. The required array spacing is the distance that matches the kernel in the FFT format to prevent leakage (47).

The resolution of an acoustical-imaging system is, in general, limited by the acoustic frequency and the size of the receiving aperture. If resolving power is required, high-resolution processing techniques can be employed, but this necessitates heavier computation than would otherwise be the case. High-resolution reconstruction can be achieved by using recursive algorithms to minimize the estimation error (48). High resolution can also be achieved by orthogonal decomposition of the wave-field which will improve the images by removing the degradation due to the limited size of the apertures (49).

Quantization is a major step in image processing. A fixed number of quantization levels are assigned to each data sample. Because the sampled wavefield is complex, two-dimensional quantization is applied and this can be done in polar or rectangular coordinates. Quantization error further degrades the image reconstruction. High bit-rates should therefore be used to minimize this type of degradation. However, for imaging systems with limited storage or transmission capabilities, low bit-rate imaging techniques are highly desirable and often necessary.

Because the phase variation of the received wavefield governs the image formation, we can reduce bit-rates by recording only the phase of the resultant wavefield. The process is then simplified to one-dimensional quantization (50-52). The complexity of the reconstruction algorithm is considerably reduced. Theoretical analysis as well as experimental results indicate that the phase-only technique can permit reconstruction with as few as 3 bits/sample and yet retain high image quality (53). Moreover, some of the image quality lost by going to low bit-rates can be restored by using enhancement algorithms without substantially increasing the computational burden.

4. POTENTIAL, LIMITATIONS AND TRADEOFFS

As we have seen in the preceding sections, there are many broad categories of operation associated with ultrasonic imaging. In addition to the broad classifications already discussed (intensity-imaging techniques, pulse-echo concepts and phase-amplitude measurements), if scanning is used, the imagers can be more narrowly classified as to whether they are C-scan which gives orthographic images, or B-scan and tomographic, which give slice image, whether the scan technique employed was mechanical or electronic, whether the scan modality was linear or compound or involved sectors of arcs, or whether the measured acoustic waves were transmitted through the object or reflected from it or its internal elements. In general, each category of operation has certain inherent advantages and disadvantages. Instruments of the above kinds are presently being used or developed for oceanic research, seismic exploration, microscopy and non-destructive evaluation as well as for medical applications.

They are the instruments of today, but what about the future? Accurate predictions are difficult to make with confidence. Nevertheless we can observe the record of the past and extrapolate from present research.

Let us now consider the problems. Today's instruments frequently suffer from slow data acquisition, inadequate resolution, excess artifact generation, inaccurate image reconstruction, poor penetration of sound into the object, low dynamic range and inferior classification of image pixels. Research is currently taking place in all of these problem areas and we can expect progress in each. The question is how much progress will be made.

Data acquisition can be speeded up substantially by using the rapid scanning techniques associated with electronic arrays and laser-beam readout. Reconstruction accuracy can be improved significantly by greater reliance on digital processing as opposed to analog methods. Resolution can be enhanced by going to higher frequencies. Perhaps the most difficult problem category is pixel quantification. Much research is going on in this area and recent development suggests that we can look for important progress even here (54-56).

But the progress will not come without tradeoffs. In some areas of concern we can expect a particular performance aspect to be improved, but only at the expense of poorer performance in some other aspect. More rapid data acquisition is technically feasible but, all other things being equal, greater rapidity means reduced dynamic range and sensitivity. This is particularly true with laser-beam readout. For example, the laser in the SLAM system is scanned at television speeds and the images are produced in a thirtieth of a second. But laser beams are noisy and sensitivity suffers whenever they are employed. The sensitivity decreases even further with increasing scanning rates. Higher resolution can be accomplished by going to higher acoustic frequencies or using techniques associated with superresolution. However, the former approach reduces penetrability and the latter increases artifact generation and reduces sensitivity.

One of the most interesting of the current research efforts is in acoustic microscopy (57). Most of the generally applicable problem areas mentioned above are pertinent in this work. Penetrability is of concern as is dynamic range, sensitivity, speed of data acquisition, blurring due to diffraction, overlapping of image elements, resolution, artifact generation, reconstruction accuracy, and inability to correctly classify the pixels (58). Progress in acoustic microscopy can be noted and generalized to indicate and illustrate the expectations and limitations involved in other fields of ultrasonic diagnostic instrumentation.

In an article of limited length such as this, it is, of course, impossible to go into much detail. However, research results to date on the tomographic microscope permit a certain amount of optimism. Perhaps it is safe to predict that slice images of complex structure in the interior of microscopic objects will be feasible in the future. We can expect data for the projections to be acquired in a matter of seconds. Digital reconstruction should be possible almost simultaneously. The question of penetrability will always be a problem and only for selected objects will it be practical to obtain useful images. Nevertheless we can confidently hope to develop non-invasive instruments to produce unambiguous and useful micrographs of the internal structure of tiny opaque objects such as integrated circuits and microscopic life.

5. CONCLUSIONS

As indicated in the previous sections, by now a wide variety of system concepts exist for acoustical imaging. Several of the systems have demonstrated astonishing capability. The techniques employed in these systems include intensity-mapping, pulse-echo operation, and phase-amplitude measurement. They involve holography, tomography, and acousto-optic interaction.

Some of the systems are active in the sense that a transmitter produces acoustic energy which is either reflected from or transmitted through the object of interest. Other systems are passive in that the object itself is the source of the acoustic energy. The field of view of some of the systems is as small as a few micrometers, in the case of acoustic microscopy, and as large as hundreds of miles, in the case of tomographic temperature mapping of the ocean (54). The frequencies employed in these systems vary from the gigahertz range to a few hertz with corresponding wavelengths from micrometers to kilometers. The applications, present and future, for acoustical imaging are as diverse as those in medical diagnosis, non-destructive evaluation, seismic exploration, microscopy and oceanic search.

For a brief moment, consider specifically the question of medical diagnosis. To be able to see inside a human body and its organs is of transendant importance to the radiologist. Tissue is generally opaque to light but relatively transparent to ultrasound. Acoustic waves at low levels appear to have little or no deleterious effect on the tissue being irradiated, which is thought not to be the case when X-rays and nuclear particles are employed. Ultrasound can be safely used for in-vivo and non-destructive imaging of fetuses, human reproductive organs, plaque formation in carotid arteries, tumors of the eye and socket, tissue irregularities in female breasts, stones in the bladder or kidney, necrotic centers in samples of liver, and valve motion in the heart. Acoustic images are unique in that they are mappings of the mechanical and elastical properties of the tissue. The use of ultrasound is therefore complementary to other diagnostic imaging.

However, it is not only in medical diagnosis that acoustical imaging is complementary to other imaging. The same is true in oceanic research, seismic exploration, microscopy and non-destructive evaluation. Much work has been done, much progress has been made, and much utility from acoustical imaging has been realized. But there remain many difficult questions to be resolved before the full capability of acoustical imaging can be applied to many of the practical problems. Much innovative work therefore remains to be done. But this work is being done. From what has already been accomplished we can reasonably expect that modified versions of the old systems will be employed in the old ways and that some of the new systems, still in the research stage, will find useful employment in entirely new ways.

REFERENCES

1. Donald I (1963). The use of ultrasonics in the diagnosis of abdominal swellings. Brit Med J 2:1154-1155.

2. Fleming JE and Hall AJ (1968). Two-dimensional compound scanning-effects of maladjustment and calibration. Ultrasonics 6:160-166.

3. Fry WJ, et al (1968). Ultrasonic visualization system employing new sacnning and presentation methods. J Acoust Soc Amer 44:1324-1388.

4. Wells PNT (1969). Physical Principles of Ultrasonic Diagnosis, Academic Press, New York.

5. Ericksen KR, et al (1974). Ultrasound in medicine - A review. IEEE Trans Sonics Ultrason SU-21:3.

6. Hellman LW (1977). Ultrasound in historical perspective. (In) Ultrasound in Obstetrics and Gynecology, (Sanders RC and James AE, eds), Appleton-Century-Crofts, New York, p 1.

7. Deter RL and Hobbins JC (1979). A survey of abdominal ultrasound scanners: The clinician's point of view. Proc IEEE 67(4):664-671.

8. Havlice JJ and Taenzer JC (1979). Medical ultrasonic imaging: An overview of principles and instrumentation. Proc IEEE 67(4):620-641.

9. Maginness MG (1979). Methods and terminology for diagnostic ultrasound imaging systems. Proc IEEE 67(4):641-653.

10. Ophir J and Maklan NF (1979). Digital scan converters in diagnostic ultrasound imaging. Proc IEEE 67(4):654-664.

11. Wade G (1976). Acoustic Imaging: Cameras, Microscopes, Phased Arrays, and Holographic Systems. (Wade G, ed), Plenum Press, New York.

12. Wade G (1975). Acoustical imaging with holography and lenses. IEEE Trans Sonics Ultrason SU-22(6).

13. Spallanzani L (1826). Litera prima. (In) Opere de Lazzaro Spallanzani, Vol. 5, Dalla Societa Tipogr De'Classici Italiani, Milano, p 209.

14. Langevin P and Chilowski MC (1916). Procedes et appareils pour la production de signaux sous-marins direges pour la localisation a distance d'obstacles sous-marins, French Patent No. 502913.

15. Langevin P. (1918) French Patent 505 703, (1921) British Patent 145 691.

16. Hartridge H (1920). The avoidance of objects by rats in their flights. J Physiol 54:54-57.

17. Keating PN, Sawatari T and Zilinskas G (1979). Signal processing in acoustic imaging. Proc IEEE 67(4):496-510.

18. Kessler LW and Yubas DE (1979). Acoustic microscopy - 1979. Proc IEEE 67(4):526-536.

19. Kino GS (1979). Acoustic imaging for non-destructive evaluation. Proc IEEE 67(4):510-525.

20. Sokolov SJ (1935). Ultrasonic oscillations and their application. Techn Phys U.S.S.R. 2:522.

21. Sokolov SJ (1935). Uber die paraktische ausnutzung de beugnung des lichtes an ultraschallwellen. Phys Z 36:142.

22. Ahmed M and Wade G (1979). Bragg-diffraction imaging. Proc IEEE 67(4):587-603.

23. Debye P and Sears FW (1932). On the scattering of light by supersonic waves. Proc National Acad Sci(US)18:409.

24. Lucas R and Biquard P (1932). Optical properties of solids and liquids under ultrasonic vibrations. J Phys Rad 7(3):464.

25. Sokolov SJ (1949). Ultrasonic microscope. Akedemia Nank U.S.S.R., Doklady (Tekniches Kaya Fizika) 64:333-335.

26. Jakowatz CV Jr and Kak AC (1976). Computerized tomography using X-rays and ultrasound. School of Engineering, Purdue University, West Lafayette, Indiana, Res Rep Tr-EE 76-EE:76-25.

27. Lemons RA and Quate CF (1973). A scanning acoustic microscope. Proc Ultrason Symp 18-21, IEEE Catalog 73 CHO 807-8 SU, (deKlerk J ed), pp 18-21.

28. Lemons RA and Quate CF (1974). Acoustic microscope - scanning version. Appl Phys Lett 24(4):163-165.

29. Erickson JJ and Brill AB (1970). Digitization of ultrasonic images. Radiology 95(3):589-593.

30. Ophir J, et al (1974). The CUPAD system: Real-time handling of randomly acquired clinical ultrasound data. (In) Ultrasound in Medicine, (White, Denis ed), Vol. 1, Plenum Press, New York.

31. Schorum S and Tidel H (1977). The Pho-sonic SM: A computer controlled image forming system. (In) Ultrasound in Medicine, Vol. 3B, Plenum Press, New York.

32. Wade G (1980). Ultrasonic imaging by reconstructive tomography. (In) Acoustical Imaging, (Yang KY ed), Plenum Press, New York, Vol. 9, pp 379-431.

33. Greenleaf JF, Johnson SA, Lee SL, Herman GT and Wood EH (1974). Algebraic reconstruction of spatial distribution of acoustic absorption within tissue from their two-dimensional acoustic projections. (In) Acoustical Holography, (Green PS, ed), Plenum Press, New York, Vol. 5, pp 591-603.

34. Carson PL, Oughton TV and Hendee WR (1976). Ultrasonic transaxial tomography by reconstruction. (In) Ultrasound in Medicine, (White D and Barns R, eds), Plenum Press, New York, Vol. 5, pp 341-350.

35. Greguss P (1965). Ultrasonic holograms. Res Film 5(4):330-337.

36. Thurstone FL (1966). Ultrasound holography and visual reconstruction. Proc Symp Biomed Eng 1:12-15.

37. Mueller RK and Sheridon NK (1966). Sound holograms and optical reconstruction. Appl Phys Lett 9:328-329.

38. Collier RJ, Burckhardt CB and Lin LH (1971). Optical Holography. Academic Press, New York.

39. Kessler LW, Palermo PR and Korpel A (1972). Practical high-resolution acoustic microscopy. (In) Acoustical Holography, (Wade G, ed), Plenum Press, New York, Vol. 4, pp 51-71.

40. Lee H, Schueler C, Flesher G and Wade G (1982). Ultrasonic planar scanned tomography. (In) Acoustical Imaging, (Powers John, ed), Plenum Press, New York, Vol. 11, pp 309-323.

41. Korpel A (1966). Visualization of the cross-section of a sound beam by Bragg-diffraction of light. Appl Phys Lett 9:425-427.

42. Hance HV, Parks JK and Tsai CS (1967). Optical imaging of a complex ultrasonic field by diffraction of a laser beam. J Appl Phys 38(4):1981-1983.

43. Wade G, Landry J and deSouza AA (1967). Acoustic Imaging: Cameras, Microscopes, Phased Arrays, and Holographic Systems. (Wade G, ed), Plenum Press, New York.

44. Robinson EA and Treitel S (1980). Geophysical Signal Analysis. Prentice-Hall, Englewood Cliffs, New Jersey.

45. Goodman JW (1968). Introduction to Fourier Optics. McGraw-Hill, New York, pp 42-54.

46. Lee H and Wade G (1982). Analysis and processing of phase-only reconstruction in acoustical imaging. Proceedings on Physics and Engineering in Medical Imaging, IEEE Computer Society Press, pp 240-246.

47. Lee H and Wade G (1982). Resolution for images from Fresnel or Fraunhofer diffraction using FFT. IEEE Trans Sonics Ultrason SU-29(3):151.

48. Lee H and Wade G. Constructing an imaging operator to enhance resolution. Submitted to Journals of Acoustical Society of America, (to be published).

49. Lee H and Wade G (1982). High-resolution imaging for systems with small apertures. Journals of the Acoustical Society of America 72(6):2033-2035.

50. Metherell AF (1969). The relative importance of phase and amplitude in acoustical holography. (In) Acoustical Holography, (Metherell AF, ed), Plenum Press, New York, Vol. 1, Chapter 14.

51. Power J, Landry J and Wade G (1969). Computed reconstructions from phase-only and amplitude-only holograms. (In) Acoustical Holography, (Metherell A, ed), Plenum Press, New York, Vol. 2, Chapter 13.

52. Oppenheim AV and Lim JS (1981). The importance of phase in signals. IEEE Proc 69(5):529-541.

53. Lee H and Wade G (1982). Evaluating quantization error in phase-only holograms. IEEE Trans Sonics Ultrason SU-29(5):251-254.

54. Behringer D, Birdsall T, Brown M, Cornuelle B, Heinmiller R, Knox R, Metzger K, Munk W, Spiesberger J, Spindel R, Webb D, Worcester P and Wunsch C (1982). A demonstration of ocean acoustic tomography. Nature 299:121-125.

55. Wang RC, Lee PPK, Persson HW, Shenk EA and Gramiak R (1982). Frequency-dependent angle scattering of ultrasound by liver. Journals of Acoustical Society of America 72(2):343-352.

56. O'Donnell M (1983). Quantitative volume backscatter imaging. IEEE Trans Sonics Ultrason 30(1):26-36.

57. Lin ZC, Lee H and Wade G (1983). Scanning tomographic acoustic microscope. ECE Technical Report #83-01, University of California at Santa Barbara.

58. Finette S, Bleier A and Swindell W (1983). Breast tissue classification using diagnostic ultrasound and pattern recognition techniques: I. Methods of pattern recognition. Ultrasonic Imaging 5(1): 55-70.

ULTRASOUND TOMOGRAPHY BY GALERKIN OR MOMENT METHODS

Steven A. Johnson and Frank Stenger

Departments of Bioengineering, Electrical
Engineering and Mathematics
University of Utah, Salt Lake City, Utah

INTRODUCTION

Ultrasound B-scan imaging is now a well established and valuable clinical tool. Improvements in transducer arrays and microprocessor controls have led to the development of real-time linear and sector scanners which produce images of remarkable clarity and resolution compared with B-scanners of only a few years ago. Further improvements in B-scan images are predicted to occur as larger transducers, apertures and improved dynamic focusing methods are employed. The use of Doppler ultrasound alone or in combination with real-time B-scan imaging is expected to increase in importance as the clinical significance of high resolution Doppler images is appreciated.

During the past decade x-ray CT (computed tomography) and recently NMR imaging have made remarkable contributions to the field of diagnostic imaging. X-ray (CT) and NMR images not only provide resolution of about 1 mm but also can be calibrated by absolute reference standards. The resulting quantitative images have proven to have valuable diagnostic value because of the greater ability thus provided to distinguish healthy and diseased tissue by their image values.

It is natural to ask if the successful B-scanner technology could be improved further by incorporating quantitative tissue characterization features. Such tissue characterization features, based on back scatter, are now being investigated for inclusion on clinical B-scanners, but the tissue characterization they provide is based on structural and statistical properties of tissues and not absolute tissue properties. The statistical properties of tissues, e.g. the spatial Fourier transform, are often correlated with the state of health or disease and are therefore valuable, but they are not easy to measure quantitatively with backscatter techniques unless special precautions are taken to model or compensate for angular dependence. These back scatter statistical properties are dependent on the gradient of the density of the tissues and thus the constant of integration has been lost (1). Thus we have partially answered our question. We have noted that some quantitative information about tissue properties can be obtained from B-scans but not the absolute mechanical properties of tissue. It is then appropriate to ask if ultrasound can in principle provide additional tissue properties. The answer is yes if the scattering into many angles is used. We next describe how such scattering has been investigated for imaging applications.

During the past decade many investigators have attempted to develop an ultrasound tomography which could provide resolution and absolute tissue characterization analogous to x-ray CT. The first ultrasound CT systems appeared at about the same time that NMR imaging was demonstrated and shortly after x-ray CT was announced (2,3). The development of ultrasound CT has not shown the rapid development into a clinical tool as have the other modes. The rapid development of NMR imaging has been surprising even to many closely associated with medical imaging. The reasons why ultrasound imaging has not followed the same pace is now becoming more clear.

Ultrasound CT has not made the rapid progress of x-ray CT or NMR imaging for several reasons: (1) the theory for x-ray CT or NMR imaging is much simpler than that of ultrasound CT; (2) neither x-ray nor NMR imaging are subject to diffraction and refraction effects, but these two effects dominate the ultrasound signal detection process; (3) x-ray detector arrays and NMR detectors are relatively simple to design and build compared to ultrasound detector arrays. This chapter will address suggestions for improving the resolution and quantitative imaging of ultrasound CT by addressing the theory of ultrasound CT and treating refraction and diffraction effects.

Initial attempts to develop ultrasound CT greatly simplified the problem by treating the propagation of ultrasound as if the acoustic energy followed straight line rays like x-rays. As a consequence, refraction and diffraction effects limited resolution to about 10 to 20 wavelengths. Correcting for refraction of the ray paths did not help because resolution is determined mostly by diffraction effects. As the diffraction problem became better understood, several investigators suggested including diffraction and refraction in the ultrasound imaging formulations by borrowing Wolf's clever perturbation theory, inverse Born, scattering algorithm from optical holography theory (4,5). Much effort has been expended in developing this approach with the greatest success coming from replacing the Born approximation with the more accurate Rytov approximations. Although this work has brought fresh and important insight to the quest, there are several serious defects in these Rytov or Born perturbation methods (6): (1) these methods must assume unrealistically low absorption; (2) the refraction is only treated approximately, e.g. the far field of a simple lens is not given correctly by either the Born or Rytov approximate methods; (3) the inclusion of density scattering has been solved but still depends on attenuation being treated correctly.

The use of an attenuating bath has been suggested to allow imaging of human tissue at higher frequencies, e.g. 1 to 3 MHz. A new algorithm to allow better treatment of refraction effects for ultrasound CT has been proposed (7). Such approaches may help, yet the fact remains that the perturbation methods are not exact except in the limit of vanishing small perturbations. Therefore, an ultrasound imaging method based upon a more exact treatment of the wave equation would have the capability of imaging objects with larger attenuation and larger variations in refractive index and density. Such a method would also have the desirable feature of acting as a standard of comparison.

We are aware of two methods which do not depend on the object producing only small perturbations to the total field. The first method is based upon extrapolating the scattered data to zero wavelength. This method looks promising for medical imaging and has potential for wider development and use if its sensitivity to noise can be reduced (8,9).

The second method is less sensitive to noise and provides "exact" inverse scattering solutions (by exact we mean the equation or theoretical model used for imaging contains no unrealistic or major approximations). The second method, although "exact" in the above sense, is nevertheless iterative in nature and may probably require a later generation of computers for cost-effective implementation. Nevertheless, the method is of academic and laboratory significance and may serve as the starting point for further improvements in ultrasonic imaging; a description follows.

2. ULTRASONIC IMAGING BY SOLUTION OF THE INVERSE SCATTERING PROBLEM

The ultrasound imaging problem in its most general and complete form is equivalent to solving the so-called inverse scattering problem. The inverse scattering problem may be defined as finding the spatial distribution γ of material properties so that the wave equation scattered field solutions $f^{(s)}$ for any incident field $f^{(i)}$ matches the given scattered field. The solution to the direct scattering problem is defined to be the solution $f^{(s)}(x)$ to a wave equation at some point x for γ given. Thus, the solution γ to the inverse scattering problem is that γ for which the solution to the direct scattering problem matches given scattered field values. In the usual ultrasound imaging environment only the scattered fields on a detector are given by measurement, but the scattered fields within the object are not known. If the fields were known within the object, then the solution to the inverse scattering problem could be obtained by a linear operation. Since both the internal field and material property are unknown, the solution to the inverse scattering problem also requires, at least implicitly, the solution to the direct scattering problem.

Because of the past difficulty of solving the inverse scattering problem, approximate solutions have been developed. One of the most widely known approximation methods is based upon using first order perturbation theory to linearize the procedure for solving the wave equation as a function of γ. This approach leads to the derivation of the well-known Born and Rytov approximations (5). For the Born approximation, equations for the perturbation in the acoustic pressure field P and perturbation in material properties γ are given by the equations

$$p = P_0 + P_1 \tag{1}$$

$$\gamma = \gamma_0 + \gamma_1 \tag{2}$$

where p is the total field when $\gamma = \gamma_0 + \gamma_1$ and P_2 is the field when $\gamma = \gamma_0$ where γ_0 is a constant and is the material property of the homogeneous background (5). The Rytov approximation is obtained by the transformations

$$p = \exp(W) \tag{3}$$

$$W = W_0 + W_1 \tag{4}$$

where W_0 is the logarithm of the incident field when $\gamma_1 = 0$, where W is the logarithm of the total field when $\gamma = \gamma_0 + \gamma_1$ and where W_1 is the perturbation in W caused by the perturbation γ_1 in γ.

For medical applications, the assumption that γ_1, P_1, and W are small with respect to γ_0, p_0, and W_0, respectively, may in many cases be erroneous because of refraction and attenuation. This may be seen by inspection of the Helmholtz wave equation

$$\nabla^2 p(\underline{x}) + \frac{\omega^2}{[c(\underline{x})]^2} p(\underline{x}) = 0 \tag{5}$$

The speed of sound $c(\underline{x})$ may be considered to be complex and we could write $c(\underline{x}) = c_R(x) + ic_I(x)$. However, it is more convenient to first expand $[c(\underline{x})]^{-2}$ by equation (2) before examining the real and imaginary parts. By adding $k_0^2 F(\underline{x}) p(\underline{x}) = \omega^2 (c_0^{-2} - [c(\underline{x})]^{-2})\, p(\underline{x})$ to both sides of equation (5) we obtain

$$\nabla^2 p(\underline{x}) + k_0^2 [1 - F(\underline{x})] p(x) = 0 \tag{6}$$

where $k_0^2 = \omega^2/c_0^2$ is the background wave number squared. Thus by inspection of equation (2) we see that $\gamma_0 = k_0^2$ and $\gamma_1 = F(\underline{x}) k_0^2$. We now consider two examples which illustrate the effect of $F(\underline{x})$ on the relative sizes of P_0 and P_1.

The first example is that of a simple lens. It is clear that a small value of $F(\underline{x})$, say five or ten percent, distributed in a lens-shaped region, can focus an incident plane wave to a local zone where the total field P may be many times larger than the unperturbed field P_0. Thus, the scattered or perturbed field P_1 cannot be small in comparison to P_0. The second example is that of a plane wave normally incident upon a block of homogeneous material with a small but finite loss corresponding to the imaginary part of $F(\underline{x})$. It is easily shown that even though the imaginary part of $F(\underline{x})$ is small, the line integral of the linear acoustical attenuation coefficient can reduce the amplitude of the internal field to a small fraction of the incident field in only 10 cm of travel. Soft tissue has an attenuation of only about $1\ \text{db cm}^{-1}\ \text{MHz}^{-1}$. For an average speed of sound of 1500 m. sec^{-1} and at 1 MHz, $1\ \text{dbMHz}^{-1}\ \text{cm}^{-1}$ corresponds to a ratio of imaginary to real part of the wave number of about 0.005. Nevertheless, at 1 MHz at a depth of 10 cm in the block the field is down 10 db relative to the field at the surface, that is, the pressure is reduced by 68 percent.

These examples illustrate the need for caution when applying the Born or Rytov theory. The situation can be improved if the background fluid is tailored to have a real and imaginary speed of sound equal to the mean respectively of the real and imaginary parts of the speed of sound for the object in the background fluid. A method for obtaining an inverse scattering solution to the exact Helmholtz wave equation or which does not require actual field to be nearly equal to the incident field with no object present, would avoid difficulties illustrated by the two examples above. We next review the background of a method for obtaining a direct solution to the exact Helmholtz wave equation which we will use to develop a robust inverse scattering solution to the exact Helmholtz wave equation.

Methods for solving the direct scattering problem by the method of moments, of which the Galerkin method is a special case, are presented in the book by Harrington (10) and the paper by Richmond (11). Recently the method of Richmond was adapted independently by Hagmann (12) and by Yoon et al (13) to solve the inverse scattering problem for the case of a single source. They found that the inverse scattering problem for a single source is an ill-posed problem and therefore very small amounts of noise in the data caused a large amount of noise in the solution for γ. As the number of points in the image domain grows, the condition number of the matrices to be inverted by the single source method grows rapidly and the solution becomes unstable and inaccurate. The stability of the solution also decreased as the detector to object distance was increased.

The ill-posed nature of the inverse scattering solution may be largely removed by solving a system of non-linear equations resulting from using multiple sources (14). The multiple source solution has demonstrated good tolerance to additive noise in the scattering data and is not sensitive to the position of the detection for over-determined cases.

3. FORMULATION OF EQUATIONS WHICH MAY BE SOLVED FOR DIRECT AND INVERSE SCATTERING SOLUTIONS

We now introduce an improved wave equation containing equation (6) as a special case and introduce an equivalent integral equation. We then transform the integral equation into a set of multivariate quadratic equations. The solution to these quadratic equations lends to the direct scattering solution and the inverse scattering solution.

The wave equation for a body immersed in a homogeneous fluid is given by (1)

$$\nabla^2 f_\emptyset(\underline{x}) + k_0^2 f_\emptyset(\underline{x}) = -k_0^2 \gamma_\kappa(\underline{x}) f_\emptyset(\underline{x}) - \nabla\cdot[\gamma_\rho(\underline{x})\nabla f_\emptyset(\underline{x})] \tag{7}$$

where $f_\emptyset(\underline{x})$ is a scalar field such as acoustic pressure for a source at position $\emptyset$, $k_0 = \omega/c_0$, ω = angular frequency of incident field, c_0 is constant speed of sound in a homogeneous background fluid, $\gamma\kappa = (\kappa - \kappa_0)/\kappa_0$ and $\gamma_\rho = (\rho - \rho_0)/\rho_0$ are the fractional change in compressibility κ and density ρ, respectively, outside the body in the fluid background. A useful integral equation which is equivalent to (7) and includes the boundary conditions is (1)

$$f_\emptyset^{(s)}(\underline{x}) \triangleq f_\emptyset(\underline{x}) - f^{(i)}(\underline{x}) = \int [k_0^2\gamma_\kappa(\underline{x}')f_\emptyset(\underline{x}')g(\underline{x}';\underline{x}) + \int \gamma_\rho \nabla' f_\emptyset(\underline{x}')\cdot\nabla' g(\underline{x}';\underline{x})]d^Q\underline{x}' \quad (8)$$

where $\underline{x} = (x_1,\ldots,x_Q)$, $\underline{x}' = (x_1',\ldots,x_Q')$ are spatial coordinates in a space of dimension $Q \leq 3$, $f_\emptyset^{(s)}(\underline{x})$ is the scattered field $f_\emptyset(\underline{x}) - f_\emptyset^{(i)}(\underline{x})$, $f_\emptyset^{(i)}(\underline{x})$ is the field $f_\emptyset(\underline{x})$ for no object present; i.e. for $\gamma_\kappa = \gamma_\rho = 0$, and where $g(\underline{x}';\underline{x})$ is a Green's function. For two-dimensional problems $g(\underline{x}';x) = H_0^{(2)}(k_0|\underline{x} - \underline{x}'|)$ where $H_0^{(2)}(\cdot)$ is the first order Hankel function of the second kind. For three-dimensional problems $g(\underline{x}';\underline{x}) = (4\pi|\underline{x}-\underline{x}'|)^{-1} \exp(ik_0|\underline{x}-\underline{x}'|)$, and where $|\cdot|$ means absolute value.

We now expand the fields $f_\emptyset, f_\emptyset^{(i)}$, and $f_\emptyset^{(s)}$ and factors containing the material properties using an accurate basis set $\{\psi_i(\underline{x})\}$. Let

$$f_\emptyset(\underline{x}) = \sum_{i=1}^{N} a_{\emptyset i}\,\psi_i(\underline{x}) \quad ; \quad f_\emptyset^{(i)}(\underline{x}) = \sum_{i=1}^{N} b_{\emptyset i}\,\psi_i(\underline{x}) \qquad (9,10)$$

$$f_\emptyset^{(s)}(\underline{x}) = \sum_{i=1}^{N} d_{\emptyset i}\,\psi_i(\underline{x}); \; k_0^2\,\gamma_\kappa(\underline{x}')f_\emptyset(\underline{x}) = \sum_{i=1}^{N} c_{\emptyset i}\,\psi_i(\underline{x}') \qquad (11,12)$$

$$\gamma_\rho(\underline{x}')\nabla f_\emptyset(\underline{x}') = \sum_{i=1}^{N}\sum_{q=1}^{Q} A_{q\emptyset i}\,\psi_i(\underline{x}')\hat{e}_q \,, \qquad (13)$$

where N is the number of basis functions and Q is the dimension of the space and corresponding Green's function, and where $\hat{e}_q$ is a unit vector along coordinate q. Upon substitution of (13) into (8) we obtain upon interchanging integration and summation,

$$\sum_{i=1}^{N}(a_{\emptyset i}-b_{\emptyset i})\psi_i(\underline{x}) = \sum_{j=1}^{N} c_{\emptyset j}g_j(\underline{x}) + \sum_{j=1}^{N}\sum_{q=1}^{Q} A_{q\emptyset j}\,G_{qj}(\underline{x}) \qquad (14)$$

where

$$g_j(\underline{x}) = \int g(\underline{x}';\underline{x})\psi_j(\underline{x}')d^Q \underline{x}' \quad ; \quad G_{qj}(\underline{x}) = \int \frac{\partial g}{\partial x_q} \psi_j \, d^Q \underline{x}' \quad . \tag{15,16}$$

Using the "expansion method" equation (14) could be developed into a system of algebraic equations describing the scattering by evaluating (14) at a set of grid points $\{\underline{x}_j\}$. Alternately, by taking the inner product of (14) with all members of a weighting or testing set of basis functions $\{W_i(\underline{x})\}$ a system of N algebraic equations may be generated by the "method of moments" (10). If $W_i(\underline{x}) \equiv \psi_i(x)$ for all i and $\{\psi_i\}$ are independent then the procedure becomes the classical "Galerkin method" (10). If $W_i(x) \equiv \delta(\underline{x}-\underline{x}_i)$ the method of moments becomes the expansion method. The choice of the expansion basis functions $\{\psi_i(\underline{x})\}$ and the testing functions $\{W_i(\underline{x})\}$ is problem dependent and often significantly effects the accuracy of solution and the number of expansion basis functions needed to obtain a solution (10). The sinc basis functions have many properties which make them especially attractive for applications using the method of moments, and in the remaining sections we will use only basis sets based on the sinc functions.

4. ALGEBRAIC SCATTERING EQUATIONS DERIVED USING SINC BASIS FUNCTIONS

The sinc functions are attractive as a basis set because when properly formulated the basis set is complete, orthogonal and retains some local or regional properties (by local or regional we mean that each coefficient of the basis expansion are related to the values of the function to be expanded in a unique local region, and this property often aids in incorporating measurements or boundary conditions) for representing spatially band limited functions.

In one-dimension a function $f(x)$ may be expanded in a basis set consisting of sinc functions shifted in integral multiples of step size h as follows

$$f(x) \approx C(f,h) \triangleq \sum_{\ell=-\infty}^{\infty} f(\ell h) S(\ell,h)(x) \tag{17}$$

where the shifted sinc functions are given by

$$s(\ell,h)\ (x) \triangleq \frac{[\sin\ (\pi/h)(x-\ell h)]}{(\pi/h)(x-\ell h)} \quad . \tag{18}$$

If f(x) is band limited to contain no Fourier components greater than

π/h then $f(x) = C(f,h)$. The orthogonality of $\{S(\ell,h)(x)\}$ is expressed by

$$\int_{-\infty}^{\infty} C(f,h)(t)S(\ell,h)(t)dt = h\, f(\ell h) \quad . \tag{19}$$

The nth derivative of function $C(f,h)$ has a sinc expansion given by

$$\left(\frac{d}{dx}\right)^n C(f,h)(x) = h^{-n} \sum_{j=-\infty}^{\infty} \{\sum_{\ell} \delta_{\ell j}^{(n)} f(\ell h)\}S(j,h)(x) \tag{20}$$

$$\delta_{\ell j}^{(n)} = \frac{d^n}{dx^n} S(\ell,1)(x)\Big|_{x=j} \quad . \tag{21}$$

In particular

$$\delta_{\ell j}^{(1)} = \begin{cases} 0 & , \quad \ell = j \\ (-1)^{j-\ell}(j-\ell)^{-1} & , \quad \ell \neq j \end{cases} \quad . \tag{22}$$

The cardinal function $C(f,h)$ may be integrated by use of the following formula

$$\int_{-\infty}^{\infty} C(f,h)(t)dt = h \sum_{\ell=-\infty}^{\infty} f(\ell h) \tag{23}$$

These and other useful properties of the sinc function for solving integral equations and for other applications are given by Stenger (15).

We take products of Q sinc functions to form basis sets for problems in Q dimensions. We illustrate this for two dimensions by setting

$$\psi_{i(m,n)}(\underline{x}) = S(m,h)(x_1)S(n,h)(x_2) \tag{24}$$

to obtain

$$f_\emptyset(x_1,x_2) = \sum_m \sum_n f_\emptyset(mh,nh)S(m,h)(x_1)S(n,h)(x_2) \quad . \tag{25}$$

Here $i = i(m,n)$ means i is an index which indexes point (m,n) in a finite two-dimensional array of grid points. Thus by (9-12), (24,25) and by property (17) we obtain relationships for the coefficients $a_{\emptyset i}$, $b_{\emptyset i}$, $c_{\emptyset i}$

as follows

$$a_{\emptyset i(m,n)} = f_{\emptyset}(mh,nh) \quad , \quad b_{\emptyset i(m,n)} = f_{\emptyset}^{(1)}(mh,nh) \tag{26,27}$$

$$d_{\emptyset i(m,n)} = f_{\emptyset}^{(s)}(mh,nh), \quad c_{\emptyset i(m,n)} = k_0^2\, \gamma_{\kappa}(mh,nh) f_{\emptyset}(mh,nh) \ . \tag{28,29}$$

The relationship between $A_{q\emptyset j}$, γ_ρ, and $f_\emptyset$ is obtained by noting that $\gamma_\rho \nabla' f \cdot \nabla' g = \gamma_\rho(\underline{x}')\sum_q (\partial f/\partial x_q')\ (\partial g/\partial x_q')$. We examine the factor $\gamma_\rho(\underline{x}')(\partial f/\partial x_q')$ and note that $(\partial f/\partial x_q')$ has a sinc expansion via application of (20,21) to (23). The result is the product of γ_ρ and a sinc expansion of $(\partial f/\partial \underline{x}_q')$. We next perform a sinc expansion on the product by use of (17). These nested expansions are simplified by noting that $S(\ell,h)\ (x = rh) = \delta_{\ell r}$ (where $\delta_{\ell r} = 0$ if $\ell \neq r$ and $\delta_{\ell r} = 1$ if $\ell = r$). Applying these steps for the terms $\gamma_\rho(\partial f/\partial x_1)$ and $\gamma_\rho(\partial f/\partial x_2)$ we obtain on setting $x_1 = x$, $x_2 = y$

$$\gamma_\rho(x,y)\ \frac{\partial f_\emptyset(x,y)}{\partial x} = h^{-1} \sum_m \sum_n \gamma_\rho\ (mh,nh)\ \sum_\ell [\delta_{\ell m}^{(1)}\ f_\emptyset(\ell h,nh)] S(m,h)(x) S(n,h)(y) \tag{30a}$$

$$\gamma_\rho(x,y)\ \frac{\partial f_\emptyset(x,y)}{\partial y} = h^{-1} \sum_m \sum_n \gamma_\rho(mh,nh)\ \sum_\ell [\delta_{\ell n}^{(1)}\ f_\emptyset(mh,\ell h)] S(m,h)(x) S(n,h)(y) \tag{30b}$$

We now have obtained expressions for $a_{\emptyset i}$, $b_{\emptyset i}$, $c_{\emptyset j}$, $d_{\emptyset j}$ and $A_{q\emptyset j}$ in terms of $f_\emptyset(mh,nh)$, k_0^2, $\gamma_\kappa(mh,nh)$ and $\gamma_\rho(mh,nh)$ as per (26-29) and (30). These expressions could be substituted into (14) to obtain a set of equations with coefficients $\psi_i(\underline{x})$, $g_j(\underline{x})$ and $G_{gj}(\underline{x})$ to be evaluated. Instead of evaluating the coefficients at all grid points (mh, nh) we use the Galerkin method to expand $g_j(\underline{x})$ and $G_{qj}(\underline{x})$ in terms of $\Psi_i(\underline{x})$. Let

$$g_j(\underline{x}) = \sum_i g_{ji}\, \Psi_i(\underline{x}) \quad \text{and} \quad G_{qj}(\underline{x}) = \sum_i G_{qji}\, \Psi_i(\underline{x}) \tag{31,32}$$

where $\Psi_i(\underline{x})$ is given by (24). Thus, by (25) the coefficients g_{ji} and G_{qji} are given by

$$g_{ji} = g_j(\underline{x}_i) \quad \text{and} \quad G_{qji} = G_{qj}(\underline{x}_i) \ . \tag{33,34}$$

On substitution of (26-29), (30) and (31,32) into (14) we obtain a set of equations involving only

$$f_\emptyset(mh,nh)\ ,\ k_0^2\ ,\ \gamma_\kappa(mh,nh),\ \gamma_\rho(mh,nh)\ ,\ g_{ji},\ G_{qji}\ ,\ \text{and } \Psi_{i(m,n)}(\underline{x}).$$

Thus on factoring, we obtain

$$\sum_i f_{\emptyset i}^{(s)}\Psi_i \triangleq \sum_i (f_{\emptyset i}-f_{\emptyset i}^{(i)})\Psi_i = \sum_i\sum_j [k_0^2\,\gamma_\kappa(\underline{x}_j)f_\emptyset(\underline{x}_j)g_{ji}+\sum_q A_{q\emptyset j}G_{qji}]\Psi_i \tag{35}$$

where the sum over $A_{q\emptyset j}$ represents the complete expressions for (30). According to the Galerkin method we may eliminate the factor Ψ_i by taking the inner product of (35) with ψ_j. Let us introduce the natural notation $f_\emptyset(mh,nh) \triangleq f_{\emptyset i(m,n)} \triangleq f_{\emptyset(m,n)}$, $\gamma_\kappa(mh,nh) \triangleq \gamma_{\kappa i(m,n)}$, $\gamma_\rho(mh,nh) \triangleq \gamma_{\rho i(m,n)}$, $g_{j(r,t)i(m,n)} \triangleq g_{ji} \triangleq g_{(r,t)(m,n)}$ and $G_{qj(r,t)i(m,n)} \triangleq G_{qji} \triangleq G_{q(r,t)(m,n)}$, where $i = i(m,n)$.

Then by (19) and (30) we obtain

$$f_{\emptyset(m,n)}^{(s)} \triangleq f_{\emptyset(m,n)} - f_{\emptyset(m,n)}^{(i)} = \sum_{r,t}\{k_0^2\,\gamma_{\kappa(r,t)}\,f_{\emptyset(r,t)}\,g_{(r,t)(m,n)} + \gamma_{\rho(r,t)}\sum_\ell[\delta_{\ell r}^{(1)}\,f_{\emptyset(\ell,t)}\,G_{1(r,t)(m,n)} + \delta_{\ell t}^{(1)}\,f_{\emptyset(r,\ell)}\,G_{2(r,t)(m,n)}]\}\ . \tag{36}$$

Equation (36) is the complete equation for describing the direct scattering problem with γ_κ and γ_ρ known and the inverse scattering problem with γ_κ and γ_ρ unknown. Note that the sums over r and t are convolutions because of the definitions of the Green's function $g(\underline{x}';\underline{x})$. The sum over ℓ is also a convolution as seen from (20-22). The sums over r,t and ℓ may be rearranged so that all coefficients of each $f_{\emptyset(r,t)}$ are summed and grouped as a total coefficient. Such a form is useful for direct scattering problems where the γ_κ and γ_ρ are known. This form is given by

$$f_{\emptyset(m,n)}^{(s)} \triangleq f_{\emptyset(m,n)} - f_{\emptyset(m,n)}^{(i)} = \sum_{r,t}\{k_0^2\,\gamma_{\kappa(r,t)}\,g_{(r,t)(m,n)} + \sum_\ell [\delta_{r\ell}^{(1)}\,\gamma_{\rho(\ell,t)}G_{1(\ell,t)(m,n)} + \delta_{t\ell}^{(1)}\,\gamma_{\rho(r,\ell)}G_{2(r,\ell)(m,n)}]\}f_{\emptyset(r,t)}\ . \tag{37}$$

We next present methods for obtaining the direct and inverse scattering solutions from (36) and (37) for the special case of $\gamma_\rho(\underline{x}') = 0$. Even with this restriction it will be clear that the general case of $\gamma_\rho(\underline{x}') \neq 0$ may be obtained by extension of the method. This conceptually simple extension requires additional programming complexity and additional memory for storing the coefficients $G_{(r,\ell)(m,n)}$ and is therefore reserved for future publications.

5. METHODS FOR SOLVING MODEL EQUATIONS FOR CASE OF $\gamma_\rho = 0$

In formulating the direct and inverse scattering solutions for the case $\gamma_\rho = 0$, it will be useful to use the single index notion i to represent a point $\underline{x}_i$ with $i = i(m,n)$, i.e. $\underline{x}_{i(m,n)} = (x_i, y_i) = (mh, nh)$ since the more explicit two index notation (m,n) is now not required. We also emphasize the major role which subscript $\emptyset$ will now play in the inverse scattering solution to obtain unique solutions. In order to obtain a sufficient number of equations to solve for $f_\emptyset^{(s)}$ it will be necessary to also solve for $f_\emptyset(\underline{x})$ inside the body. Thus we must consider two sets of equations, namely, one set relating $f_\emptyset^{(s)}$ anywhere to $f_\emptyset$ inside the body and a second set relating $f_\emptyset$ at point x_i in the body to all other points $\underline{x}_j$ in the body. It will also be useful to write these equations in terms of residuals $r_{1\ell\emptyset}$ and $r_{2m\emptyset}$. Thus we write

$$r_{1\ell\emptyset} \triangleq f_{\emptyset\ell} - \sum_{j=1}^{N} \gamma_{\kappa j} f_{\emptyset j} C_{\ell j} - f_{\emptyset\ell}^{(i)} = 0 \qquad \begin{array}{l} \ell = 1,\ldots,N \\ \emptyset = 1,\ldots,\Phi \end{array} \tag{38}$$

$$r_{2m\emptyset} \triangleq f_{\emptyset m}^{(s)} - \sum_{j=1}^{N} \gamma_{\kappa j} f_{\emptyset j} D_{mj} = 0 \qquad \begin{array}{l} m = 1,\ldots,M \\ \emptyset = 1,\ldots,\Phi \end{array} \tag{39}$$

where $C_{\ell j} = k_0^2 \, g_{j\ell}$ and $D_{mj} = k_0^2 g_{jm}$. The subscript m in (39) refers to a measurement position on a detector. The subscript ℓ in (38) refers to any grid point in a region including the body. By inspection a necessary condition for existence of a common solution to (38) and (39) in that the number of equations exceed the number of unknowns, i.e. that $M\,\Phi \geq N$.

In his study of the direct scattering problem, Richmond (11) obtained a similar set of equations to (38) and (39) above using pulse basis functions. The advantage of the sinc basic functions lies in their economy. Richmond recommends 10 sample points per incident wavelength for his evaluation of $C_{\ell j}$ and D_{mj} to be accurate for problems of low contrast, i.e. $|\gamma_\kappa| \leq .10$. We find that 4 samples per incident wavelength is sufficient for accurate reconstruction of γ_κ for objects whose spatial frequency spectrum of γ_κ does not exceed the reciprocal of the incident wavelength. For solving the direct scattering problem, i.e. finding $f_\emptyset^{(s)}$, we follow the steps in Richmond's method, namely: first, for given γ_κ solve the linear system (38) for $f_{\emptyset j}$, second, for $f_{\emptyset j}$ from step one, and for given γ_κ compute $f_{\emptyset m}^{(s)}$ by evaluating the sum in (39).

Solving the inverse scattering problem is more difficult since both γ_j and $f_{\emptyset j}$ are unknown and are present in (38) and (39) as quadratic terms or products $\gamma_j\, f_{\emptyset j}$. We have investigated (16,17) three methods for solving this non-linear problem: (1) iteratively solving alternate linear systems method, i.e. solving of (38) for $f_{\emptyset j}$ with $\gamma_{\kappa j}$ fixed then solving (39) for $\gamma_{\kappa j}$ with $f_{\emptyset j}$ fixed, (2) non-linear row action methods (18), (3) and optimization techniques (19,20). The last two methods are complex and have not been computationally faster than the first method (17) so we only describe the alternate iterative method (or AIM method for short).

The AIM method for iteratively solving alternate linear systems may be described by the following algorithm: Let $f_\emptyset^{(k)}$ or $\gamma^{(k)}$ be the kth iterate of $f_\emptyset$ or γ respectively. Then

1. Pick a trial value for $\gamma^{(k)}(\underline{x})$ and $f_\emptyset^{(k)}(\underline{x})$. Pick accuracy δ.
2. Solve (38) for $f_\emptyset^{(k+1)}(\underline{x})$ with $\gamma^{(k)}(x)$ fixed.
3. Solve (39) for $\gamma^{(k+1)}$ with $f_\emptyset^{(k+1)}(\underline{x})$ fixed.
4. Test for convergence by forming the objective function.

$$F_{OBJ} = \sum_\emptyset \sum_\ell |r_{1\ell\emptyset}| + \sum_\emptyset \sum_m |r_{2m\emptyset}|. \tag{40}$$

   If $F_{OBJ} < \delta$, go to step 5, else go to step 2.
5. Stop. Print results, make file, display image, etc.

Various methods for solving the linear systems may be employed such as Gauss elimination, fixed point methods (applicable only to equation (38)), e.g.

$$f_{\emptyset \ell}^{(k+1)} = \sum_{j=1}^{N} \gamma_{kj}^{(k)} f_{\emptyset j}^{(k)} C_{\ell j} - f_{\emptyset \ell}^{(i)} \qquad (41)$$

or the Kaczmarz method (21,11).

6. RESULTS OF COMPUTER SIMULATION STUDIES

The performance of the 5-step algorithm for obtaining γ_κ for the case $\gamma_\rho = 0$ was investigated using an 11 by 11 pixel test object. Simulated scattering data was obtained by the following two steps: First solve equation (38) for $f_{\emptyset \ell}$, with $\gamma_{\kappa j}$ given by the values of the test object. Second using the $f_{\emptyset \ell}$ values so obtained and the given $\gamma_{\kappa j}$ values, evaluate $f_{\emptyset m}^{(s)}$ from equation (39). The values of $f_{\emptyset m}^{(s)}$ thus obtained represent noise free data or $f_{TRUE}^{(s)}$. The distribution is normalized such that $||\Delta f^{(s)}||_1/||f_{TRUE}^{(s)}||_1 = \beta$. Here $||\cdot||_1$ means the "L one norm" (also called L_1 norm, $||f||_1 = \Sigma|f_i|$) . Thus noisy data $f^{(s)} = f_{TRUE}^{(s)} + \Delta f^{(s)}$ is obtained for algorithm evaluation.

Figure 1 shows the 11 by 11 test object. Figures 2-3 show the inverse scattering solution obtained using noisy scattering data with $\beta = .10$, the data overdetermined such that $M\Phi = 2.4N$, and the image band limited to have no spatial frequencies higher than the reciprocal of the incident field's wavelength. The detector radius was 10 incident field wavelengths and the pixel separation was one-fourth incident field wavelengths. If a picture quality index or "one norm error rate" $\alpha = ||\gamma_{picture} - \gamma_{TRUE}||_1/||\gamma_{TRUE}||_1$, is computed it is seen that $\alpha = .06$. Without the spatial filtering $\alpha = .17$. These results suggest that with overdetermination of the data, with small contrast, i.e. $|\gamma| \leq .10$, with a small detector radius, and with a small number of pixels (here N = 11 x 11 = 121) the problem is very well posed.

The question of the effect of increasing the number of pixels N and the contrast in γ on the noise properties of the solution image was investigated. We also investigated the effect of increasing the detector radius. This investigation was conducted to determine the inverse scattering solution when the internal fields were known with certainty. An image computed using the 5-step algorithm, could not have less noise than the (a) method using the correct internal fields. The solution to the inverse scattering problem with the correct internal fields given, i.e. evaluating (39) for γ, we call the "pseudo inverse scattering problem" (23). The simultaneous solution of equations (38) and (39) we call the "complete inverse scattering problem" (23).

0	0	1	4	8	10	8	4	1	0	0
0	0	0	0	0	-1	0	0	0	0	0
0	2	10	25	43	52	43	25	10	2	0
0	0	-1	-2	-4	-5	-1	-2	-1	0	0
1	10	36	91	158	190	158	91	36	10	1
0	-1	-3	-9	-15	-19	-15	-9	-3	-1	0
4	25	91	229	398	478	398	229	91	25	4
0	-2	-9	-22	-39	-47	-39	-22	-9	-2	0
8	43	158	398	691	831	691	398	158	43	8
0	-1	-15	-39	-69	-83	-69	-39	-15	-4	0
10	52	190	478	831	1000	831	478	190	52	10
-1	-5	-19	-47	-83	-100	-83	-47	-19	-5	-1
8	43	158	398	691	831	691	398	158	43	8
0	-4	-15	-39	-69	-83	-69	-39	-15	-4	0
4	25	91	229	398	478	398	229	91	25	4
0	-2	-9	-22	-39	-47	-39	-22	-9	-2	0
1	10	36	91	158	190	158	91	36	10	1
0	-1	-3	-9	-15	-19	-15	-9	-3	-1	0
0	2	10	25	43	52	43	25	10	2	0
0	0	-1	-2	-4	-5	-4	-2	-1	0	0
0	0	1	4	8	10	8	4	1	0	0
0	0	0	0	0	-1	0	0	0	0	0

Fig. 1. An 11 by 11 image of a Gaussian test object. The pixels, i.e., grid points, are separated by 1/4 wavelength. Thus, the image is 11/4 by 11/4 wavelengths. Each pair of upper and lower numbers corresponds to a scaled value of γ_k at each grid point. The top number in each pair is 10,000 times the value of the real part of γ_k and the bottom number is 10,000 times the imaginary part of γ_k. Using this test object, simulated scattering data was obtained using the method of moments to solve the direct solution. Ten percent Gaussian random noise was then added to the simulated scattering data, i.e. the ratio of the one norm of the noise to the one norm of the scattering data is ten percent. The simulated scattering data with noise thus generated was used as input for the method of moments inverse scattering solution method. The results of the calculations to retrieve the original object are shown in the next figure.

-16 12	6 0	-2 -4	-2 -2	20 -35	8 43	-9 -22	2 19	20 -2	-29 -14	38 -21
-3 -8	33 12	-17 -6	20 -10	40 45	68 -33	69 42	12 -30	-3 0	49 5	-27 -10
3 30	-18 -15	52 7	91 -14	164 -49	205 1	118 -71	100 22	21 0	-19 -2	21 14
3 -12	38 30	85 -16	215 -25	377 -65	454 -51	458 -20	200 -34	97 -13	26 14	-4 -14
-8 19	37 -39	186 -4	365 -80	706 -61	830 -81	635 -60	103 -24	186 7	18 -16	10 13
18 -5	39 21	204 -43	479 -37	834 -90	1004 -86	839 -72	185 -34	192 -34	54 -13	-5 -25
-2 -21	34 26	157 -30	411 -43	695 -71	825 -98	701 -91	383 -18	163 -14	28 -25	-5 0
8 14	24 -42	131 3	225 -34	373 -33	495 -52	393 -25	234 -29	68 5	61 12	18 0
11 -8	5 16	34 -40	72 9	162 -25	191 -23	179 4	79 1	37 16	-13 -38	-22 28
-18 -9	12 -18	26 41	-11 -13	74 -2	47 6	41 -11	16 -3	23 -31	9 27	15 -31
14 -20	9 15	-2 -25	13 38	15 -11	22 16	-1 -12	20 16	5 24	24 0	-4 23

Fig. 2. Reconstructed image of object shown in Figure 1 using the method of moments without spatial filtering. The image was made from simulated scattered data with ten percent added noise using the test object of Figure 1 and the iterative reconstruction algorithm described in the text. The above results after 20 iterations shows about 17 percent noise, i.e., $\alpha = 0.17$ from equation (48).

0	-2	0	8	17	18	11	3	2	5	6
2	0	-4	-3	1	6	6	1	-4	-8	-8
1	0	4	20	41	49	37	15	2	2	5
4	1	-2	-4	-3	-2	-2	-2	-1	-1	-2
3	8	31	86	153	186	157	87	27	3	3
6	4	-3	-15	-22	-23	-17	-8	0	4	3
5	21	89	234	400	477	406	210	90	18	4
6	1	-11	-36	-53	-52	-36	-18	-7	-2	0
7	33	156	409	688	811	688	112	162	38	8
2	0	-20	-53	-80	-80	-57	-31	-17	-14	-9
8	39	188	491	823	967	817	488	192	45	10
-1	-4	-23	-56	-88	-93	-70	-39	-21	-17	-13
8	36	161	416	698	822	695	110	155	32	8
-1	-4	-18	-45	-73	-82	-61	-34	-14	-9	-8
7	25	95	241	108	187	114	240	83	13	4
0	0	-9	-27	-45	-51	-39	-19	-5	0	0
6	13	35	86	152	187	160	91	30	4	3
0	1	0	-9	-17	-18	-11	-3	0	1	2
4	4	6	18	36	44	37	22	11	7	4
-3	0	4	3	0	0	2	3	2	0	0
3	1	2	9	17	17	12	9	10	10	4
-5	-2	4	7	5	1	0	2	3	2	0

Fig. 3. Reconstruction of image shown in Figure 1 using the method of moments and spatial filtering. This image is the spatial filtered variation of Figure 2. Since Figure 2 is sampled at 4 samples per wavelength of the incident field the image contains noise frequencies up to twice the reciprocal of the incident fields wavelength. Let λ_0 be the incident field's wavelength. Then spatial frequencies in the range from $1/\lambda_0$ to $2/\lambda_0$ may be removed from Figure 2 without affecting the practical resolution limit of the image of γ_k. Figure 3 is the image in Figure 2 filtered in this manner. The noise is reduced considerably since α as defined by equation (48) is reduced from 0.17 in Figure 2 to 0.05 here.

r = 10 λ γ < 0.1 - j0.01 5% noise	50% Over		100% Over		200% Over		300% Over	
	B_U	α	B_U	α	B_U	α	B_U	α
3 x 3	---	---	45.0	.70	.27	.083	.22	.093
5 x 5	220	.91	8.7	.12	.74	.057	.84	.049
7 x 7	$4x10^4$	4.90	220.0	.47	15.00	.140	6.40	.060

TABLE 1.

Upper bound on the norm error B_U and actual norm error α for various size images as function of percent overdetermination of the problem. Constant grid spacing h = 1/4 λ, problem size is NxN, detector radius is at 10 wavelengths. Five percent Gaussian noise was added to the scattered field. A two-dimensional Gaussian test object similar to that in Figure 1 was used.

Our study was composed of the following 6 steps which provide two different bounds or estimates on the noise present in the solution to the complete inverse scattering problem.

1. Generate the scattered field data which would be observed at a detector using a Gaussian object distribution. Define $\underline{f}^{(s)}$ as the complete column vector of ΦM components given by

$$\underline{f}^{(s)} \triangleq \{\underline{f}_1^{(s)}, \underline{f}_2^{(s)}, \ldots, \underline{f}_\emptyset^{(s)}, \ldots, \underline{f}_\Phi^{(s)}\}^T \tag{42a}$$

where

$$\underline{f}_\emptyset^{(s)} \triangleq \{f_\emptyset^{(sc)}(\underline{x}_1), \ldots, f_\emptyset^{(sc)}(\underline{x}_j), \ldots, f_\emptyset^{(sc)}(\underline{x}_M)\} . \tag{42b}$$

2. Generate a Gaussian random noise vector $\Delta\underline{f}^{(s)}$ with a mean of zero and normalized such that

$$\frac{\|\Delta\underline{f}^{(s)}\|_1}{\|\underline{f}^{(s)}\|_1} = \beta . \tag{43}$$

3. Starting with the overdetermined linear system (39)

$$A\underline{\gamma} = \underline{f}^{(s)} \tag{44}$$

the least squares solution for $\underline{\gamma}$ which we accept as "the best solution", is obtained by solving:

$$A^H A\underline{\gamma} = A^H \underline{f}^{(s)} \quad . \tag{45}$$

Here A^H is the complex conjugate transpose of A.

4. Compute the "upper bound on the norm error ratio in $\Delta\underline{\gamma}$," B_U, using the condition number inequality (23, page 431).

$$\frac{\|\Delta\underline{\gamma}\|_1}{\|\gamma\|_1} \leq K_1(A^H A) \frac{\|A^H \Delta\underline{f}^{(s)}\|_2}{\|A^H \underline{f}^{(s)}\|_1} \triangleq B_U \quad . \tag{46}$$

Here $K_1(A^H A)$ is the one norm condition number of $A^H A$.

5. Determine the actual norm error from the true value of $\gamma(\underline{x}_j)$ i.e. $\underline{\gamma}_{TRUE}$, by solving for $(\underline{\gamma}_{TRUE} + \Delta\underline{\gamma}) = \underline{\gamma}_N$ from the equation

$$A^H A(\underline{\gamma}_{TRUE} + \Delta\underline{\gamma}) = A^H(\underline{f}^s + \Delta\underline{f}^s) \quad . \tag{47}$$

6. Determine $\Delta\underline{\gamma}$ by solving (47) for $(\underline{\gamma}_{TRUE} + \Delta\underline{\gamma})$ and compute the "one norm error ratio of the image" or α by

$$\frac{\|\Delta\underline{\gamma}\|_1}{\|\underline{\gamma}_{TRUE}\|_1} \triangleq \alpha \quad . \tag{48}$$

From this type of analysis of the pseudo-inverse problem, we have found the following trends.

1. For a given grid spacing h, the condition number $K_1(A^H A)$ is an exponentially increasing function of the edge dimension $n = N^{1/2}$ of the square imaging grid.

2. Although the condition number and, hence, the error bound in (46) are exponentially increasing with increasing edge dimension, the actual error as computed by (48) does <u>not</u> show this increase with edge dimension.

3. The amount of overdetermination required to make the norm error

in the image as low as the norm error in the scattered field is extensive, as much as 300 percent.

4. For weakly overdetermined systems, increasing the detector radius significantly increases the condition number, but for strongly overdetermined systems, it makes very little difference.

5. Increasing the percentage overdetermination improves (decreases) both the upper bound B_U and the actual norm error α in the image.

6. For a fixed radius and number of pixels and for fixed noise in the scattered field, the condition number and noise in the computed image of γ both decrease as the contrast in the test object is increased. For example, for detector radius R = 10 and 300 percent overdetermined data and for a 7 by 7 image, the condition number and noise are 860 and 6 percent, respectively, for a test object consisting of a centered Gaussian radial distribution having a peak complex value γ of 0.1-i 0.01, but the condition number and noise are 200 and 4 percent, respectively, for the same test object with Gaussian peak of 1.0-i 0.1.

7. SUMMARY

Presently clinical ultrasound imaging is done almost exclusively with the B-scan mode (a 2-D echo strength to brightness mode) with A-scan (an amplitude of echo graph along a single line) or with Doppler scanning (for measuring blood flow velocity). Ultrasound transmission computed tomography has not found clinical application yet because of the low resolution (no better than 5 to 10 mm) of the time of flight and attenuation images. An improved transmission technique called diffraction tomography is based on perturbation solutions to the Helmholtz wave equation and in principle provide improved resolution for very low attenuating objects, but has as yet not been implemented due to significantly large attenuation in tissues. Diffraction tomography is based on the use of the so-called Born or Rytov approximation and therefore does not provide a complete description of wave phenomena.

We have developed a method for solving an exact wave integral wave equation for an inverse scattering solution for both the compressibility and density distributions in a body. Our method depends on the use of multiple incident fields and the representation of the total fields and material properties in a basis function expansion. These techniques are modifications of the moment method or Galerkin method. The solutions obtained by this method, when redundant data are used, are unique and are stable even with noisy data. We illustrate the robustness of our method by presenting a solution image made with 10 percent noisy data. We also provide tabulated data from experiments which show the effect of picture size (number of pixels) and degree of overdetermination on the image quality. The present algorithms used for obtaining a solution require an enormous amount of computation and therefore are too slow and expensive for clinical applications (24). Improvements in algorithm speed (24) and anticipated advances in computer technology could open applications for laboratory scanning and eventually perhaps even for clinical scanning for Galerkin or moment method solutions. The demonstration of the accuracy and robustness of these new methods for a small number of pixels should act as an incentive for further research into method of moment or other more exact approaches to inverse scattering.

ACKNOWLEDGEMENTS

We acknowledge the work of Michael L. Tracy in writing the FORTRAN programs which provided the images in Figures 1 and 2 and for the conducting the study which produced Table 1. A more complete description of this study has been made (17).

Discussion with T.H. Yoon, S.Y. Shin, J.W. Ra, Z.H. Cho, S.B. Park, and G.H. Kim of the Department of Electrical Engineering, Korea Advanced Institute of Science and Technology, Seoul, Korea, are appreciated. Exchanges of information with M.J. Hagmann, now at the National Institute of Health, Bethesda, Maryland, are also appreciated. We appreciate discussion on iterative methods for solving non-linear equations with C. Wilcox, Department of Mathematics, University of Utah. The support of our colleagues M.J. Berggren and D.A. Christensen is appreciated. The editing and typing help of Sylvia Morris in the Department of Mathematics, University of Utah is also appreciated. Support in part from grants PDP-110B from the American Cancer Society and R01-CA1-29728 from the National Cancer Institute are appreciated.

REFERENCES

1. Morse PM and Ingard KV (1968). Theoretical Acoustics. McGraw-Hill Book Company, New York.

2. Greenleaf JF, Johnson SA, Lee SL, Herman GT and Wood EH (1974). Algebraic reconstruction of spatial distributions of acoustic absorption within tissue from their two-dimensional acoustic projections. (In) Acoustical Holography, (Green PS, ed), Plenum Press, New York, Vol. 5, pp 591-603.

3. Greenleaf JF, Johnson SA, Samayoa WF and Duck FA (1975). Algebraic reconstruction of spatial distributions of acoustic velocities in tissue from their time-of-flight profiles. (In) Acoustical Holography, (Booth, Newell ed), Plenum Press, New York, Vol. 6, pp 71-90.

4. Wolf, E (1969). Three-dimensional structure determination of semi-transparent objects from holographic data. Optics Communicator 1 (4):153-156.

5. Mueller RK, Kaveh M and Wade G (1979). Reconstructive tomography and applications to ultrasonics. Proc IEEE 67:567.

6. Kaveh M, Soumekh M, Lu ZQ, Mueller RK (1982). Further results on diffraction tomography. (In) Acoustical Imaging, Plenum Press, New York, Vol. 12, pp 599-608.

7. Norton SJ and Linzer M (1982). Correcting for ray refraction in velocity and attenuation tomography: A perturbation approach. Ultrasound Imaging 4:201-233.

8. Stenger F (1982). Asymptotic ultrasonic inversion based on using more than one frequency. (In) Acoustical Imaging, Plenum Press, New York, Vol. 11, pp 425-444.

9. Stenger F, Berggren MJ, Johnson SA and Wilcox CH (January 1983). Rational function frequency extrapolation in ultrasonic tomography. Submitted to Ultrasonic Imaging.

10. Harrington RF. Field Computation by Moment Methods. The Macmillan Company, Inc., New York; copies may also be obtained from the author in the Electrical Engineering Department, Syracuse University, New York.

11. Richmond J (May 1965). Scattering by a dielectric cylinder of arbitrary cross-sectional shape. IEEE Transactions on Antennas and Propagation, pp 334-341.

12. Hagmann MJ, Gandhi OP and Ghodgaonkar DK (1981). Application of moment methods to electromagnetic biological imaging. MIT's International Microwave Symposium Digest, p 432.

13. Yoon TH, Kim SY and Ra JW (June 1982). Reconstruction of distributed dielectric objects using low-frequency waves. Conference Proceedings of the 1982 International GeoScience and Remote Sensing Symposium, Munich, Germany; IEEE Transactions on GeoScience and Remote Sensing, IEEE Symposium Digest 3:3.1-4.

14. Johnson SA, Yoon TH and Ra JW (Feb 1983). Inverse scattering solutions of the scalar Helmholtz wave equation by a multiple source moment method. Electronics Letters 19(4):130.

15. Stenger F (April 1981). Numerical methods on Whittaker cardinal, or sinc functions. SIAM Review 23(2).

16. Johnson SA and Tracy ML. Inverse scattering solutions by a sinc basic, multiple source, moment method-Part I: Theory. Submitted to Acoustical Imaging.

17. Tracy ML and Johnson SA (May 1983). Inverse scattering solutions by a sinc basic, multiple source, moment method - Part II: Numerical evaluations. Submitted to Acoustical Imaging.

18. Censor Y (Oct 1981). Raw-action methods for huge and sparce systems and their applications. SIAM Review 23(4):446-466.

19. Fletcher R (1980). Practical methods of optimization. (In) Unconstrained Optimization, Vol. 1, John Wiley & Sons, New York.

20. Gill PE, Murray W and Wright MH (1981). Practical Optimization. Academic Press, New York.

21. Kaczmarz S (1935). Angenaherte Auflosung von System Linearer Gleichungen. Bull Akad Polon Science Lett, A. 35, pp 355-357.

22. Herman GT (1980). Image Reconstruction from Projections, the Fundamentals of Computerized Tomography. Academic Press, New York.

23. Ralson A and Rabinowitz P (1978). A First Course in Numerical Analysis, McGraw-Hill Book Company, New York.

24. Fast imaging algorithms based upon Fast Fourier Transform (FFT) implementation of the back projection algorithm

$$\gamma_\ell \quad FFT^{-1}\,\{FFT[\gamma_\ell']/FFT[\mathrm{sinc}^2(k_0|\underline{x}|)]\}$$

where

$$\sum_{i}^{N} \gamma_\ell\, \mathrm{sinc}^2(k_0|\underline{x}_\ell - \underline{x}_i|) \qquad \gamma_\ell' = \frac{2\pi}{\Phi} \sum_{\emptyset}^{\Phi}(1/f_{\emptyset\ell}) \sum_{m}^{M} S_{\Phi m}^{(s)}\, D_{m\ell}^{*}$$

for solving Equation (39) and Fast Fourier Transform implementation of conjugate gradient methods for solving Equation (38) look promising for reducing computational time dependence from $N^{5/2}$ of this paper to $N^{3/2} \log N$. These suggestions were made by S.A. Johnson and D.T. Borup, respectively, Department of Electrical Engineeering, University of Utah, prior to April 1, 1983.

NMR IMAGING: PRINCIPLES, ALGORITHMS AND SYSTEMS

Z.H. Cho, H.S. Kim, C.H. Oh, H.W. Park and S.W. Lee

Department of Electrical Sciences,
Korea Advanced Institute of Science, Seoul, Korea;
Department of Radiology, Columbia University, New York

1. INTRODUCTION

Recently Nuclear Magnetic Resonance (NMR) became a reality and extensive research efforts are underway both at academic institutions and industrial research and development centers all over the world (1-6). It is also becoming apparent, now, that the recently obtained image quality appears to be close to X-ray CT images in resolution and superior in many other aspects, such as contrast sensitivity.

Although similarity seems to exist between X-CT and NMR-CT, basic principles underlying the NMR-CT differ substantially compared with X-CT and complexity involved appears much more with NMR-CT. It appears that our present understanding of NMR-CT imaging is at best pedestrian and nearly unexplored in terms of full utilization of the imaging parameters involved in NMR, such as exact evaluation of the T_1 and T_2 and imaging of these values in a variety of different conditions. Further exploration of new imaging methods, related instrumentation and algorithms appears to be unlimited. In this chapter, we will attempt to integrate basic image formation in NMR imaging and algorithms associated with it so that a coherent approach to the full understanding of NMR imaging can be realized.

From the image reconstruction point of view, NMR images can be formed either by direct mapping or by projection reconstruction. To date, potentially useful direct mapping method is the Fourier imaging (7) while the line-integral (8) and plane-integral (9) projection techniques belong to projection reconstruction.

An interesting aspect of NMR imaging is its diversity in image formation, data collection, and reconstruction. Many different imaging and data processing methods are presently known, and some of them will be discussed in this chapter.

Great advances have been made during the last several years in the area of instrumentation development for whole-body NMR tomography. A magnet, gradient coils, an RF coil, and associated electronics are required to form an NMR tomographic system. A main-field strength of 500 Gauss to 15 KGauss is expected to be used with field gradients of 0.01 to 1 Gauss/cm, formed by x, y and z gradient coils. With these strengths of magnetic fields and gradients, the required RF range is approximately 2 to 70 MHz for proton resonance or density (hydrogen-density or water-content) imaging. Expected resolution in a conventional NMR imaging system depends on the field homogeneity and available strength of field gradients. For example, for a magnet with a 5 KGauss static magnetic field and a homogeneity, ΔH, of 4 parts per million (ppm) (0.02 G), the expected resolution limit is approximately 0.5 mm if one employs gradients of $G_{x,y,}$ or $_z$ = 0.4 Gauss/cm.

A typical NMR tomographic imaging system designed for human imaging is shown in Fig. 1. In this figure, a four-part air-core magnet with a bore hole diameter of 60-70 cm surrounds two pairs of field gradient coils (x and y), one pair of Helmholtz-type z gradient coils, and an RF coil, respectively. Although the configuration of magnet and gradient coil structure may differ depending on the design and type of magnet, i.e., whether magnet is resistive, superconductive, or permanent magnet, the basic concept will remain similar to this for the majority of NMR imaging systems in the future.

2. PRINCIPLES OF NMR TOMOGRAPHY (10,22)

2.A Principles of NMR Physics

NMR was discovered more than thirty years ago (11) and has become an indispensable analytical method and tool in chemistry and physics. Although the basic physics of NMR is well founded and discussions can easily be found elsewhere, we will describe a few topics necessary for the understanding of NMR as an imaging tool.

Most of the materials contain nuclei that are either protons or neutrons or a combination of both (12). Nuclei containing an odd number of protons or neutrons, or both in combination, possess a nuclear "spin" and a "magnetic moment". This situation is equivalent to an aggregation of many small magnets (see Fig. 2). In the real world, many materials are composed of several nuclei, and the most common nuclei are ^{1}H, ^{2}H, ^{7}Li, ^{13}C, ^{31}P, and ^{27}I. Although materials may be composed of nuclei with an even number of protons and neutrons which possess no spin or magnetic

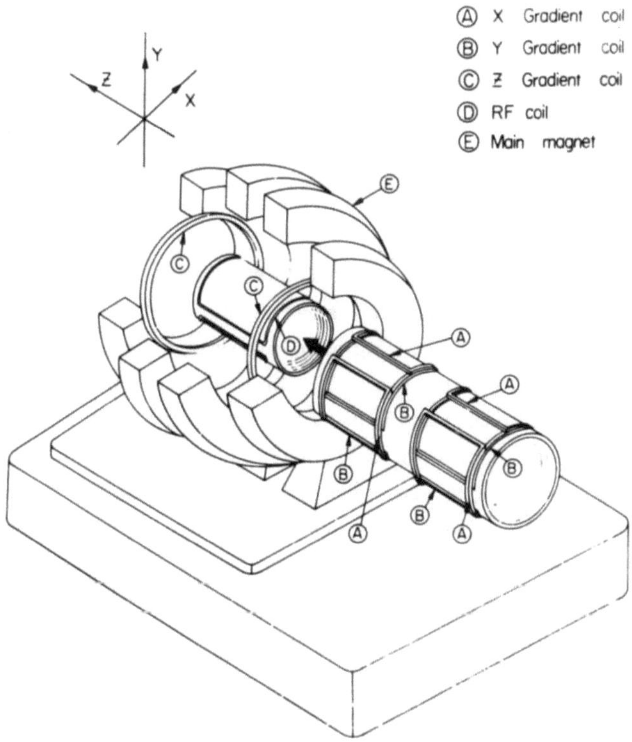

Fig. 1. Sketch of actual NMR tomograph. Locations of gradient and RF coils are shown to illustrate the physical arrangement of an NMR imaging system.

moment, they usually contain some nuclei with an odd number of protons or neutrons as well. Therefore, NMR is practically applicable to meet solid- and liquid-phase materials.

When a material is placed in a magnetic field, some of the randomly oriented nuclei experience external magnetic torque which tends to align the nuclei parallel to the direction of applied magnetic field. The fraction of magnetized nuclei is limited by thermal agitation, and this small probability of magnetization of nuclei at room temperature is related to imaging sensitivity and has been a difficulty in NMR imaging.

The spinning nucleus responds to the external magnetic field like a gyroscope precessing around the direction of the field. The rotating or precession frequency of the spins is usually called the Larmor precession frequency $\vec{\omega}_0$ (12,13). To visualize the behavior of nuclei, three distinct nuclear-spin states associated with surrounding magnetic field situations are illustrated in Fig. 2.

Another important phenomenon of NMR is that the applied external magnetic field creates an energy "absorption state" from a statistical point of view. The proton has intrinsic angular momentum or spin of $h/2$, where h is Planck's constant divided by 2π. When proton nuclei are placed in a magnetic field, the nuclei will be in two energy states (12,13), namely, $+\mu H_0$ and $-\mu H_0$ states (Zeeman splitting), where μ and H_0 are the nuclear magnetic moment and applied magnetic field, respectively (see Fig. 3). At a room temperature with a given magnetic field, number of nuclei or protons having $-\mu H_0$ energy state or low energy state are slightly larger than $+\mu H_0$ or high energy state nuclei, and for those nuclei or protons, irradiation of electromagnetic radiation of energy E equivalent to $2\mu H_0$ will raise the proton energy state up to $+\mu H_0$. This energy is usually supplied by an RF magnetic field H_1. The excited proton then tends to return or relax to its low energy state which produces a "free induction decay" (FID) signal that is the fundamental form of the nuclear signal obtainable from an NMR system.

Two relaxation mechanisms are associated with these excited nuclear spins, namely, transverse or spin-spin relaxation and longitudinal or spin-lattice relaxation, respectively. Transverse relaxation is faster than longitudinal relaxation so that the spin-spin relaxation time constant T_2 is always smaller than the spin-lattice relaxation time constant T_1. It is interesting to note that both of these relaxation times (T_1 and T_2) are sensitive to the molecular bonding and environments surrounding the nuclei. For instance, the mean T_1 values of normal tissues compared with many malignant tissues are substantially different so that it allows us to differentiate malignant tissues from normal ones in many instances.

A similar tendency is observed for T_2 values. Imaging capabilities of these two important parameters, T_1 and T_2, together with the spin densities of the objects thus make NMR imaging a unique, versatile, and powerful technique in medical imaging. Let us now review the few fundamental processes involved in NMR tomographic imaging in a little more detail.

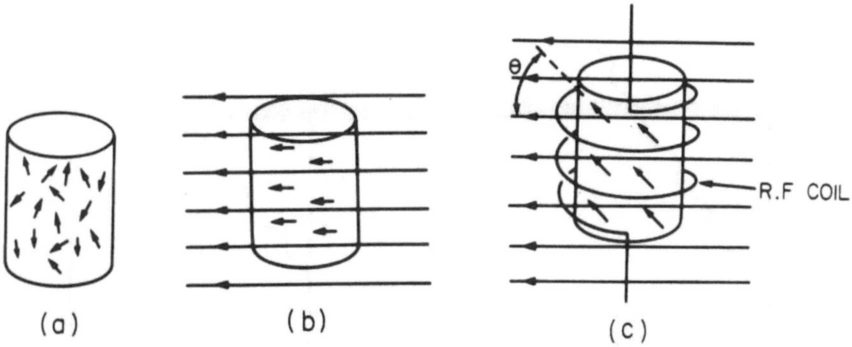

Fig. 2. Behavior of nuclear spins in various situations. (a) Without magnetic field. (b) With strong magnetic field. (c) Additional application of RF pulse with Larmor frequency.

For convenience, a coordinate system is chosen as in Fig. 4, where the magnetic field vector $\vec{H}_o$ is taken in the z axis direction. When the net magnetization vector $\vec{M}$ is at an angle θ to $\vec{H}_o$, the net energy of the system is (12,13)

$$E = -\vec{M} \cdot \vec{H}_o = -MH_o \cos\theta, \tag{1}$$

and the equilibrium magnitude of the net magnetization is

$$M_o = N(-\gamma h)^2 H_o I(I+1)/3KT_o, \tag{2}$$

where N is the number of spins, γ is the gyromagnetic ratio, I is the spin quantum number, K is Boltzmann's constant, and T_o is object temperature. The signal strength which is proportional to M_o may be increased by increasing the field strength H_o. Lowering T_o, which is usually not possible in medical applications, would also improve the equilibrium magnetization.

2.B Bloch Equation (24)

The magnetization, $\vec{M}$, placed in a magnetic field $\vec{H}$ is governed by Bloch equation as follows

$$\frac{d\vec{M}}{dt} = \gamma\vec{M} \times \vec{H} - (M_x\hat{x}+M_y\hat{y})/T_2 - (M_z-M_o)\hat{z}/T_1, \tag{3}$$

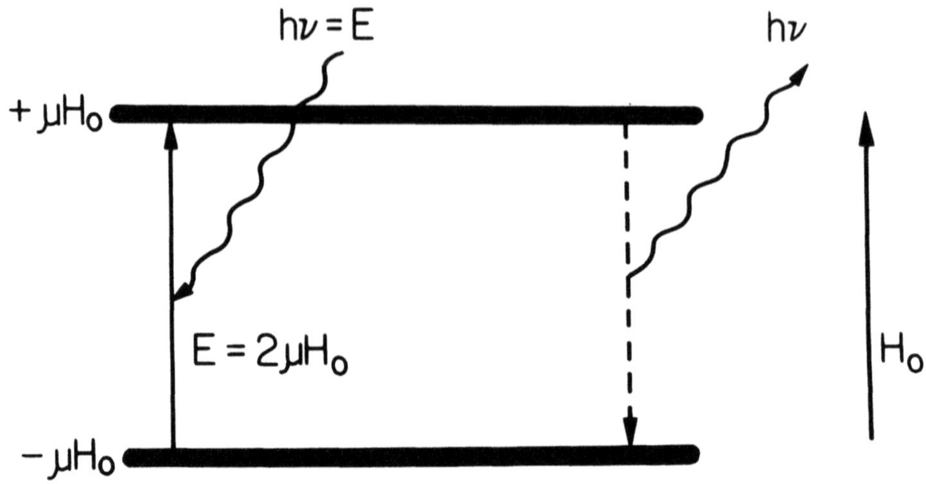

Fig. 3. Quantum mechanical description of NMR spin excitation and eventual emission of nuclear signal. Energy $\hbar\nu$ is supplied by RF pulse through RF coil.

where T_1 and T_2 are the relaxation time constants which represent the spin-lattice and spin-spin relaxation mentioned earlier.

The term $\vec{M} \times \vec{H}$ gives rise to the precessional motion mentioned earlier. To see how this comes about, we take $\vec{H} = H_o \hat{z}$, and temporarily ignore the T_1 and T_2 terms of Eq. (3). Initially, the magnetization has components M_x^o, M_y^o, and M_z^o. Then

$$\frac{d\vec{M}}{dt} = \gamma\vec{M} \times H_o\hat{z}. \tag{4}$$

In x,y,z Cartesian coordinates, Eq. (4) appears as (left) and their corresponding solutions as (right)

$$\frac{dM_x}{dt} = \gamma H_o M_y \qquad M_x(t) = M_x^o \cos\omega_o t - M_y^o \sin\omega_o t$$

$$\frac{dM_y}{dt} = \gamma H_o M_x \qquad M_y(t) = M_x^o \sin\omega_o t + M_y^o \cos\omega_o t$$

$$\frac{dM_z}{dt} = 0. \qquad M_z(t) = M_z^o , \tag{5}$$

where ω_o is the Larmor frequency which plays a key role in NMR imaging and is a strong function of given static magnetic field H_o for a given nuclei whose gyromagnetic ratio is γ, i.e.,

$$\omega_o = -\gamma H_o \ . \tag{6}$$

Including relaxation times the solution becomes

$$\begin{aligned}
\tilde{M}_x(t) &= e^{-t/T_2} \quad M_x(A) \\
\tilde{M}_y(t) &= e^{-t/T_2} \quad M_y(A) \\
\tilde{M}_z(t) &= M_z^o \ e^{-t/T_1} + M_o(1-e^{-t/T_1}).
\end{aligned} \tag{7}$$

where M_z^o is the z-component magnetization of the spin at the moment of excitation.

To observe the behavior of magnetization in RF field, it is convenient to see it in a (uniformly) rotating frame. Let $M_{x'}$ and $M_{y'}$ as the x' and y' components of magnetization in the rotating frame which rotates around z axis with frequency ω, then

$$\begin{aligned}
M_x(t) &= M_{x'}\cos\omega t - M_{y'}\sin\omega t \\
M_y(t) &= M_{x'}\sin\omega t + M_{y'}\cos\omega t.
\end{aligned} \tag{8}$$

Then the differential equations satisfied by $M_{x'}$ and $M_{y'}$ are

$$\begin{aligned}
\frac{dM_{x'}}{dt} &= (\gamma H_o + \omega)M_{y'} \\
\frac{dM_{y'}}{dt} &= -(\gamma H_o + \omega)M_{x'} + \gamma H_1 M_z \\
\frac{dM_z}{dt} &= -\gamma H_1 M_{y'} \ .
\end{aligned} \tag{9}$$

A conventional value to take for ω is ω_o, the Larmor frequency of the static field. Then, using the Larmor equation $\omega_o = -\gamma H_o$, the solution of Eq. (9) becomes

$$\begin{aligned}
M_{x'}(t) &= M_{x'}(0) \\
M_{y'}(t) &= M_{y'}(0) \cos\omega_1 t - M_z(0) \sin\omega_1 t \\
M_z(t) &= M_{y'}(0) \sin\omega_1 t + M_z(0) \cos\omega_1 t,
\end{aligned} \tag{10}$$

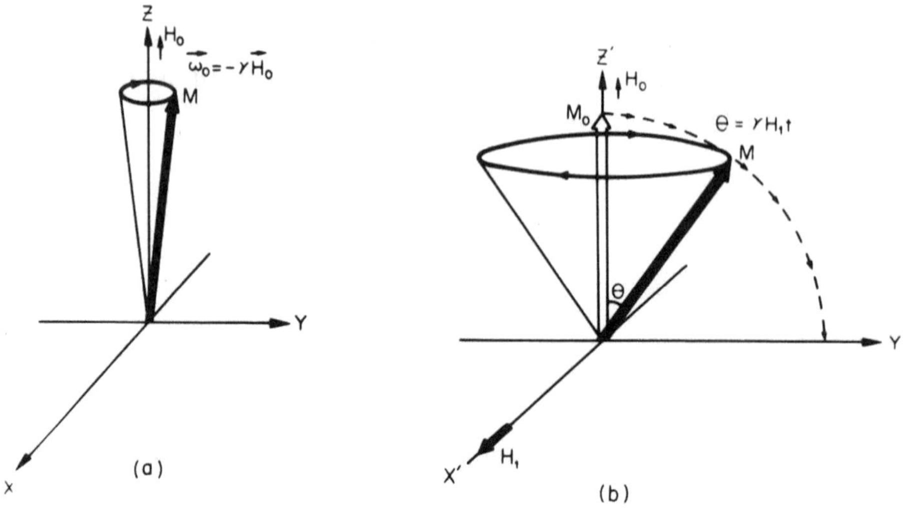

Fig. 4. Spin orientations with and without RF pulse.
(a) Spin in the absence of RF pulse. (b) Spin with an application of suitable size of RF pulse. Usually RF pulse which could tilt the spins to 90° is used for the selection of a line or a slice (or slices) in NMR tomography.

which says that the magnetization vector rotates around the z axis with angular frequency $\omega_1 = -\gamma H_1$.

The classical flipping angle or precession angle θ is given by

$$\theta = \gamma H_1 t_p, \tag{11}$$

where t_p is the RF pulse duration. Application of an RF pulse which tips the magnetization $\vec{M}$ into the x-y plane results in excitation of the spin system (see Fig. 4). When H_1 is applied along the x' axis for a pulse period of t_p, the spin rotates or flips by an angle θ from the z axis toward the y' axis. In general, θ is set to $\pi/2$ or π depending on the mode of excitation and type of NMR experiments. For the simplest case, $\theta = \pi/2$ is used to observe the maximum transverse component of magnetization.

2.C Relaxation Times

After H_1 is turned off, the rotating magnetization induces a current in the pickup coil surrounding the object. The magnetization then relaxes, through interactions with the environment, to its thermal equilibrium so that the spins realign with the original H_o field direction. At the same time, the transverse component of magnetization which is related to the entropy of the system also decays by a dephasing effect of spin-spin interaction.

In addition to the spin-spin relaxation, there are other dephasing effects due to the magnetic field inhomogeneity and field gradients, the latter is deliberately added in NMR imaging to resolve the spatial distribution of spin density. In fact, they produce shifts in the Larmor frequencies throughout the sample, which results in a phase incoherency that eventually makes the composite sinusoidal signal decay faster than the inherent transverse relaxation time T_2. The effective transverse relaxation time resulting from this field inhomogeneity effect alone is given by (13)

$$\frac{1}{T_2^*} = \frac{1}{T_2} + \frac{\gamma \Delta H}{2}, \tag{12}$$

where ΔH is the field inhomogeneity. When a field gradient is added to resolve the spatial distribution of spin density, T_2^* is further reduced and Eq. (12) becomes (22)

$$\frac{1}{T_2^{**}} = \frac{1}{T_2^*} + \gamma GR, \tag{13}$$

where G (in Gausses per centimeter) is a field gradient and R (in centimeters) is the object radius. The composite sinusoidal signal decaying with an effective transverse relaxation time T_2^{**}, in NMR imaging is known as FID, which is usually detected with a phase sensitive detector. These signals look similar to a decaying demodulated AM signal.

Concurrently, longitudinal or spin-lattice relaxation causes the spins to realign in the $\vec{H}_o$ (or z) direction since it is the lowest energy state or thermal-equilibrium state. Since it involves energy dissipation through the lattice, the longitudinal relaxation time T_1 is usually larger than T_2 and related to the z component of magnetization as

$$M_z(t) = M_o[1-(1-M_z^o/M_o)e^{-t/T_1}] , \tag{14}$$

where M_z^o is the z component of magnetization at the starting time of relaxation.

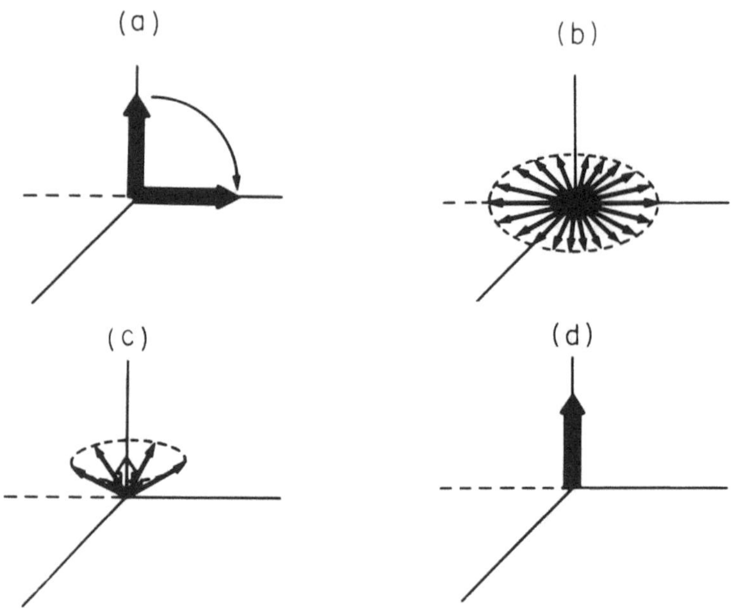

Fig. 5. Sequential illustrations of the spin relaxation process.

The two processes work simultaneously and vary greatly depending on the characteristics of the material. In the case of human tissues, for instance, T_1 and T_2 are in the order of 0.5 s and 50 ms, respectively, in the magnetic field of about 1 kG. T_1 cannot be smaller than T_2, since if all the spins are aligned in the z direction, there is no transverse component. In Fig. 5, sequential behaviors of the relaxation processes are shown. In general, T_1, T_2^*, T_2^{**} have the following relation:

$$T_2^{**} \leq T_2^* \leq T_2 \leq T_1. \tag{15}$$

Another important topic to be introduced is the spin-echo process which plays an important role in NMR imaging as readers will find in the following.

There are two basic forms of the spin-echo techniques applicable to NMR imaging, namely, the Hahn spin-echo technique and the Carr-Purcell Meiboom-Gill (CPMG) technique (12,13). The principle of the Hahn spin-echo technique is as follows (see Fig. 6): First a 90° RF pulse with suitable pulse length is applied along the x' axis and consequently magnetization vectors flip to the y' axis. The spin magnetizations then dephase in time due to field inhomogeneity or added field gradients. Subsequent application of a 180° RF pulse again along the x' axis rotates the spins as shown in Fig. 6(b). The spin magnetizations now continue processing but start rephasing. This rephasing process is equivalent to a regrowth of the FID signal and signal will grow close to the original amplitude at twice the dephasing time depending on the T_2 decay. At this instant, all the spins are completely rephased along the -y' axis. Alternatively, a 180° pulse can be added along the y' axis which will result in spin flips as shown in Fig. 6(f). This method is called CPMG technique and often used for non-negative spin-echo sequence.

3. IMAGE FORMATION ALGORITHMS (22-24)

3.A Direct Fourier Transform Imaging (7)

First direct Fourier imaging method was proposed by Kumar, Welti, and Ernst (KWE) (7). In this case, imaging can be proceeded by total 3-D excitation of an object in three series of time sequences and the result of 3-D Fourier transform of those data is considered as the 3-D spin density function.

In Fig. 7, a basic form of RF pulsing and field gradient sequences of the 3-D KWE method is illustrated. In this case, three orthogonal field gradients G_x, G_y, and G_z are applied in sequence. The z component of the local magnetic fields are, then, given as

$$H_z(x,y,z) = \begin{cases} H_o + G_z z, & \text{for} \quad 0 < t < t_z \\ H_o + G_y y, & \text{for} \quad t_z < t < t_z + t_y \\ H_o + G_x x, & \text{for} \quad t_z + t_y < t \, , \end{cases} \tag{16}$$

where each time scale t_z and t_y is varied according to preassigned

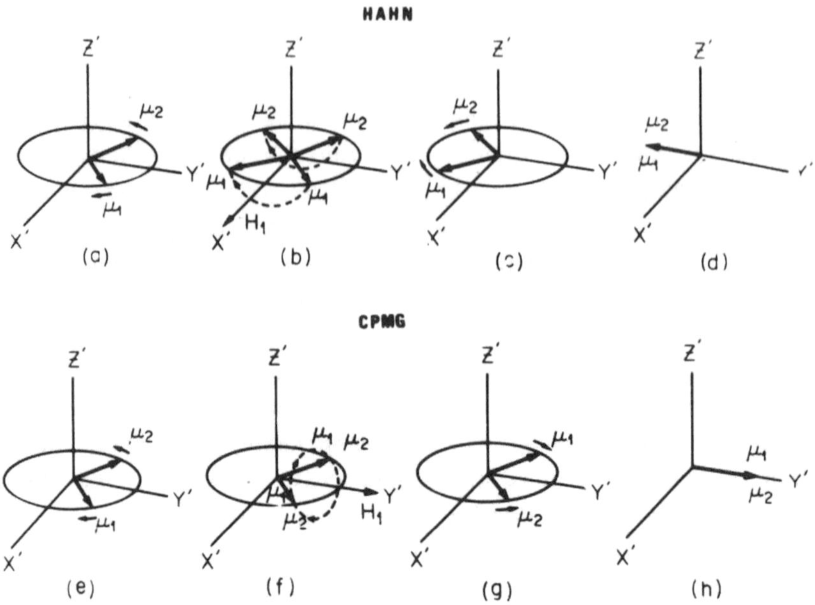

Fig. 6. Hahn spin-echo; (a) spin magnetizations dephase, (b) pulse is applied along x' axis, (c) the spins are being (rephased)-refocused, (d) spin-echo is generated along y' axis. Carr-Purcell and Meiboom-Gill (CPMG) spin-echo; (e) spins are dephased, (f) 180° pulse is applied along y' axis, (g) the spins are being refocused, (h) spin-echo is generated along y' axis.

sequences, i.e., $t_z \simeq 0 \sim t_a$, $t_y \simeq t_a \sim 2t_a$, and t_a is the optimal observation time of FID. The series of the FID signals obtained in t_x periods, then, will form a full 3-D FID signal set sufficient for reconstruction of the entire volume spin density.

Observed FID signal $s(t_x, t_y, t_z)$ is, neglecting relaxation process, given by

$$s(t_x,t_y,t_z) = M_o \int_x \int_y \int_z \rho(x,y,z)\cdot \exp\,[i(G_x x t_x + G_y y t_y + G_z z t_z)]\, dz dy dx. \tag{17}$$

Fourier transform of Eq. (17) results in spatial spin density function

$\rho(x,y,z)$, i.e.,

$$f(\omega_x,\omega_y,\omega_z) = \int\limits_{t_x}\int\limits_{t_y}\int\limits_{t_z} s(t_x,t_y,t_z)\exp[-i(\omega_x t_x+\omega_y t_y+\omega_z t_z)]\, dt_z dt_y dt_x$$
$$\equiv \rho\ (x,y,z). \tag{18}$$

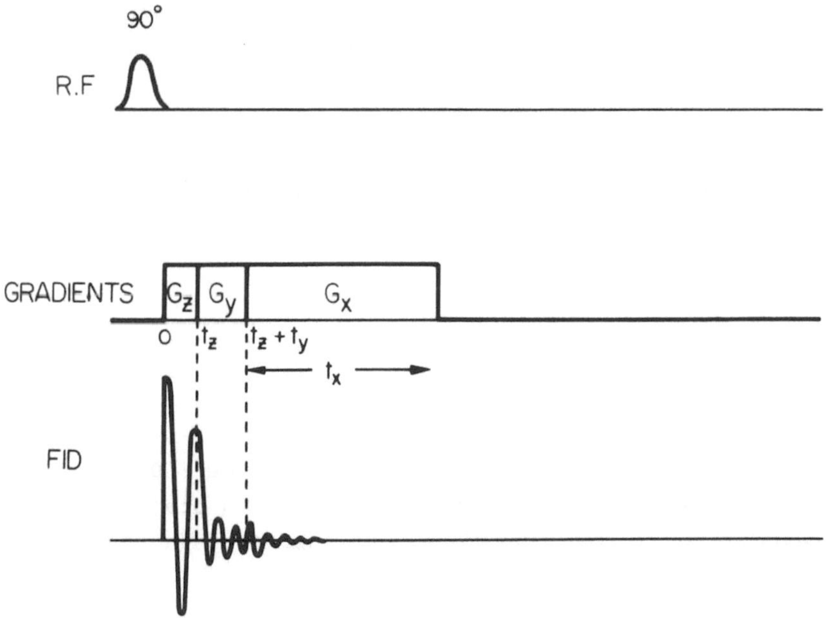

Fig. 7. Imaging sequences of original 3-D KWE direct Fourier imaging $n\Delta t_z$ and $n\Delta t_y$ sequences are required for a complete 3-D volume imaging.

3.B Line-integral Projection Reconstruction (LPR)

(a) <u>Basic principles of LPR</u>: Projection reconstruction using a 2-D and 3-D image reconstruction algorithm is well developed, especially in the areas of X-CT and radionuclide emission tomography (14-21). Although there are several different ways of reconstructing the image, the basic form of data collection is similar, i.e., line-integral projection data of parallel or fan mode are obtained in angular steps, by rotating the object around 180° or 360°. The most familiar and convenient method of reconstructing 2-D or 3-D images is the Fourier convolution method (14). This Fourier convolution method may be summarized as follows. The reconstructed 2-D image $f(x,y)$ is given by

$$f(x,y) = \int_0^{\pi} d\theta \, [p_\ell(x',\theta) * \xi(x')], \tag{19}$$

where $p(x',\theta)$ is projection data, $\xi(x')$ is the filter kernel which corrects for 1/r blurring that results from circular symmetric linear superposition (14,17), and (x',y') is the coordinate system rotated by θ from (x,y).

In Fourier transform NMR, the nuclear signal (FID) can be considered as the inverse Fourier transform of the spatial domain spin density function by its own physical process, i.e., FID signal at a given angular view θ is given by (10,22)

$$s_\theta(t) = M_o \int_{x'} \int_{y'} \rho(x',y';z_o) \exp[i\gamma x' G_{x'} t] \, dy' dx', \tag{20}$$

provided a plane at $z = z_o$ is selected. Although the FID signal $s_\theta(t)$ appears in time domain, it represents Fourier domain projection data. Projection data $p_\ell(x',\theta)$ is, therefore, obtained through the Fourier transform of the FID signal, i.e.,

$$p_\ell(x',\theta) = F[s_\theta(t); t \rightarrow x']. \tag{21}$$

The basic form of projection data obtainable in Fourier transform NMR is similar to the data obtained by X-CT. In Fig. 8, FID signals obtained at different angular views are shown together with the field gradient and RF excitation sequences. The RF signal applied (see Fig. 8) contains only one frequency component in this case and satisfies the 90° spin flip. After $0 \sim \pi$ rotation of projection in an appropriate step by adjustment of the field gradients G_x and G_y, a complete projection data set sufficient for reconstruction of a slice at a given plane z_o is obtained. At this point, 2-D image reconstruction can proceed according to Eq. (19).

Each FID signal $s_\theta(t)$ is Fourier transformed, convolved with a filter kernel, and backprojected.

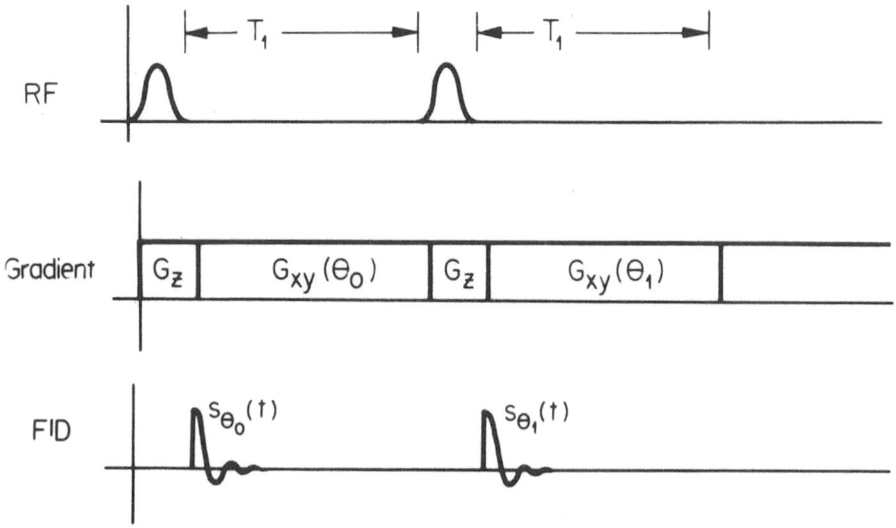

Fig. 8. FID, RF pulse timing, and field gradient sequences in the basic single-slice line-integral projection reconstruction.

(b) Multiplane line-integral projection reconstruction by plane encoding: This method is a combination of simple line-integral projection reconstruction and plane-encoding technique. The principle of this method is illustrated in Fig. 9. As is shown in Fig. 9, for the data set at a view θ_i, the same G_{xy} and G_z are applied n times each with a different frequency composition of RF pulses. To obtain a complete view data of n planes, data acquisition is repeated n times.

The key to this method is the coding and encoding of the signals according to the composite RF pulse sequence. Simple illustration of the encoding procedure using the Hadamard coding matrix can be given as follows. (See also Fig. 9). Let FID's obtained at each 90° composite RF pulse sequence be $S^1_{\theta_o}(t)$, $S^2_{\theta_o}(t)$, ..., $S^n_{\theta_o}(t)$. Each FID is then a composite line-integral projection set which includes data from several planes at an angular view θ_o, i.e., $S_{\theta_o z_o}(t)$, $S_{\theta_o z_1}(t)$, etc. Composite FID's, $S^1_{\theta_o}(t)$, $S^2_{\theta_o}(t)$,.., $S^n_{\theta_o}(t)$ are, therefore, given as

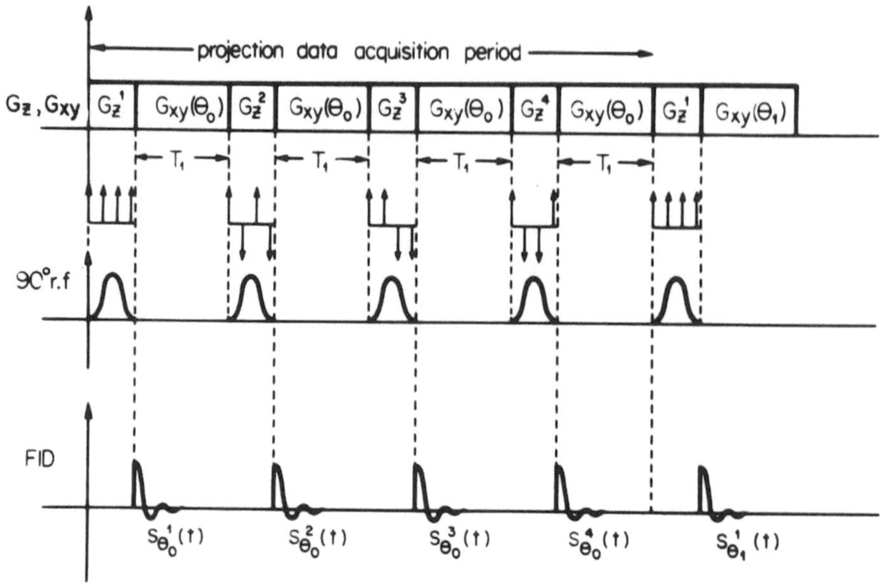

Fig. 9. Pulse and gradient sequences of the multi-plane line integral projection reconstruction using plane-encoding (four-plane encoding).

$$\underline{s}_{\theta_o}(t) = \begin{bmatrix} s^1_{\theta_o}(t) \\ s^2_{\theta_o}(t) \\ \vdots \\ s^n_{\theta_o}(t) \end{bmatrix} = \begin{bmatrix} H_n \end{bmatrix} \cdot \begin{bmatrix} s_{\theta_o z_o}(t) \\ s_{\theta_o z_1}(t) \\ \vdots \\ s_{\theta_o z_{n-1}}(t) \end{bmatrix} . \qquad (22)$$

From Eq. (22), a desired FID signal $s_{\theta_i z_j}$ which corresponds to each slice z_j can be obtained by matrix inversion. The advantge of this method is again the statistical improvement of the image with increase of total scanning time. This encoding technique can also be extended to other encoding matrices, such as Fourier matrix. More detailed analysis of the encoding method will be published in a separate paper (28).

3.C Plane-integral Projection Reconstruction (PPR) (9)

(a) Basic principles of the total volume excitation and PPR: RF pulse and field gradient sequences of the total volume PPR imaging are shown in Fig. 10. The FID signal of total volume plane-integral projection data in z' direction can be expressed as

$$s(t) = M_o \int_{x'}\int_{y'}\int_{z'} \rho(x',y',z')\exp[i\gamma(H_o+z'G_{z'})t]dz'dy'dx'$$

$$= M_o \int_{z'} \left[\int_{y'}\int_{x'} \rho(x', y', z')dx'dy' \right] \exp[i\gamma(H_o+z'G_{z'})t]dz' \quad (23)$$

where (x',y',z') is the coordinate system rotated from (x,y,z) so that z' direction coincides with the projection direction. With phase sensitive detection, the FID signal obtained is

$$s(t) = M_o \int_{z'} p_p(z')\exp[i\gamma z'G_{z'}t]dz', \quad (24)$$

where $p_p(z')$ is a plane-integral.

In Eq. (24) s(t) are the projection data in the Fourier domain, and the Fourier transform of s(t) are, therefore, spatial domain plane-integral projection data with which reconstruction can be performed. This plane-integral method first proposed by L.A. Shepp is an efficient and powerful total volume imaging technique in NMR (9).

(b) Algorithm for total volume image reconstructions with plane-integral projection data (9): 2-D image reconstruction from 1-D line-integral projection data is a well-known technique and has been successfully applied to X-CT. True 3-D volume image reconstruction of an object with 1-D plane-integral projection data is, however, relatively new (9).

From Fourier transform theory, it can be shown that the 3-D volume image $f(x,y,z)$ can be obtained if the object function $O(\omega_T, \theta, \phi)$ is known, i.e.,

$$f(x,y,z) = \frac{1}{(2\pi)^3} \int_0^{\pi} d\phi \int_0^{\pi} d\theta \int_{\infty}^{\infty} d\omega_T O(\omega_T,\theta,\phi) \exp(-i\omega_T T)\omega_T^2 \sin\theta \quad (25)$$

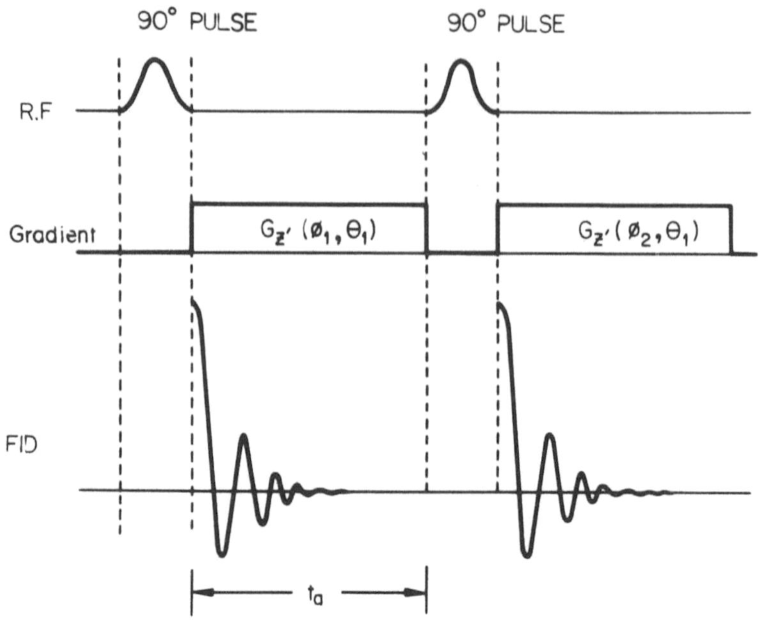

Fig. 10. RF pulse and field-gradient sequences applied to total volume plane-integral projection reconstruction.

where $T = z\cos\theta + (x\cos\phi + y\sin\phi)\sin\theta = z'$.

FID signal obtainable in NMR (Eq. 24) and the basic form of the plane-integral projection data are related as

$$s_{\theta,\phi}(t) = F^{-1}[p_p(T,\theta,\phi);\ T\to\omega_T]$$
$$\equiv \tilde{p}_p(\omega_T,\theta,\phi), \quad \text{for} \quad t = \omega_T, \tag{26}$$

where $p_p(T,\theta,\phi)$ is the plane-integral along T.

From Eq. (26), it is also easy to see that $p_p(T,\theta,\phi)$ is related to $\tilde{p}_p(\omega_T,\theta,\phi)$ as

$$p_p(T,\theta,\phi) = \frac{1}{2\pi}\int_{-\infty}^{\infty}\tilde{p}_p(\omega_T,\theta,\phi)\exp(-i\omega_T T)\,d\omega_T. \tag{27}$$

Double differentiation of $p_p(T,\theta,\phi)$ with respect to T by the Fourier differentiation theorem gives

$$-p''_p(T,\theta,\phi) = \frac{1}{2\pi} \int_{-\infty}^{\infty} \tilde{P}_p(\omega_T,\theta,\phi)\exp(-i\omega_T T)\omega_T^2 d\omega_T. \tag{28}$$

If we assume that $O(\omega_T, \theta, \phi)$ equals $\tilde{P}_p(\omega_T, \theta, \phi)$ in NMR, from Eq. (25) the reconstructed image function can be obtained by

$$f(x,y,z) = -\frac{1}{(2\pi)^2} \int_0^{\pi} d\phi \int_0^{\pi} d\theta \; p''_p(T,\theta,\phi)\sin\theta \tag{29}$$

Although basic principles are the same, there are other plane-integral projection reconstruction methods which can be implemented in a slightly different way (30). One of the methods is direct two-stage method which has substantial computational advantage. In this two-stage method (29), first blurred 2-D superimposed image is reconstructed and final image is derived from this blurred 2-D image by use of conventional 2-D Fourier convolution technique. This latest version indirect method will be published shortly (30).

3.D Imaging Modes

The spin density $\rho(x,y,z)$ obtained in the various imaging methods described above is not a real spin density but rather weighted by T_1 or T_2. T_1 varies greatly between normal and abnormal tissues and the image of spin density weighted by T_1 is found to be clinically useful. With this in mind there have been several attempts to extract T_1 information as well as spin density and T_2. Following are the typical imaging modes and corresponding terminology currently in use.

(a) Saturation recovery: The saturation recovery method involves simply repeating the pulse sequence at regular intervals T. The earlier equations are unchanged except for the replacement of $\rho(x,y,z)$ with $\rho'(x, y, z)$, i.e.,

$$\rho'(x,y,z) = \rho(x,y,z)\,[1-\exp(-T/T_1(x,y,z))] \tag{30}$$

Now ρ' is a function of both T_1 and ρ.

(b) Inversion Recovery: Inversion recovery is similar to the saturation recovery technique except that the 180° RF pulse is added first, then

90° RF pulse follows at time T_I. $\rho'(x,y,z)$ is then related with $\rho(x,y,z)$ as,

$$\rho'(x,y,z) = \rho(x,y,z)\ [1-2\exp(-T_I/T_1(x,y,z))]. \tag{31}$$

It is easy to see that increased dependency of the image upon T_1 over that obtained in the saturation recovery.

(c) Spin echo: By application of 180° pulse following the first 90° pulse at time $t = T_s$, signal regrowth occurs at $t = 2T_s$ by spin echo. $\rho(x,y,z)$ is then replaced with $\rho'(x,y,z)$ which is given by,

$$\rho'(x,y,z) = \rho(x,y,z)\exp[-2T_s/T_2(x,y,z)]. \tag{32}$$

As seen, the image is now weighted with T_2 as is indicated.

4. SOME INSTRUMENTATION PROBLEMS AND REQUIREMENTS (27)

In this section, we will discuss a few key problems arising in NMR imaging due to the pulse shape, finite pulse rise time, and stability factor of the instrument, namely, selective excitation by variety of RF pulses and effects of their pulse shapes, gradient pulse and its finite pulse rise time, and finally the random phase fluctuation effect due to the magnetic field instability. Each topic is analyzed, problems and solutions are indicated. A new scheme is proposed with which the problems can be overcome and both experimental as well as theoretical studies of this new proposed method were carried out.

Finally, experimentally obtained human images with KAIS-NMR tomograph using the proposed technique are shown.

Although the method is applied to line-integral projection reconstruction, its basic concept can be applied to other NMR imaging methods, such as the plane-integral projection reconstruction discussed earlier.

4.A Selection of Slices by 90° RF Pulses

Ideal selective 90° RF pulse which is applied along x' direction flips the spins only in a given frequency band into y' direction and does not affect the spins which are located outside of the given frequency band (12,13). After the RF pulse, the magnetization $\vec{M}$ starts precessing around z axis, and the FID signal obtained becomes,

$$s(t) = \int_{\omega_o - \frac{\Delta\omega}{2}}^{\omega_o + \frac{\Delta\omega}{2}} M(\omega)\exp(i\omega t)d\omega, \tag{33}$$

where $\Delta\omega$ is a frequency band, and $M(\omega)$ is the spatially dependent magnetization which is sometimes noted as spin density function $\rho(\omega)$. By Fourier transform of $s(t)$, $M(\omega)$ is obtained.

In an actual system, however, an ideal selective pulse which provides rectangular frequency band selection does not exist. In other words, the spins in the selected frequency band are not all flipped by the same amount along y' axis so that they possess a unique set of $M_{x'}$ and $M_{y'}$ (25,26). In addition, this non-ideal rectangular RF pulse causes excitation of the spins outside the rectangular frequency band giving spurious unwanted signals.

If we define a complex magnetization $\tilde{M}(\omega)$ as

$$\tilde{M}(\omega) = M_{y'}(\omega) + iM_{x'}(\omega), \tag{34}$$

then, FID signal becomes

$$s(t) = \int_{\omega_o - \frac{\Delta\omega}{2}}^{\omega_o + \frac{\Delta\omega}{2}} \tilde{M}(\omega)\, e^{i\omega t}\, d\omega. \tag{35}$$

Therefore, by Fourier transform of Eq. (35), complex magnetization $\tilde{M}(\omega)$ is obtained.

For the search of an optimum RF pulse shape, spin responses to the different types of RF pulse have been measured and corresponding computer simulations were performed. In the experiment, a small amount of water contained in a vessel was excited by RF pulses of different shapes at near the Larmor frequency. The inhomogeneity within the volume of water was less than 100 Hz and was relatively small compared to the bandwidth of the RF pulse. By Fourier transform of FID, the complex magnetization $\tilde{M}(\omega)$ was obtained and its magnitude and phase were calculated as

$$|\tilde{M}(\omega)| = \left[M_{y'}^{2}(\omega) + M_{x'}^{2}(\omega) \right]^{\frac{1}{2}}$$

and

$$\phi(\omega) = \tan^{-1}[M_{x'}(\omega)/M_{y'}(\omega)] \ . \tag{36}$$

The experiment was repeated changing the frequency for each pulse shape, and the distributions of $|\tilde{M}(\omega)|$ and $\phi(\omega)$ versus frequency were obtained. The experiment was carried out for several pulse shapes and the corresponding results are shown in Fig. 11. At each figure, the RF pulse shape, the magnitude of magnetization v.s. frequency, and the calculated and experimentally obtained phase v.s. frequency are shown at the first, second, and third and fourth column, respectively.

Observing the distributions of the magnitudes and phases of the magnetization v.s. frequency, two problems can be noticed, namely, selectivity definition and phase distribution. In terms of selectivity, full truncated sinc and full Hanning appear most optimum. The phase distribution, however, is very widely spread in both cases.

The experimentally observed results were confirmed by compuer simulations (by solving the Bloch equations numerically, generating FID signals, and then performing the Fourier transform). These simulation results are also shown in Fig. 11 side-by-side with the experimental results. As is seen, the close agreement between the experimental results and simulations is found. It has been suggested that the dephasing effect which results in wide distribution of phase can be corrected by application of gradient reversal (26).

The effects of rephasing by gradient reversal obtained by computer simulation are shown in Fig. 12 for two cases, i.e., for the full Hanning and full truncated sinc pulse, respectively. Gradient reversal with full truncated sinc or full Hanning could provide accurate spin rephasing in an ideal condition, i.e., uniform excitation of the spins in a selected region. As is shown, gradient reversal requires a negative gradient (see Fig. 12) with the same area shown in A and B.

4.B Rise Time Effect of Gradient Pulse on Image Quality

Right after the RF excitation, if a gradient pulse is applied, e.g., along x-direction, an FID signal is obtained during the gradient pulse application which is given by

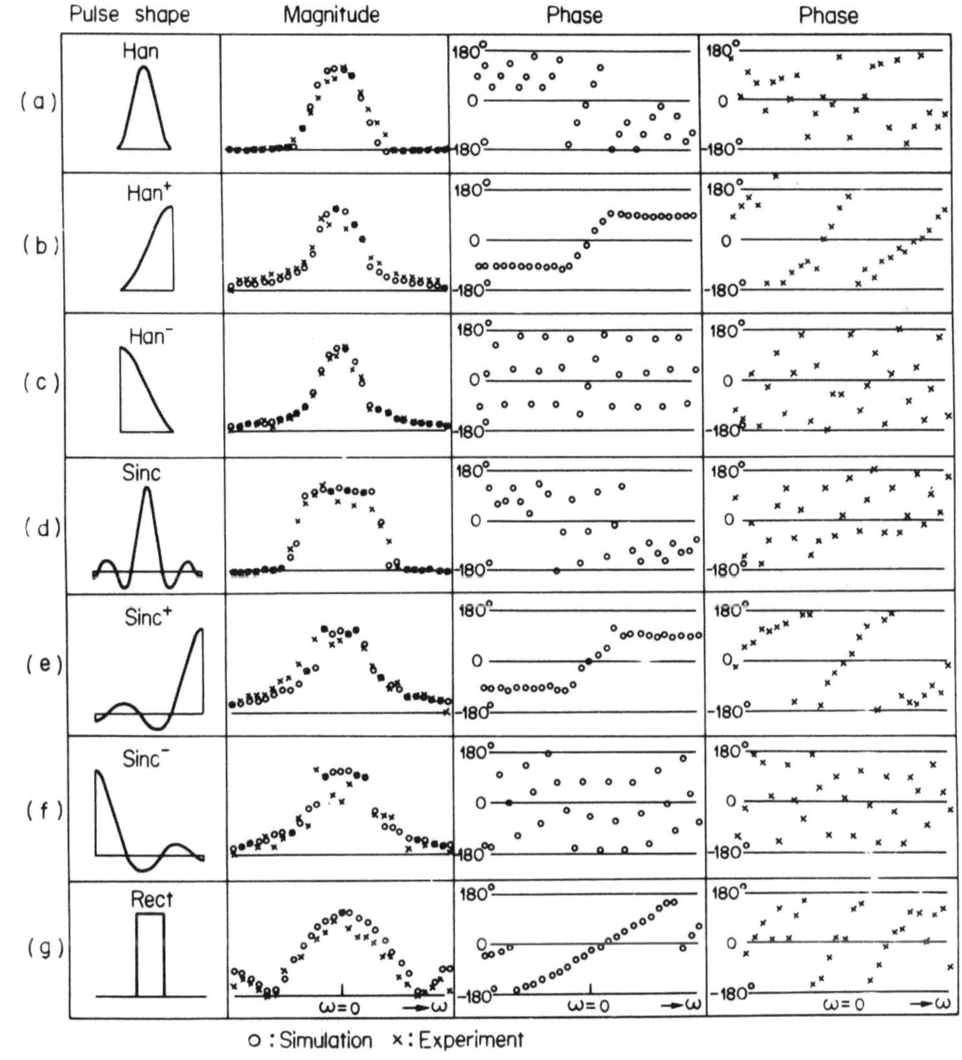

Fig. 11. Frequency domain magnitude and phase distributions as a function of frequency are shown for the several different RF pulse shapes: (a) Hanning, (b) Hanning+ , (c) Hanning- , (d) Sinc, (e) Sinc+ , (f) Sinc- , and (g) Rectangular.

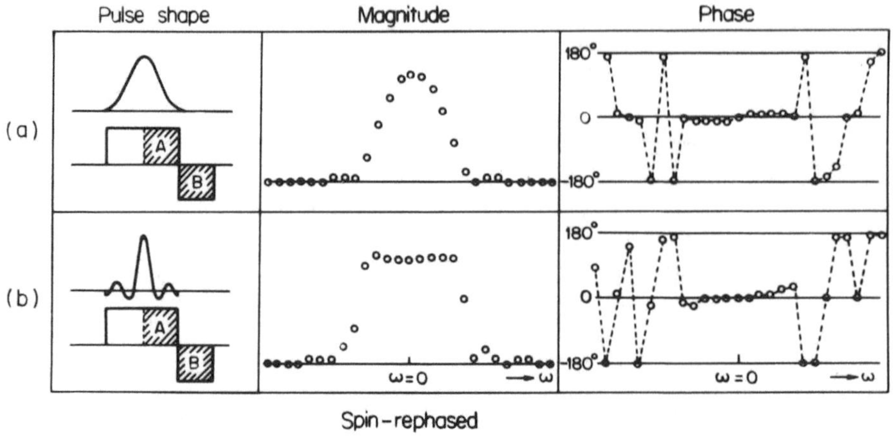

Fig. 12. Frequency domain magnitude and phase distributions as a function of frequency for the full Hanning and full truncated sinc shapes with spin rephase. Spin rephase is obtained with the gradient reversal.

$$s(t) = M_o \int_y \int_x \rho(x,y) \exp[i\gamma x \int_o^t G_x(t')dt'] dxdy$$

$$= M_o \int_x P_\ell(x) \exp[i\gamma x \int_o^t G_x(t')dt']dx, \qquad (37)$$

where $G_x(t')$ is a time varying gradient, $\rho(x,y)$ is spin density and $P_\ell(x)$ is a line-integral projection data along x axis, respectively. If gradient pulse is constant during the integration time t, Eq. (37) can be written as

$$s(t) = M_o \int_x P_\ell(x) \exp(i\gamma x G_x t) dx$$

$$= \frac{M_o}{\gamma G_x} \int_\omega P_\ell(x) \exp(i\omega t) d\omega \qquad (38)$$

where G_x is a constant gradient and $\omega = \gamma x G_x$. Projection data is then obtained via Fourier transform of FID signal. In an actual system, however, gradient pulse has a finite rise time and its effect on image is detrimental unless proper measures are taken. This effect could be eliminated by use of the pulse scheme proposed in the following section.

4.C Correction of Phase Instability

In a practical system, magnetic field fluctuates both in short as well as long-term period. Among others, main magnetic field fluctuation in the FID signal (thereby) degrades the resolution as well as the uniformity of the reconstructed image.

Let $s^*(t)$ be an inverse Fourier transform of projection data $p(\omega)$ which is a real value. Since $s^*(t)$ consists of a complex conjugate pair with both at $t \geq 0$ and $t \leq 0$, it may be given as,

$$s^*(t) = \begin{cases} s_R(t) + is_I(t), & \text{for } t \geq 0 \\ s_R(-t) - is_I(-t), & \text{for } t < 0, \end{cases} \tag{39}$$

where $s_R(\pm t)$ and $s_I(\pm t)$ are real and imaginary components of $s^*(t)$, respectively.

In a conventional NMR system, however, data is collected only for a time period $t \geq 0$ (see Fig. 8) and actual FID signal obtained is then given by

$$s(t) = \begin{cases} s_R(t) + is_I(t), & \text{for } t \geq 0 \\ 0 \quad , & \text{for } t \leq 0, \end{cases} \tag{40}$$

where $s_R(t)$ and $s_I(t)$ are usually obtained by quadrature phase sensitive detection. The Fourier transform of Eq. (40) can be written as

$$F[s(t)] = \frac{1}{2} [p(\omega) + iq(\omega)] \tag{41}$$

where both $p(\omega)$ and $q(\omega)$ are the real values. Therefore, from Eqs. (40) and (41), projection data can be obtained by Fourier transform of FID signal and by taking the real part only.

In most of NMR systems, however, due to finite power supply stability (10 parts per million), center frequency fluctuates at the rate of $10 \sim 100$ Hz.

Because of this center frequency instability, phase delay between FID signal and reference signal exists. Therefore, Eq. (41) may be written as

$$F[s(t)] = \frac{1}{2}[p(\omega) + iq(\omega)]e^{i\phi} \tag{42}$$

If ϕ is in continuous fluctuation (experiencing both long and short term instability), it is difficullt to obtain the projection data $p(\omega)$ alone from Eq. (42).

On the other hand, if full data, i.e., FID's of both $t \geq 0$ and $t \leq 0$ are obtained, it should, theoretically, be possible to extract only real value. In the case of spin-echo technique, the measured FID $s(t)$ phase delay (or change) ϕ can be written as

$$s(t) = \begin{cases} [s_R(t) + is_I(t)]\, e^{i\phi}, & \text{for } t \geq 0 \\ [s_R(-t) - is_I(-t)]\, e^{i\phi}, & \text{for } t < 0. \end{cases} \tag{43}$$

Since Fourier transform of a complex conjugate pair results real value only, Fourier transform of Eq. (43) will result real value with phase delay $e^{i\phi}$, i.e.,

$$F[s(t)] = p(\omega)\, e^{i\phi}. \tag{44}$$

Eq. (44) will directly lead to real value by simply taking absolute value or magnitude of Eq. (44), i.e.,

$$|F[s(t)]| = |p(\omega)|\ . \tag{45}$$

This full data set can be obtained by 180° spin-echo technique that will be described in the following.

5. PROPOSED SCHEME: AN EXAMPLE APPLIED TO THE PROJECTION RECONSTRUCTION

In the earlier sections we have examined; (a) necessity of rephasing the spins using gradient reversal for 90° RF selection pulse, (b) effect of gradient rise time and its correction, and (c) use of possibility of correcting field instability effect by taking FID signals of both $t \geq 0$ and $t < 0$. A new method which could solve the above three problems simultaneously is proposed and its pulse scheme is shown in Fig. 13. This technique is similar to the conventionally known NMR pulse scheme, e.g., pulse scheme shown in GE NMR INTRODUCTION (31). Note, however, the subtle difference which leads to the novel technique described in this chapter. In this scheme the whole object is excited by a 90° RF pulse without application of any gradient pulse. Immediately after the end of 90° r.f. pulse, x-y gradient pulse follows. FID signal obtained during this period is distorted by the rise time of the gradient pulse, hence it is discarded. After t_{r_1} time delay, a 180° pulse is applied together with z-selection gradient pulse and subsequent reading pulse allows us to collect spin echo FID signal. In this way, RF pulse selectively flips the spins 180° within the frequency band and spins are refocused while all the spins outside the frequency band decay away. The first half of the spin-echo pulse corresponds to FID signal of $t < 0$, while the second half is for $t \geq 0$, allowing us to obtain real value only by taking magnitude of the Fourier transformed value. In addition, the signals are obtained at the region where gradient pulse is constant.

Experiments were performed with this new proposed pulse scheme for the line-integral projection reconstruction. In Fig. 14, human head images obtained by the new method are shown. Images shown in Fig. 14 are obtained by line-integral projection reconstruction with slice-encoding method discussed in section 3.B(b). With this method, multiple-slice images are obtained in a simultaneous way and shown are the images of 4-sagittal slices which were obtained 4 x 4 Hadamard-like matrix coding (28). In Fig. 15, images obtained by the multislice total volume plane-integral projection technique are shown (30). Note that the spin-echo technique (magnitude method) described can equally be applied to the plane-integral projection method.

6. CONCLUSIONS

Although diversity and complexity involved in NMR imaging is extensive, it is apparent that along with the progress in the understanding of basic principles of NMR and image formation (reconstruction) related to NMR imaging, a more unified approach will be possible. Thus the choice and direction of NMR imaging, depending on the specific application will be more clear.

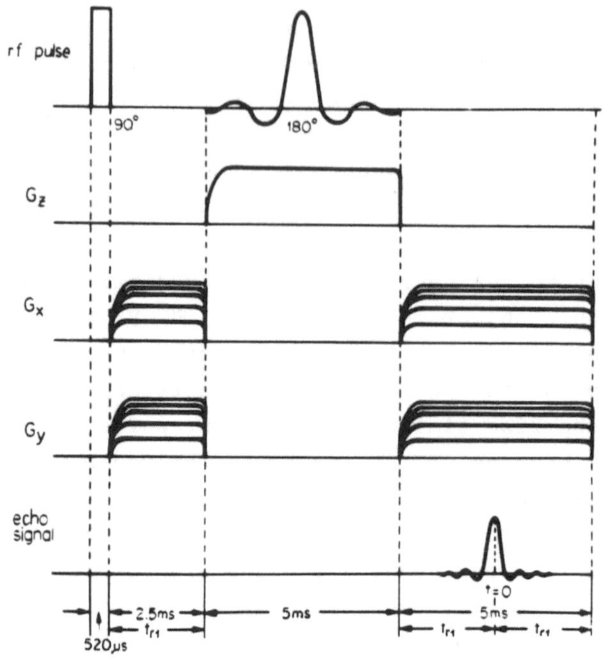

Fig. 13. Pulse scheme for selective spin-echo pulse sequence. In this scheme, a slice (or slices) is selected by the 180° spin-echo pulse.

In this chapter, we summarized the NMR principles briefly to the extent which is necessary for the understanding of NMR imaging, image reconstruction problems in connection to NMR, and physical, instrumental problems associated with NMR imaging.

Even though specific pulse sequences and image reconstruction methods described in this chapter may soon be outmoded, the basic principle underlying the methods will remain. Future developments will be more in the direction of finer methodologic instrumental development such as automatic tuning of the RF coil, precise alignment of the gradient coil, homogeneity correction, and more accurate calculation of the T_1 and T_2 values in association with image reconstruction and data acquisition modes.

Development of NMR imaging at present (1983) is indeed fast and dynamic. No one knows what direction the field will take, but it is certain that

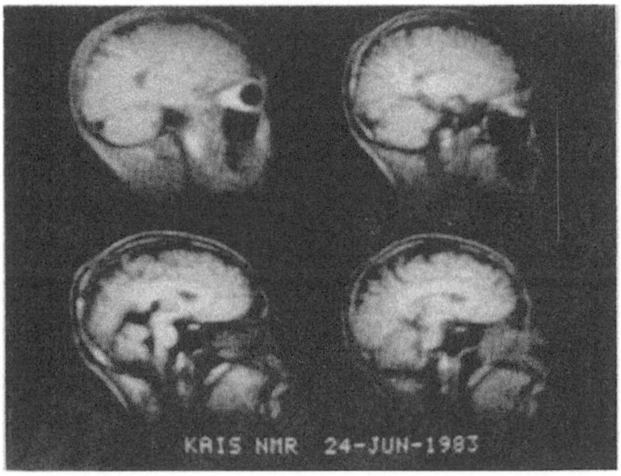

Fig. 14. Four sagittal images obtained by the encoding line-integral projection reconstruction.

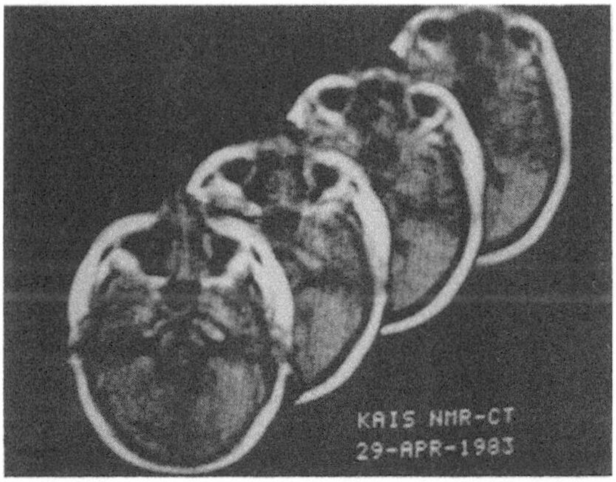

Fig. 15. Four transaxial images of human head obtained by plane-integral projection reconstruction.

the width and depth of NMR imaging is unprecedented with any other imaging modality we have ever experienced.

REFERENCES

1. Lauterbur P (1973). Image formation by induced local interactions: Examples employing nuclear magnetic resonance. Nature 242:190.

2. Damadian R (1971). Tumor detection by NMR. Science 171:1151.

3. Andrew E, Bottomley P, Hinshaw W, Holland G and Moore W (1972). Nature 270:cover.

4. Moore W, Holland G and Kreel L (1980). The NMR CAT scanner: A new look at the brain. CT 4(2):1.

5. Holland G, Hawkes R and Moore W (1980). Nuclear magnetic resonance (NMR) tomography of the brain: Coronal and sagittal sections. J Comput Assist Tomogr 4:429.

6. Hounsfield G and Maudsley A (1976). Fast scan photon density imaging by NMR. J Phys E Sci Instrum 9:271.

7. Kumar A, Welti D and Ernst R (1975). NMR Fourier zeugmatography. J Mag Res 18:69.

8. Lai C-M, House W and Lauterbur P (May 1978). Nuclear magnetic resonance zeugmatography for medical imaging. Presented at Conference Technology for Non-invasive Monitoring of Physiological Phenomena, Section 30, IEEE and ERA Electro/78, Boston, Massachusetts.

9. Shepp L (1980). Computerized tomography and nuclear magnetic resonance. J Comput Assist Tomogr 4:94.

10. Cho ZH, Hilal SK, Kim HS and Song HB (1983). Computer modeling and simulation of Fourier transform NMR tomography. (In) NMR Imaging, Saunders, Philadelphia, Pennsylvania, p 453.

11. Bloch F (1946). Nuclear induction. Phys Rev 70:460.

12. Shaw D (1971). Fourier Transform NMR Spectroscopy. Elsevier, New York.

13. Farrar T and Becker E (1971). Pulse and Fourier Transform NMR. Academic Press, New York.

14. Shepp L and Logan B (1974). The Fourier reconstruction of a head section. IEEE Trans Nucl Sci NS-21:21.

15. Hounsfield G, Ambrose J, Perry J, et al (1973). Computerized transverse axial scanning. Brit J Radiol 46:1016.

16. Cormack A (1973). Reconstruction of densities from their projections, with applications in radiological physics. Phys Med Biol 18:195.

17. Cho ZH (1974). General views on 3-D image reconstruction and computerized transverse axial tomography. IEEE Trans Nucl Sci NS-21: 44.

18. Brooks R and DiChiro G (1976). Principles of computer assisted tomography (CAT) in radiographic and radioisotopic imaging. Phys Med Biol 21:689.

19. Ter-Pogossian MM (1977). Basic principle of computed axial tomography. Sem Nucl Med 7:109.

20. Cho ZH, Chan JK and Erikson L (1976). Circular ring transverse axial positron camera for 3-D reconstruction of radionuclides distribution. IEEE Trans Nucl Sci NS-23:613.

21. Ramachandran G and Lakshiminarayanan A (1971). Three-dimensional reconstruction from radiographs and electron micrographs. Proc Nat Acad Sci(US) 48:2236.

22. Cho ZH, Kim HS, Song HB and Cumming J (1982). Fourier transform nuclear magnetic resonance tomographic imaging. Proc IEEE 70:1152.

23. Cho ZH, et al. Development of methods and algorithms for Fourier transform nuclear magnetic (NMR) imaging. Imaging System Science Lab, Department of Electrical Science, Korea Advanced Institute of Science, Seoul, Korea.

24. Hinshaw W and Lent A (1983). An introduction to NMR imaging: From the Bloch equation to the imaging equation. Proc IEEE 71:338.

25. Hoult DI (1979). The solution of Bloch equations in the presence of a varying RF field - An approach to selective pulse analysis. J Magn Res 35:69.

26. Sutherland R and Hutchison J (1978). Three-dimensional NMR imaging using selective excitation. J Phys E Sci Instrum 11:1979.

27. Oh CH, Kim HS, Park HW, Kim WS, Lee SW and Cho ZH (June 1983). Study of RF, gradient pulse and magnet instability effect in NMR tomography. IEEE Trans Nucl Sci NS-30:1899-1904.

28. Cho ZH, Park HW, Ph CH. To be published.

29. Marr R, Chen C-N and Lauterbur P (Feb 10-16,1980). On two approaches to 3-D reconstruction in NMR zeugmatography. (In) Proc Conf en Mathematics Aspects of Computerized Tomography (Mathematicsche Forschungsinstitute, Oberwolfach, West Germany.

30. Cho ZH and Lee SW. To be published.

31. GE NMR INTRODUCTION (1982).

Lecture Notes in Medical Informatics

Vol. 1: Medical Informatics Europe 78. Proceedings 1978. Edited by J. Anderson. XI, 822 pages. 1978.

Vol. 2: D. Fenna, S. Abrahamsson, S.O. Lööw and H. Peterson, The Stockholm County Medical Information System. VI, 163 pages. 1978.

Vol. 3: Long-Term Studies on Side-Effects of Contraception – State and Planning. Symposium 1977. Edited by U. Kellhammer and K. Überla. VI, 240 pages. 1978.

Vol. 4: Clinical Trials in 'Early' Breast Cancer. Proceedings 1978. Edited by H.R. Scheurlen, G. Weckesser and I. Armbruster. VI, 283 pages. 1979.

Vol. 5: Medical Informatics Berlin 1979. Proceedings 1979. Edited by B. Barber, F. Grémy, K. Überla and G. Wagner. XXIII, 970 pages. 1979.

Vol. 6: Health Care Technology Evaluation. Proceedings, 1978. Edited by J. Goldman, VII, 118 pages. 1979.

Vol. 7: Technology and Health: Man and his World. Proceedings, 1978. Edited by G. Wagner P.L. Reichertz and E. Masè. VI, 243 pages. 1980.

Vol. 8: Mathematical Aspects of Computerized Tomography. Proceedings 1980. Edited by F. Natterer. VIII, 309 pages. 1981.

Vol. 9: Computers and Mathematical Models in Medicine. Proceedings, 1977. Edited by D. Cardús and C. Vallbona. VIII, 315 pages. 1981.

Vol. 10: B. Fries, Applications of Operations Research to Health Care Delivery Systems. V, 107 pages. 1981.

Vol. 11: Medical Informatics Europe 81. Proceedings 1981. Edited by F. Grémy, P. Degoulet, B. Barber, and R. Salamon. XXIII, 975 pages. 1981.

Vol. 12: G. Wiederhold, Databases for Health Care. VI, 75 pages. 1981.

Vol. 13: The Computer and Blood Banking. Proceedings, 1981. Edited by J.R. Möhr and A. Kluge. X, 275 pages. 1981.

Vol. 14: F. Wingert, Medical Informatics. X, 247 pages. 1981.

Vol. 15: Digital Image Processing in Medicine. Proceedings 1981. Edited by K.H. Höhne. VIII, 197 pages. 1981.

Vol. 16: Medical Informatics Europe 82. Proceedings, 1982. Edited by R.R. O'Moore, B. Barber, P.L. Reichertz, and F. Roger. XX, 938 pages. 1982.

Vol. 17: Biomedical Images and Computers. Proceedings, 1980. Edited by J. Sklansky and J.-C. Bisconte. VII, 332 pages. 1982.

Vol. 18: D.P. Pretschner, Engymetry and Personal Computing in Nuclear Medicine. V, 129 pages. 1982.

Vol. 19: R.L. Blum, Discovery and Representation of Causal Relationships from a Large Time-Oriented Clinical Database: The RX Project. XIX, 242 pages. 1982.

Vol. 20: S. Biefang, W. Köpcke, M.A. Schreiber, Manual for the Planning and Implementation of Therapeutic Studies. V, 100 pages. 1983.

Vol. 21: Influence of Economic Instability on Health. Proceedings, 1981. Edited by J. John, D. Schwefel, and H. Zöllner. VIII, 528 pages. 1983.

Vol. 22: Objective Medical Decision-making; Systems Approach in Acute Disease. Proceedings, 1983. Edited by J.E.W. Beneken and S.M. Lavelle. VIII, 243 pages. 1983.

Vol. 23: Selected Topics in Image Science. Edited by Z.-H. Cho and O. Nalcioglu. IX, 308 pages. 1984.

Medizinische Informatik und Statistik

Band 20: Robuste Verfahren. 25. Biometrisches Kolloquium der Deutschen Region der Internationalen Biometrischen Gesellschaft, Bad Nauheim, März 1979. Herausgegeben von H. Nowak und R. Zentgraf. V, 121 Seiten. 1980

Band 21: Betriebsärztliche Informationssysteme. Frühjahrstagung, München, 1980. Herausgegeben von J. R. Möhr und C. O. Köhler. XI, 183 Seiten. 1980.

Band 22: Modelle in der Medizin. Theorie und Praxis. Herausgegeben von H. J. Jesdinsky und V. Weidtman. XIX, 786 Seiten. 1980.

Band 23: Th. Kriedel, Effizienzanalysen von Gesundheitsprojekten. Diskussion und Anwendung auf Epilepsieambulanzen. XI, 287 Seiten. 1980.

Band 24: G.K. Wolf, Klinische Forschung mittels verteilungsunabhängiger Methoden. X, 141 Seiten. 1980.

Band 25: Ausbildung in Medizinischer Dokumentation, Statistik und Datenverarbeitung. Herausgegeben von W. Gaus. X, 122 Seiten. 1980.

Band 26: Explorative Datenanalyse. Frühjahrstagung, München, 1980. Herausgegeben von N. Victor, W. Lehmacher und W. van Eimeren. V, 211 Seiten. 1980.

Band 27: Systeme und Signalverarbeitung in der Nuklearmedizin. Proceedings. Herausgegeben von S. J. Pöppl und D. P. Pretschner. IX, 317 Seiten. 1981.

Band 28: Nachsorge und Krankheitsverlaufsanalyse. 25. Jahrestagung der GMDS, Erlangen, September 1980. Herausgegeben von L. Horbach und C. Duhme. XII, 697 Seiten. 1981.

Band 29: Datenquellen für Sozialmedizin und Epidemiologie. Herausgegeben von Ralph Brennecke, Eberhard Greiser, Helmut A. Paul und Elisabeth Schach. VIII, 277 Seiten. 1981.

Band 30: D. Möller, Ein geschlossenes nichtlineares Modell zur Simulation des Kurzzeitverhaltens des Kreislaufsystems und seine Anwendung zur Identifikation. XV, 225 Seiten. 1981.

Band 31: Qualitätssicherung in der Medizin. Probleme und Lösungsansätze. GMDS-Frühjahrstagung, Tübingen, 1981. Herausgegeben von H. K. Selbmann, F. W. Schwartz und W. van Eimeren. VII, 199 Seiten. 1981.

Band 32: Otto Richter, Mathematische Modelle für die klinische Forschung: enzymatische und pharmakokinetische Prozesse. IX, 196 Seiten. 1981.

Band 33: Therapiestudien. 26. Jahrestagung der GMDS, Gießen, September 1981. Herausgegeben von N. Victor, J. Dudeck und E. P. Broszio. VII, 600 Seiten. 1981.

Band 34: C. E. M. Dietrich, P. Walleitner, Warteschlangen-Theorie und Gesundheitswesen. VIII, 96 Seiten. 1982.

Band 35: H.-J. Seelos, Prinzipien des Projektmanagements im Gesundheitswesen. V, 143 Seiten. 1982.

Band 36: C. O. Köhler, Ziele, Aufgaben, Realisation eines Krankenhausinformationssystems. II, (1–8), 216 Seiten. 1982.

Band 37: Bernd Page, Methoden der Modellbildung in der Gesundheitssystemforschung. X, 378 Seiten. 1982.

Band 38: Arztgeheimnis – Datenbanken – Datenschutz. Arbeitstagung, Bad Homburg, 1982. Herausgegeben von P. L. Reichertz und W. Kilian. VIII, 224 Seiten. 1982.

Band 39: Ausbildung in der Medizinischen Informatik. Proceedings, 1982. Herausgegeben von P. L. Reichertz und P. Koeppe. VIII, 248 Seiten. 1982.

Band 40: Methoden der Statistik und Informatik in Epidemiologie und Diagnostik. Proceedings, 1982. Herausgegeben von J. Berger und K. H. Höhne. XI, 451 Seiten. 1983.

Band 41: G. Henrich, Bildverarbeitung von Computer-Tomogrammen zur Unterstützung der neuroradiologischen Diagnostik. VIII, 203 Seiten. 1983.

Band 42: K. Boehnke, Der Einfluß verschiedener Stichprobencharakteristika auf die Effizienz der parametrischen und nichtparametrischen Varianzanalyse. II, 6, 173 Seiten. 1983.

Band 43: W. Rehpenning, Multivariate Datenbeurteilung. IX, 89 Seiten. 1983.

Band 44: B. Camphausen, Auswirkungen demographischer Prozesse auf die Berufe und die Kosten im Gesundheitswesen. XII, 292 Seiten. 1983.

Band 45: W. Lordieck, P. L. Reichertz, Die EDV in den Krankenhäusern der Bundesrepublik Deutschland. XV, 190 Seiten. 1983.

Band 46: K. Heidenberger, Stategische Analyse der sekundären Hypertonieprävention. VII, 274 Seiten. 1983.

Band 47: H.-J. Seelos, Computerunterstützte Screeninganamnese. IX, 221 Seiten. 1983.

MIX
Papier aus verantwortungsvollen Quellen
Paper from responsible sources
FSC® C105338

If you have any concerns about our products,
you can contact us on
ProductSafety@springernature.com

In case Publisher is established outside the EU,
the EU authorized representative is:
Springer Nature Customer Service Center GmbH
Europaplatz 3, 69115 Heidelberg, Germany

Printed by Libri Plureos GmbH
in Hamburg, Germany